The Local Turn in Tourism

ASPECTS OF TOURISM

Series Editors: **Chris Cooper** (Leeds Beckett University, UK), **C. Michael Hall** (University of Canterbury, New Zealand) and **Dallen J. Timothy** (Arizona State University, USA)

Aspects of Tourism is an innovative, multifaceted series, which comprises authoritative reference handbooks on global tourism regions, research volumes, texts and monographs. It is designed to provide readers with the latest thinking on tourism worldwide and in so doing will push back the frontiers of tourism knowledge. The series also introduces a new generation of international tourism authors writing on leading edge topics.

The volumes are authoritative, readable and user-friendly, providing accessible sources for further research. Books in the series are commissioned to probe the relationship between tourism and cognate subject areas such as strategy, development, retailing, sport and environmental studies. The publisher and series editors welcome proposals from writers with projects on the above topics.

All books in this series are externally peer-reviewed.

Full details of all the books in this series and of all our other publications can be found on http://www.channelviewpublications.com, or by writing to Channel View Publications, St Nicholas House, 31-34 High Street, Bristol, BS1 2AW, UK.

ASPECTS OF TOURISM: 95

The Local Turn in Tourism

Empowering Communities

Edited by
**Freya Higgins-Desbiolles and
Bobbie Chew Bigby**

CHANNEL VIEW PUBLICATIONS
Bristol • Jackson

DOI https://doi.org/10.21832/HIGGIN8793
Library of Congress Cataloging in Publication Data
A catalog record for this book is available from the Library of Congress.
Names: Higgins-Desbiolles, Freya, editor. | Bigby, Bobbie Chew, editor.
Title: The Local Turn in Tourism: Empowering Communities/Edited by
  Freya Higgins-Desbiolles and Bobbie Chew Bigby.
Description: Bristol, UK; Jackson, TN: Channel View Publications, 2022. |
  Series: Aspects of Tourism: 95 | Includes bibliographical references
  and index. | Summary: "This book considers the vital importance of local
  communities to just and sustainable tourism futures. The contributors
  examine how tourism can be reoriented to better connect people, place
  and planet. This local turn starts by centring local communities at the
  heart of tourism and identifies ways to ensure local community rights
  and benefits"— Provided by publisher.
Identifiers: LCCN 2022029333 (print) | LCCN 2022029334 (ebook) |
  ISBN 9781845418793 (hardback) | ISBN 9781845418786 (paperback) |
  ISBN 9781845418816 (epub) | ISBN 9781845418809 (pdf)
Subjects: LCSH: Tourism—Social aspects. | Sustainable tourism.
Classification: LCC G156.5.S63 L64 2022 (print) | LCC G156.5.S63 (ebook)
  | DDC 306.4/819—dc23/eng20221013
LC record available at https://lccn.loc.gov/2022029333
LC ebook record available at https://lccn.loc.gov/2022029334

British Library Cataloguing in Publication Data
A catalogue entry for this book is available from the British Library.

ISBN-13: 978-1-84541-879-3 (hbk)
ISBN-13: 978-1-84541-878-6 (pbk)

Channel View Publications
UK: St Nicholas House, 31–34 High Street, Bristol, BS1 2AW, UK.
USA: Ingram, Jackson, TN, USA.

Website: www.channelviewpublications.com
Twitter: Channel_View
Facebook: https://www.facebook.com/channelviewpublications
Blog: www.channelviewpublications.wordpress.com

Copyright © 2023 Freya Higgins-Desbiolles, Bobbie Chew Bigby and the authors of individual chapters.

All rights reserved. No part of this work may be reproduced in any form or by any means without permission in writing from the publisher.

The policy of Multilingual Matters/Channel View Publications is to use papers that are natural, renewable and recyclable products, made from wood grown in sustainable forests. In the manufacturing process of our books, and to further support our policy, preference is given to printers that have FSC and PEFC Chain of Custody certification. The FSC and/or PEFC logos will appear on those books where full certification has been granted to the printer concerned.

Typeset by Nova Techset Private Limited, Bengaluru and Chennai, India.

Contents

Figures and Tables	xi
Contributors	xv
Dedication	xxi
Preface	xxiii
Foreword	xxvii
Helena Norberg-Hodge	
Local Roots	xxxi
vesper tjukonai	
Introduction	1
Freya Higgins-Desbiolles and Bobbie Chew Bigby	
Introduction	1
Going Local	2
Localising Tourism and a Degrowth Approach	10
Social Movements for Defence of the Local	12
Key Terminologies for Localising Tourism	16
This Volume	22

Part 1 Theorising Local Communities in Tourism Anew

1 Place-based Governance in Tourism: Placing Local Communities at the Centre of Tourism	31
Bobbie Chew Bigby, Joseph Edgar and Freya Higgins-Desbiolles	
Introduction	31
Place and Placelessness	32
Case Study: Karajarri People, Country and Tourism	35
Degrowth+Decolonising	47
Conclusion	50
2 Circular *Oikonomia*, Posthumanism and Local Space to Socialise Tourism	54
Lucia Tomassini and Elena Cavagnaro	
Introduction	54
The Circular Economy as a Circular Oikonomia	56

	Posthumanism	59
	A Theoretical Framework of Circular Oikonomia and Posthumanism	61
	Conclusion	64
3	Travel Boycotts, Ethical Consumption and Destination Communities: Expanding the Morality of Neighbourliness *Siamak Seyfi and C. Michael Hall*	69
	Introduction	69
	Boycotts in Tourism	71
	Boycott and Social Justice	73
	Boycott or Not: Is Boycotting Always the Best Practice?	74
	The Debates over Boycotts and the Problem of Community	76
	Conclusion	78
4	The Local Turn in Tourism: Place-based Realities, Dangers and Opportunities *Can-Seng Ooi*	82
	Introduction	82
	Two Moral Limits of the Market	83
	Localisation in Popular Approaches of Making Tourism More Sustainable	85
	The Local Turn and Dealing with Local Realities	90
	Conclusions	95

Part 2 Case Studies of Local Community (Dis)/(Re)/Empowerment

5	Unheard Voices: Youth Activism for Social and Environmental Justice *Antonia Canosa*	103
	Introduction	103
	Background	104
	Youth Participatory Filmmaking	106
	Youth4Sea	107
	Our Home Holiday Town	108
	Discussion and Conclusion	111
6	Enhanced Food Security Through Localised Community Cryptocurrency: Experiences of a Costa Rican Tourism Town *Mary Little*	116
	Introduction	116
	Community Currencies as a Resilience-Building Strategy	118
	Conceptual Framework – The Doughnut Economy	120
	Research Site: Monteverde, Costa Rica	121
	COVID-19 and Monteverde's Food Security Responses	123
	Discussion	128
	Conclusion	129

7   An Ethnographical Study of Community Tourism:
    Seeking Alternative Tourism Options for Malta Through
    'Meet the Locals'   133
    Andrew Jones and Julian Zarb
        Introduction   133
        Community Tourism: The Research Context   135
        Tourism in Malta: Promoting Alternative Approaches to
            Traditional Tourism   137
        The Research Approach and Project Delivery   139
        Current Research Outcomes   141
        Conclusions   148

Part 3 Practitioners' Views and Insights

8   Localhood   155
    Signe Jungersted
        The Epilogue   155
        Local Huh?   156
        Strategic Determinants of Localhood   156
        The End Was Just the Beginning…   158
        Easier Said than Done: Implementing Localhood   160
        From Localhood to Local Good   164

    Case Study: Transforming Relations Between DMOs
        and Communities: NAO Launching: Time for *DMOcracy*   166
        What We Want To Achieve   166
        Why DMOcracy? Why Now?   167
        Liveability Over Visitability   168

9   The Story of Cambodian Children's Trust: Evolving
    Development Practice From 'Doing For' Communities To
    'Doing With' Communities   169
    Tara Winkler
        Introduction   169
        My Start in Development: 'Doing to' Families   170
        Understanding Development: Investing in Downstream
            Systems   172
        CCT First Evolution: 'Doing for' Families   175
        CCT Second Evolution: 'Doing with' Families and
            Communities   178
        CCT's Development Insight: Empowerment is a Shift
            in Power   179
        Why Fund Downstream?   181
        CCT's Shift to an All-Khmer Leadership Team In-Country   183
        CCT's Next Evolution: Shifting the Power by Working
            Within Local Systems   186

	Case Study in Practice of Working with Communities: Freire's 'Pedagogy of the Oppressed' and Working in Mutuality: Pathways for Communities to Take Control of Their Futures	191
10	The Neighbourhood Where History, Community, Tourism and Truth-Telling Meet: A Tourism Practitioner Case Study from the Greenwood Cultural Center of Tulsa, Oklahoma *Bobbie Chew Bigby and Michelle Brown-Burdex*	194
	Introduction	194
	The Story of Greenwood	196
	The Greenwood Cultural Center and 2021 Centennial Commemoration	201
	Discussion and Reflections with Michelle, a Tourism Practitioner and Storyteller in Greenwood	205
	Conclusion	210

Part 4 Imagining New Futures

11	Convivial Tourism in Proximity *Nora Müller, Robert Fletcher and Macià Blázquez-Salom*	215
	Introduction	215
	COVID-19 in the Context of the Ecological Crisis	216
	Tourism Debates in the Pandemic Setting	216
	The Proposal of Convivial Conservation in the Realm of Tourism	218
	Conservation for the Elite	220
	The Case of La Trapa (Mallorca)	221
	Conclusion	228
12	Towards a 'More-than-Tourism' Perspective for Localising Tourism *Phoebe Everingham and Sinéad Francis-Coan*	232
	An Autoethnographic Approach for a More-than-Tourism Localised Tourism Agenda	234
	Newcastle Afoot: The Politics of Public Space and Being (Re)Enchanted by Local Place	236
	Contested Notions of Local Recreation and Public Space	242
	Reclaiming Local Space, Centring Local Community as Custodians	242
	Conclusion: (Re)Enchanting the Local for Living in More Inclusive Regenerative Futures	245

13	Reclaiming the City: Social Movements and the Local Impacts of the Global Tourism Industry	250
	Alexander Araya López	
	Introduction: The Worst Year of the Industry	250
	The Political Uses of the 'Right to the City': Challenging the Narratives of Global Tourism	251
	Methods	253
	Barcelona: Mapping the Conflicts Related to Global Mass Tourism	253
	Not just Barcelona: Tourism Conflicts in Venice, Lisbon and Amsterdam	259
	Discussion	262
	Conclusion	264
14	Conclusion: What is to be Done?	269
	Freya Higgins-Desbiolles and Bobbie Chew Bigby	
	Introduction	269
	Themes Emerging from these Chapters	270
	There is More Work to be Done	275
	Conclusion	277
	Index	280

Figures and Tables

Figures

Figure 0.1	(Left) Tourists read the sign asking visitors not to climb Uluru (Central Australia) and explaining the reasons the Anangu Traditional Owners ask this. (Right) Tourists climb Uluru, circa 2011. Credit: Co-author Freya Higgins-Desbiolles.	4
Figure 0.2	Community-centred tourism framework. Figure adapted with permission from Higgins-Desbiolles *et al.* (2019: 1937). Taylor and Francis Inc. http://tandfonline.com.	6
Figure 0.3	Applying subsidiarity to our travel circuits A framework for proximity travel	21
Figure 1.1	Uncle Joe Edgar shows visitors a *makapala* (bush banana) growing on a tree near the Karajarri Tourism and Cultural Hub. He explains the importance of these traditional bush foods and how many of them grow plentifully during the wet season. Credit: Co-author Bobbie Chew Bigby.	37
Figure 1.2	Stepping into tourism. Adapted from: Stepwise Heritage and Tourism (n.d.)	40
Figure 1.3	Karajarri tour leader Wynston Shovellor (centre) speaks to a group of students at a lagoon on Karajarri Country. He explains the significance of this waterscape to students, then demonstrates the Karajarri practice of *kuwiyinpijala*, where a person takes water in their hand and then blows it out, allowing the Country to know that person better. Credit: Co-author Bobbie Chew Bigby.	41
Figure 1.4	Principles of place-based approaches to tourism governance	46
Figure 2.1	Visual model of the proposed theoretical framework for a local turn in tourism	62
Figure 5.1	Cigarette 'butt-feeder' containers employed by young co-researchers in the Youth4Sea project. Credit: Author	108

Figure 6.1	The doughnut economy model has helped the Monteverde community envision and enact strategies to meet human needs without overshooting the ecological ceiling. Doughnut model imposed over coordinators planning a community garden site in a disused bullring. Photo source: Paula Vargas, used with permission	120
Figure 6.2	Participants in the first Verdes market selling ceviche made of green plantains from a home garden (left, photo source: Irene G. Chen, used with permission) and the Verdes market's future home, a defunct bullring the market will share with a community garden (right, photo credit: Author)	127
Figure 7.1	Discovering Malta and Gozo through its people and culture. Source: Malta Tourism Society 2018 (used with permission)	134
Figure 7.2	Community Tourism Malta: 'Meet the Locals'. Source: Malta Tourism Society 2018 (used with permission)	143
Figure 7.3	Community Tourism Malta: Opportunities	148
Figure 7.4	Community Tourism Malta: Challenges	149
Figure 9.1	A downstream approach versus an upstream approach to child protection in development	174
Figure 10.1	The destruction and burning of the Greenwood neighbourhood. In the foreground is the building containing the Village Blacksmith Shop. The printed caption reads, 'Runing [sic] the negro out of Tulsa, June 1, 1921'. Credit: Tulsa Historical Society & Museum (used with permission)	198
Figure 10.2	A group being led to the Convention Hall during the 1921 Tulsa Race Massacre. Credit: Tulsa Historical Society & Museum (used with permission)	198
Figure 10.3	Blocks of destroyed homes following the Tulsa Race Massacre. Credit: Tulsa Historical Society & Museum (used with permission)	199
Figure 10.4	Black Tulsans detained during the 1921 Tulsa Race Massacre. Credit: Tulsa Historical Society & Museum (used with permission)	199
Figure 10.5	The front entrance of the Greenwood Cultural Center in Tulsa, Oklahoma. Credit: Co-author Bobbie Chew Bigby	201
Figure 10.6	A plaque commemorating the 1921 Tulsa Race Massacre. Behind the plaque stands the historic Vernon AME Church, one of only two churches to	

	have survived the Massacre. The church stands facing the Greenwood Cultural Center. Credit: Co-author Bobbie Chew Bigby	202
Figure 10.7	A large mural of 'Black Wall St.' that is painted onto the side of the highway that intersects the historic Greenwood neighbourhood and faces the Greenwood Cultural Center. Credit: Co-author Bobbie Chew Bigby	203
Figure 10.8	Michelle Brown-Burdex speaking with President Joseph Biden on his visit to the Greenwood Cultural Center during the Centennial Commemoration on 1 June 2021. Credit: Christopher Creese, Creeseworks (used with permission)	208
Figure 11.1	La Trapa. Credit: Co-author Macià Blázquez-Salom	222
Figure 11.2	Visitors at the viewpoint of La Trapa. Credit: Co-author Macià Blázquez-Salom	225
Figure 12.1	The city arcade as it looks today. Credit: Author	239
Figure 12.2	Homage to the secret baths by local graffiti artist painted on power box outside the arcade. Credit: Author	240
Figure 12.3	Detail of the secret baths' history presented in public graffiti art. Credit: Author	241
Figure 13.1	Banners against evictions near the Carmel's Bunkers, Turó de la Rovira. Credit: Author, 2019	255
Figure 13.2	Banners against evictions at the Carrer de Mallorca, next to La Sagrada Familia. Credit: Author, 2020	256
Figure 13.3	Banner against evictions near the MACBA, El Raval. Credit: Author, 2019	257
Figure 13.4	Banner against Airbnb, noise, skateboarding and tourists near the MACBA, El Raval. Credit: Author, 2019	257
Figure 13.5	Map of selected tourism-related conflicts in Barcelona. Map created by the author with uMap	260
Figure 14.1	A preliminary schema to advance the local turn in tourism studies. Reprinted from *Annals of Tourism Research*, Vol 92, F. Higgins-Desbiolles and B.C. Bigby, 'A local turn in tourism', 103291, with permission of Elsevier	276

Tables

Table 4.1	Society is complex, heterogeneous and evolves: Implications for a local-led change strategy	93

Table 6.1	The creation of community cryptocurrencies begins by determining objectives and developing specific actions to support those objectives Community members complete actions to earn Verdes currency	124
Table 14.1	Tentative framework for community involvement	277

Contributors

Editors

Freya Higgins-Desbiolles is Adjunct Associate Professor with the Department of Recreation and Leisure, University of Waterloo, Canada; Visiting Professor with the Centre for Research and Innovation in Tourism, Taylor's University of Malaysia; and Adjunct with UniSA Business, University of South Australia. She has worked with communities, non-governmental organisations and businesses on research into social justice, human rights and sustainability issues in tourism. She is also a co-founder of the Tourism Alert and Action Forum, which advocates for community rights in tourism.

Bobbie Chew Bigby is an enrolled member of the Cherokee Nation of Oklahoma and a native of Tulsa, Oklahoma. She is currently based between Oklahoma, US and Broome, Australia where she is completing a PhD at the University of Notre Dame Australia, Nulungu Research Insitute focused on comparative Indigenous-led tourism and cultural resurgence. Bobbie has engaged in research focused on Indigenous peoples, tourism and connections to traditional culture in China, India, Cambodia, Myanmar, Australia and at home in Oklahoma Indian Country.

Authors

Alexander Araya López concluded his PhD in Sociology in 2014 at the Lateinamerika-Institut, Freie Universität Berlin. His research interests include contemporary global discussions on protest, radical politics, civil/democratic disobedience, digital activism and media representation of acts of dissent, particularly cross-comparative analysis both within Europe and in Latin America. He was also former Marie Skłodowska-Curie Fellow at Ca' Foscari, University of Venice (2018–2020).

Macià Blázquez-Salom is a Professor in the Department of Geography at the University of the Balearic Islands. He teaches and researches on tourism, sustainability and land use planning. He has been a visiting scholar in universities of Mexico (Toluca and La Paz), Nicaragua (UNAN), Dominican Republic (INTEC), Austria (Salzburg), Germany (Rurh-Bochum), Sweden (Mid-Sweden) and the Netherlands (Wageningen). As a way to link activism and research within the framework of Radical Geography and Political Ecology, he collaborates with social movements in Spain, particularly in the Balearic Islands (https://www.gobmallorca.com/), but also in Latin America (https://www.albasud.org/).

Michelle Brown-Burdex is a native of Tulsa, Oklahoma and has worked at the Greenwood Cultural Center (GCC) for the past 25 years. She currently serves as the Center's Program Coordinator and has implemented and managed numerous award-winning educational and arts programmes. Michelle is a committed storyteller and tourism practitioner, having been the lead Tour Guide at GCC where she educates people about the tragedy of the 1921 Tulsa Race Massacre and the triumphant history of Black Wall Street. Michelle is a member of the Tulsa Mass Graves Historical Narrative Committee and the American Folklore Society and recently founded the Black Wall Street Historical Society. Above all, Michelle is a proud wife, mother and grandmother.

Antonia Canosa is a Research Fellow with the Centre for Children and Young People, Southern Cross University, Australia. Antonia's work focuses on children's rights, participation and well-being in the tourism industry. Her research interests include the ethical dimensions of research with children and ethnographic, participatory and visual methodologies. She also researches in the areas of identity, belonging and connection to nature among children and young people growing up in tourist destinations.

Elena Cavagnaro is Professor of Sustainability in hospitality and tourism at NHL Stenden University of Applied Sciences and Associate Professor at the University of Groningen. Following her understanding of sustainability as a multi-dimensional and multi-layered concept, her research focuses on issues that run across and connect the social, organisational and individual layers of sustainability.

Mowandi Joe Edgar is a Parrjarri man, a descendant of the Karajarri and Yawuru people from the Kimberley region of Western Australia. He is a member and director of the Karajarri Traditional Lands Association Registered Native Title Body Corporates (RNTBC) and Nyamba Buru Yawuru (RNTBC). Joe currently serves as the Senior Cultural Advisor/Elder-in-Residence for the Broome public school cluster. In his career that has spanned over 35 years, Joe has had extensive experience in Aboriginal affairs across numerous sectors, including with Indigenous tourism, arts, culture, governance, education and environmental issues.

Phoebe Everingham is an early career researcher and sessional staff member at the University of Newcastle, Australia. She draws on multidisciplinary perspectives such as human geography, sociology/anthropology and tourism management studies. Phoebe is committed to ensuring tourism and recreation can be enjoyed by all and alongside Sinéad Francis-Coan is working to working to establish a local regenerative tourism network in Newcastle where both are based.

Robert Fletcher is Associate Professor in the Sociology of Development and Change group at Wageningen University in the Netherlands. A former ecotourism guide, he is an environmental anthropologist with research interests in conservation, development, tourism, globalisation, climate change, human–wildlife interaction, social and resistance movements and non-state forms of governance. His publications include the books *The Conservation Revolution: Radical Ideas for Saving Nature beyond the Anthropocene,* co-authored with Bram Büscher (Verso, 2020) and *Romancing the Wild: Cultural Dimensions of Ecotourism* (Duke University Press, 2014).

Sinéad Francis-Coan is a postgraduate in Leisure and Tourism studies from the University of Newcastle. Sinéad is politically active in her community and involved in advocacy across a range of issues. Sinéad has presented her research at multiple conferences and has been published as a freelance writer in Global Hobo online travel magazine. Along with Dr Phoebe Everingham, Sinéad has been working to establish a local regenerative tourism network in Newcastle where both are based.

C. Michael Hall is a Professor in the Department of Management, Marketing and Entrepreneurship at the University of Canterbury, New Zealand; Docent, Department of Geography, University of Oulu, Finland; and Visiting Professor, School of Business and Economics, Linnaeus University, Kalmar. He publishes widely on tourism, sustainability, global environmental change and regional development.

Andrew Jones is currently an Associate Professor at The University of Malta but has held positions at the University of Wales, Swansea Business School (UK), and the University of Brunei. Andrew has professional experience in regional planning, environmental management and tourism at both international, regional and local levels and has been an enthusiastic contributor, for over 35 years, to the academic tourism community within Malta, the UK and internationally. As well as Malta, Andrew has conducted teaching and research assignments in Japan, China, India, Hong Kong, Singapore, Uzbekistan, Malaysia, Japan, Germany, Greece, Italy and the UK.

Signe Jungersted is founding partner in Group NAO that works with strategic transformation and innovation within the broader experience and visitor economy. Signe is formerly Development Director with Wonderful Copenhagen (DMO), where she led and authored the 2020-strategy that declared 'The End of Tourism' and welcomed a new era of people-centric tourism development and localhood. Signe holds a master's degree in Political Science from Copenhagen University, studied Mandarin in Shanghai and has pursued management education at MIT Boston, Hyper Island Stockholm and is currently completing an EMBA in Creative Leadership in Berlin.

Mary Little LLM is an Associate Professor at the Center for Sustainable Development Studies, School for Field Studies (SFS), in Costa Rica. She teaches courses on food security and sustainable tourism at the United Nations-mandated University for Peace in Costa Rica. Before teaching, Mary advocated for the legal rights of refugees and women experiencing domestic violence. These experiences with advocacy have informed her social justice approach to research and teaching. Her research explores community-driven waste solutions, responsible tourism and regenerative food initiatives. Currently, she is focused on agrotourism as a mechanism for climate adaption and is examining food security strategies in tourism areas for her PhD candidacy at the University for Peace.

Nora Müller is a PhD student in Tourism Geographies at the University of the Balearic Islands (Spain) working on a project entitled 'Private protected areas and the touristification and commodification of nature through neoliberal conservationism'. Previously, she obtained a BSc in Environmental Sciences from the Leuphana University (Germany) and a MA in Human Geography from the Eberhard Karls University of Tübingen. From the perspective of critical geography, she is interested in the relations between human and non-human natures, especially in the realm of tourism.

Helena Norberg-Hodge is a pioneer of the new economy movement, and has been promoting an economics of personal, social and ecological well-being for more than 40 years. She is Director of Local Futures; author of *Ancient Futures* and *Local Is Our Future*; and Producer of the film *The Economics of Happiness*. Helena is a recipient of the Alternative Nobel prize, the Arthur Morgan Award and the Goi Peace Prize. Her websites are www.localfutures.org and www.helenanorberghodge.com.

Can-Seng Ooi is a sociologist and Professor of Cultural and Heritage Tourism at the University of Tasmania. Educated in Singapore and Denmark, he has conducted research in many countries including Singapore, Denmark, Australia, China and Malaysia. His research interests and expertise include tourism-in-society perspectives, tourism development strategies, place branding and cross-cultural interaction and understanding. His website is www.cansengooi.com.

Siamak Seyfi is an Assistant Professor at the Geography Research Unit of the University of Oulu, Finland. Using a multi-/interdisciplinary approach, his research interests focus on critical tourism geographies, mobilities, ethical and political consumerism as well as qualitative sociological and ethnographic research methods in tourism.

vesper tjukonai guest of Mother Earth, Kaurna Country brung up, Ngarrindjeri Country heart-held, an anglo-gael who, very grateful for elders' generous welcome, loves to tour word stories.

Lucia Tomassini is Research Lecturer in Sustainability in Hospitality and Tourism at NHL Stenden University of Applied Sciences and University of Groeningen. She has a background in Architecture and Urban Planning and an MSc in Urban Development and Reconstruction in developing countries. Her academic research embraces studies on the relationship between tourism and development agenda, social-geography, sustainability, circular economy and posthuman studies as well as uses of narrative approach in tourism and hospitality studies.

Tara Winkler is the co-founder of Cambodian Children's Trust (CCT), which she established in 2007 with Pon Jedtha to help 14 children escape an abusive orphanage. Today, CCT is a grassroots, community-led organisation that empowers families and prevents children from ending up in orphanages. Winkler's story has been featured on ABC's Australian Story and 60 Minutes Australia. She has also taken to the TED stage and the Australian Parliamentary Inquiry into Modern Slavery to raise awareness about the orphanage industry, fuelled by white saviourism. Winkler is currently co-writing her first feature film based on her book, *How (NOT) to Start an Orphanage*.

Julian Zarb has had 40 years' experience within the tourism industry in the UK, Europe and Malta. In 2003 he moved into the public service in Malta where he was appointed Director Tourism within the same Ministry for Tourism and Culture between 2010 and 2013. Today he is a board director of the Tourism Society (UK) and a Postgraduate fellow of the Royal Geographical Society. He is also a lecturer and researcher with the Institute for Travel, Tourism and Culture at the University of Malta and focuses his work on local tourism planning and community-based tourism implementation.

This book is dedicated to all the beautiful places and
the people who care for them.

Freya

Certain lands have captured my heart – Long Beach, North Carolina, USA; Ngarrindjeri Country and Kaurna Country, Australia.

Bobbie

I dedicate this book to the Greenwood community of Tulsa, Oklahoma and in particular the victims, survivors and descendants of the 1921 Tulsa Race Massacre. I acknowledge that the historic Greenwood neighbourhood sits within the Tribal reservation boundaries of the Muscogee (Creek) Nation and Cherokee Nation, the two Native Nations who were granted stewardship of these lands from the time of their forced removal. I also respect that these lands were the traditional homelands of the Wichita, Caddo and Osage Nations who were the original peoples of this land and the first to be dispossessed. I have gratitude and deep love for Tulsa, Oklahoma, this wounded and special place where I was born and raised.

Preface

Listening to Bruce Springsteen's 'My Hometown', we might recognise an almost universal nostalgia for the places we come from, that many of us have been compelled to leave because livelihoods cannot be sustained there anymore. Many songs sing the joys of being free, roaming the world's glorious places and meeting new people along the way. But can we hear the laments for places lost to tourism and tourists, as local people feel what they love being slowly and not so slowly stolen from them, paved over and then packaged and sold as something they cannot even recognise?

In examining the local turn in tourism, we are reflecting on a world that is subject to mind-spinning change through globalisation and asking what it is about local communities that might ground us and make us more whole as people, in relatedness to others and to loved places. Arguably, it is this rootless mobility enjoyed by the global elite that in part has enabled the ecological damage that comes from consuming beyond the Earth's capacity. Such people are careless because they do not stay to live with the consequences of these actions.

International tourism has played a significant role in globalisation and global change and in doing so has impacted local communities in profound ways, too often without their permission or support or benefit. It is an offense to reason that tourism is defined and enacted with myopic concentration on the tourists and the tourism industry, pushing to the periphery local communities, local peoples and local places and the claim they have on the spaces that are labelled 'tourism destinations'. This book is intended as a direct challenge to this order of affairs. Local communities should be the centre of our definition and enacting of tourism because it is they who live with the impacts of tourism, too often to their detriment.

It is not an inadvertent mistake that the conventional definition of tourism omits the local community as a central player. This is a powerplay which allows the tourism industry and tourists to profit from tourism too often at the expense of local communities. This is witnessed in Venice where local people have been displaced from the old city, as day-trippers and cruise tourists crowd them out and ruin their quality of life. In Barcelona, cheap airlines bring in weekend party tourists who book

Airbnb accommodation in the heart of residential areas and disrupt the lives of the local residents. In Bali, the central government makes the island the focus of major tourism development and local residents find subsistence living under pressure as hotels, golf courses and gated communities take fertile land and attract young people to tourism and hospitality jobs, viewed as far easier than traditional agriculture and fishing work.

These changes have profound impacts on communities around the world. COVID-19 may have offered a circuit-breaker where we have paused to reflect on the nature of this living and its impacts at personal, familial, community, societal, national and global levels. Nostalgia has been invoked for cleaner waters, peaceful landscapes, community connections, healing contact with nature, wildlife enjoying calmer conditions – largely due to a sudden stopping of the relentless on-the-go pressures of modern living and also due to the shutting down of tourism. Blue skies were not scarred by airplane vapour trails. Coastal waters were not stirred up by mega-cruise ships. Communities including Venice were able to walk their plazas without having to dodge day-trippers and lounging backpackers.

As this book was being compiled and edited, Russia invaded Ukraine on the 24 February 2022. Just as people's thoughts were turning to tourism recovery, another strike against globalisation and global order that facilitated ease of global tourism was struck. Analysts have put forward many ideas on the meaning of these events, including a clear rise in authoritarian politics and a challenge to global geopolitics, but consensus has agreed that it marks a new era in human affairs. From the perspective offered in this book, it could be seen as one of the more visible manifestations of the growing resource wars for scarcer resources as the outcomes from global climate change begin to overwhelm communities around the world. In terms of tourism, it highlights the insight that tourism dependency is not advisable and that a dedicated localisation strategy is growing ever more vital.

In this work, we are not advancing a simplistic argument that tourism is a blight but that over-dependency on it is; we are saying it needs to be rethought and reformed. With this edited volume, we commend the idea of localising tourism through undertaking a decidedly local turn in tourism. We think this begins by defining tourism by the local community and situating them at the centre of the phenomenon. We have gathered a group of scholars and practitioners to write their thoughts on why this idea might be commendable, what this might mean, how we might go about it and what are the challenges and limits to such a programme of action. Our authors come from a variety of countries and contexts. Their explorations in the pages that follow offer up an exciting agenda for transformation in our conduct of tourism. It is a journey with people to understand the meaning of love of place and how this might be better shared through forms of tourism that are in respectful relatedness. This will take us from the heavily touristed cities of Europe, to a nature gem hidden in the European countryside, to a beachside town in Australia, to Cambodian communities subjected to orphanage tourism, to a North American community using tourism for truth-telling and to an Aboriginal community in Western Australia that uses tourism to change your mind about Country and what it means to live on and with *our* place. We owe our sincere thanks to Sarah Williams and the whole team at Channel View Publications for helping us to build this community of exploration. We could not have found a better publisher and we are deeply honoured that they have supported this work.

This is a journey and like all good journeys it will take you somewhere unexpected, maybe reveal some hidden wisdoms and hopefully change you in ways that help you to grow and learn. We certainly did.

Foreword

Helena Norberg-Hodge

This book comes at an important juncture, as global tourism again begins to pick up after the interregnum brought about by the COVID pandemic. The great pause that accompanied the early phase of the pandemic especially offered the planet a moment to breathe as globalisation's incessant busyness, acceleration and movement of material and people – including for global tourism – ground to a standstill. For communities and regions that have grown heavily dependent on tourism for livelihood security, this episode exposed in the harshest terms how perilous and brittle such economic dependence is. Neglect of a real, productive, local economy, including food production for local needs, is possible as long as cash income and a global supply chain is maintained, but its sudden removal lays bare the folly of hitching a whole economic wagon to this mirage.

In Ladakh, India, where I first had my eyes opened to the basic lessons of local self-reliance and the hazards of a commodity-consumerist 'development' model back in the 1970s, I have been able to witness in real time the steady substitution of a localised agrarian economy by a tourism-based cash economy and the many cultural, psychological and ecological consequences that shift has entailed. Despite the manic pace of change there, however, it remains largely agrarian, with villages that – however atrophied compared to the pre-development era – maintain an agricultural base, and many people returned to this comparative security and safety during lockdown. The reigning ideology that endlessly valorises the urban office, the so-called service economy while neglecting the rural and bemoaning its 'backwardness' was completely turned on its head.

The great pause also offered us – or should have done – an opportunity to critically reassess every aspect of pre-pandemic 'normal', to recognise the planetary-scale catastrophe this normal was producing and to imagine profoundly different directions. This is as true of tourism as anything. Mass global tourism has never been and cannot be sustainable in any circumstance, but its democratisation especially in the form of 'cheap' air travel, heralds a hitherto unimaginable ecological impact. Like many other aspects of the global economy, this too needs to be reduced rather than expanded, and what remains must be radically re-imagined and reoriented towards the 'local turn' that this book so admirably advances.

With regard to the first point about reducing tourism, I am increasingly convinced that in a more localised world which allows for genuine biological and cultural diversity, there will be much less need to travel in the first place. In such a world, people would be working far fewer hours at a more relaxed pace. The beauty of nature would be closer to home. In other words, the soul-crushing alienation and stress that compels so much of escapist tourism today would be less. It can sound unrealistic and utopian, but my experience in both localised traditional cultures and in newer re-localised hubs around the world has affirmed to me that this change is beginning to happen.

Of course, none of this is to gainsay the incomparable benefits and joys of travel. For those from heavily industrialised contexts, there is a critical role for travel to less-industrialised places where land-based old cultures can still be experienced, especially for young people. Over many years my organisation Local Futures facilitated such immersive experiences for young Westerners (and, increasingly, urbanites from various parts of Asia) with Ladakhi farming families. Needless to say, such immersive experience with a local culture – paired with deep facilitated learning programmes – can be profoundly influential on people from the overdeveloped/techno-industrial realm, sometimes effecting changes of perspective and later of life that hardly any other experience could match. This was our experience, with many participants having been inspired to undertake relocalisation initiatives like farmer's markets when they returned to their home countries. In the other direction, it is equally important to facilitate travel to the urban-industrial contexts for people – particularly young people from rural backgrounds – in the so-called developing countries, in order to counter the shameless glorification of the West, the urban, the 'modern' that is ubiquitous in mass media and education systems around the world. This process, which we undertook by sponsoring community leaders from Ladakh to travel to the West on 'reality tours' for many years, helps in what I call positive counter-development. It replaces the myths of progress and development with a more balanced understanding of the previously unknown ecological and social realities of the conventional models of development. These reality tours also provide experience of the sincere and inspiring efforts of so many people in the industrialised world to shift direction towards slower, simpler, more localised and sufficient lives and economies. This in turn reverses the arrow of development, showing the land-based village cultures to be lighthouses of sustainable futures newly aspired to by the 'over-developed'.

For the meaningful, thoughtful international travel that will remain, air travel is still an unsustainable mode, and needs to eventually – sooner than later – be replaced by slower, far more pleasant, modes like boats and trains, as well as walking and cycling holidays, often combined with a more spiritual approach, for example the 'pilgrimage' walks.

In addition, by way of gentle critique, my impression is that too many of the 'responsible tourism' initiatives that are becoming increasingly common, while having the best of intentions, tend to be reduced to the individual level – how the individual tourist, or maybe their tour group, can do differently, can spend their money better. This is indeed welcome, yet, there can be no responsible or sustainable travel or tourism within a congenitally unsustainable economic context. Put differently, if economies were localised, tourists would automatically have less destructive impact, and this is a political project, compared to ultimately ineffective efforts to cajole them into individually seeking out the few 'green' options within a corporate-dominated economy.

There can be an important role for tourism being directed in a way to support the local economy, for example via local food cafes, farmers markets, sensitively done homestays using local produce and appropriate technology, and so on. A scaled-down tourism, in collaboration with local organisations and sensitive governments, can also do much to boost both the esteem and prestige of the local and artisanal economy, as well as its economic viability. In Ladakh, the presence of environmentally sensitive tourists has helped raise awareness of and demand for zero waste/plastic-free local products, as well as spurred the opening of a number of local and traditional food cafes that source from a network of nearby farmers.

On the other hand, a reality check would quickly show that this segment of tourists and this proportion of the total tourist-directed food and craft spend that is local is negligible compared to the overall scale of tourism that has ballooned beyond any reasonable optimum, causing a cascade of problems from farmland conversion to hotels, to mountains of plastic waste, to dangerous water depletion and contamination. Hence, the primacy of the need for establishing hard limits on tourism, no matter how green, since beyond critical thresholds it will overrun and overwhelm a place and create dangerous levels of economic dependency. Clearly, relying on conscious consumption by sensitive tourists to sustain a local food system is precarious to say the least. Tourism should supplement and encourage the local economy, but not become its central pillar. This important collection helps point the way toward a tourism that serves local cultures and economies, rather than the other way round.

Local Roots

vesper tjukonai

~~~~~~~~~~~~~~~~~~~~*What do you see, when you look up and gaze about?*

In Ngarrindjeri Country, one rare morning, Wallaby Mother was munching, a sliver of sweet green goodness in her mouth. Then Wallaby Junior jumped headfirst to safety, legs awry stuck out of the pouch. Too many days, however, the sick smell of chemical sprays clings like suffocating plastic wrap.

Yet.

W.E.I.R.D.[1] tourism entices:

> 'Soak up the peaceful atmosphere' of 'unspoilt naturally beautiful landscape.'
> 'Trust us. You will quickly fall under its peaceful and magical spell.'

W.E.I.R.D. tourism sells any locality using much the same words. But they don't fool the locals who really know the place. A loco locomotive local may be madly on the move, travelling from place to place, but, rootwise, a local is not a person. It's a place.[2] Inanimate. Unfeeling.

And yet.

Perhaps sensing some skin-tingling presence, ancient Romans believed in a *genius loci,* spirit guardian of place. They'd make shrines where the *genius* was often seen as a good serpent demon, twisting, coiling around altar offerings.[3] O noble Serpent! Symbol of wise custodianship, regeneration and renewal.

Besides.

Snakes are good for the place. They're 'middle-order predators', keeping ecological balance.[4]

But W.E.I.R.D. religion has turned Guardian Serpent into an evil, cursed thing. Their holy book *Genesis* says:

> 'Thou art cursed above all cattle, and above every beast of the field; upon thy belly shalt thou go, and dust shalt thou eat all the days of thy life'.[5]

Do those who preach death to the 'pagan' *genius loci* not know? *Genesis* is a word cousin of *genius!*[6] And when locality lost its genius, it became much the same as *region*. And *district*.

By the way.

A *region* is not any old stretch of land. Rootwise, it's 'ruled' land.[7] And rootwise, a *district* is 'seized' land.[8] The medieval *districtus* was forcefully seized and fiercely ruled by the feudal lord. The sovereign. The empire-building British king with his pompous periwig, weasel furs, frills, flounces and terrifying addiction to pillage and plunder was, etymologically, a *super-ānus*.[9]

It was not in their 'sovereign domains', that Australien anthropologist, A.W. Howitt, carefully placed 'native tribes', but in 'localities'.[10] Why locality? It doesn't have any sovereign rights.

'No matter what Act or policy is placed upon us, the First Peoples of this Land, we are here, we won't go away', Wiradjuri writer, Kerry Reed-Gilbert, set the record straight. 'Our existence is Our ownership of this land. Our spirituality. Our connection. Our sovereignty'.[11]

Knowing the Indigenous genius that's in a location, Kombu-merri philosopher, Mary Graham, observed:

> 'We [First Peoples] achieve the fullest expression of our human identity in a location in land'.

This wise Kombu-merri thinker pointed out:

> 'The custodial ethic/Aboriginal Law thus cannot be idealogised: it is a locus of identity for human beings, not a focus of identity'.[12]

~What do you hear from the land just now, when you pause and pay heed?

One moonlight bathed night, a barn owl screeched close by my ear, startling the heck outa me. And then I laughed. Delighted. Thrilled by the nocturnal creature with whom I share this space. Too many times, I hear staccatos of gunfire that send furious flurries of feathers sky-high.

W.E.I.R.D hunting tourism is supposedly good for the local economy. Not in the home of the Black-winged Stilt! Or Red-necked Stint! These migratory waders may be protected, but, I suspect, they too want to rest and recuperate in calm and quiet, before it's their turn to tour the world again.

Turn and *tour* are word twins. Rootwise, both do the rounds, but a *tour* first mainly meant military men taking turns at killing.[13] Both are products of conquest.

'From the time of William the Conqueror', notes US historian Mary Hill Cole, control was kept by moving court in the Royal Tour.[14] Hosting a Royal Tour conferred high status and much social reward. It also incurred much cost. Even debt. More dependency on the goodwill of the ruler.

The first W.E.I.R.D. royal to tour Astraylia was Prince Alfred, Duke of Edinburgh, son of Queen Victoria. The duke's processional entry to Melbourne, in 1867, was 'as impressive as possible' with 'all the military force in the colony', wrote Australien historian Cindy McCreery.[15]

Even the tiny settlement of Narrung was on the royal itinerary. Local Mr Hacket recalled:

'The Duke of Edinburgh landed at Wommeran and a kangaroo hunt was organised for his amusement'.[16]

A kangaroo 'hunt'? Hardly. The animals were herded into a pit for royal slaughter.[17] After the 'hunt', the duke was 'entertained on the Malcolm's Australian estates'.

Malcolm was none other than Neill Malcolm, 13th Laird of Poltalloch, Scotland; a British MP and slave profiteer.[18] He lived in London, never setting foot in Narrung. Too many Ngarrindjeri disappeared on Malcolm's Narrung estates. Managers relocated, by force and under the cover of darkness, those who refused to work like slaves. They were shipped out to Malcolm's Jamaican estates, where they died like slaves.[19] Nevertheless, pioneer-revering locals named a landmark after the absent luxury-loving laird. Point Malcolm Lighthouse is now a tourist attraction.

Meanwhile.

A delegation of Ngarrindjeri elders, led by *Ru:puli* Peter Pulami, waited to petition the touring duke. They waited in vain for hours in the hot sun.[20] Australien anthropologists, Ronald and Catherine Berndt, recorded:

'As a result... he [Pulami] was declared temporarily insane and spent almost two and a half years in the Lunatic Asylum in Adelaide'.[21]

On Tour, the Duke never knew.

Could it be that W.E.I.R.D. tourists see themselves as royals? Or as loyal subjects, rewarding themselves for long labours? Could it be that, tripping *en masse* to their chosen getaway utopias, W.E.I.R.D. tourists create dystopian horrors for the locals?

Ironic.

Utopia was invented as an imaginary 'no-place' island.[22] The English thinker, Sir Thomas More, was a bit of a wag, I suspect. Punning on the Greek *οὐ*: not, and *εὐ*: good, he also coined *eutopia,* a 'region of ideal happiness or good order' *Utopia* and *eutopia* sound the same to the W.E.I.R.D. ear.[23]

Utopia Station is the W.E.I.R.D. name for the 'Urapuntja' homelands where Arrernte Anmatjere elder, Rosalie Kunoth-Monks, was born. She taught:

> 'Country owns or holds you, not you holding the country and becoming master of the land. The land was your mother, your father and everything else'.[24]

~~~~~~~~~~~~~~~~~~~~*When you walk the land, what do you feel?*
~~~~~~~~~~~~~~~~~~~~~~~~~~~~*How does the land hold you?*

Sadly, just walking down the road, I feel sick. It's hazardous for my health. Travel is a no-no. The best I can do is a daily sunwise saunter. It's an old English tradition. Going *sunn-ganges,* pronounced 'sern-yangas' maybe, was thought to bring the sun's blessings dancing round kith and kin.

Going *unsunn-ganges,* or against the sun, was something done only by the *scín-lǽce,* sheen-leech, or shine-doctor. This spirit-healer knew how to tour the realms of the dead, and come back alive.

Hallowing by sunshine, however, was free for every one. During the all-in spring revels of renewal, called *Gangdæges*, Gangdays, locals would walk around their homelands 'beating the bounds'.[25] It was a chance to make sure Mother Earth was awake and rising, ready for new crops. And along the way, young ones would get to hear the old stories of the land.

And then.

Holier-than-thou monks denounced Gangdays as heathen, and turned them into pious penitent prayerful processions. Three days of shouldering a heavy cross round parish boundaries was meant to uphold their saviour's ascension to the heavens.[26] It wasn't meant to be fun.

Is it really fun being a W.E.I.R.D. tourist in this broken, exploited world of inequality?

Narungga writer, Natalie Harkin, warns:

>   please
> prepare well   take time out   start with a trip   to the French
> and American zoned   Pacific-paradise   sail through French Polynesia
> there will be restricted places   colonial-bases   explore the Tuamotu
> Archipelago   visit Moruroa   this great chain of atolls   detonated   cracked

This Narungga truth-teller asks:

> What would it take   to listen to the Traditional Owners   to learn from the lessons
> of the land   to respect voices that refuse to be   bought   buried   sold?

What would it take to turn W.E.I.R.D. tourism into local get-togethers, sunwise walks, listening to those who've gone before, preparing for those who come after us?

Wallaby welcome. Credit: vesper tjukonai

Notes

(1) W.E.I.R.D. My version: Western Egocentric Indulgent Ruthlessly Dollarised. Apologies to Canadian behavioural scientists, Henrich, J., Heine, S. and Norenzayan, A. (2010) The weirdest people in the world?. *Behavioral and Brain Sciences* 33(2-3), 63.
(2) From the Latin *locālis*, belonging to a place, formed on *locus*. Cousin of *lieu* and *locomotive*. Ayto, J. (2001) *Dictionary of Word Origins* (p. 327). London: Bloomsbury.
(3) (1907) *New International Encyclopedia*, VIII. 532/1.
(4) Prakash, K.B.D. (2021) *Environmental Echoes, Wakeup Calls for the Environment Events,* Sankalp Publication, 82.
(5) *Genesis* 3, 14-15. King James Version.
(6) *Genius* and *Genesis* both grew from the Latin verb *gigner*, to beget. So did genital, genetic, gender, generate, indigenous and genocide. Ayto, J. (2001) *Dictionary of Word Origins* (p. 252). London: Bloomsbury.
(7) From the Latin verb *regere*. Source also of reign, regulate and regimen. Hoad, T.F. (ed.) (1986) *The Concise Oxford Dictionary of English Etymology* (p. 395). Oxford: Oxford University Press.
(8) From the Latin verb *distringĕre*. Source also of distress, strain and strict. Ayto, J. (2001) *Dictionary of Word Origins* (p. 177). London: Bloomsbury.
(9) From the Latin *super*: above, plus *-ānus*. Ayto, J. (2001) *Dictionary of Word Origins* (p. 491). London: Bloomsbury.
(10) Howitt, A.W. (1904) *The Native Tribes of South-East Australia*, 1996 facsimile edition (pp. 61, 67, 90, 127, 134, 142, 146, 249, 252, 257, 258, 262, 269, 271, 272, 273, 282, 283, 284, 776, 794. Phew!) Canberra: Aboriginal Studies Press.
(11) Gilbert-Reed, K., compiler (1997) *Message Stick: Contemporary Aboriginal Writing* (p. xi). Alice Springs: IAD Press.
(12) Graham, M. (2008) Some thoughts about the philosophical underpinnings of Aboriginal worldviews. *Australian Humanities Review* (Online). http://australianhumanitiesreview.org/2008/11/01/some-thoughts-about-the-philosophical-underpinnings-of-aboriginal-worldviews/ (accessed 21 March 2022).
(13) Via the Latin *tornus:* lathe; from the Greek τόρνος, *tórnos:* circular movement, lathe. Hoad, T.F. (ed.) (1986) *The Concise Oxford Dictionary of English Etymology* (p. 509). Oxford: Oxford University Press.
(14) 'Rulers visited scattered royal castles and fortifications to coerce obedience from subjects in a time when royal authority depended upon face-to-face meetings'. Cole, M.H. (1999) *The Portable Queen: Elizabeth I and the Politics of Ceremony* (p. 13). Boston: University of Massachusetts Press.
(15) McCreery, C. (2008) A British prince and a transnational life: Alfred, Duke of Edinburgh's visit to Australia, 1867–68. In D. Deacon, P. Russell and A. Woollacott (eds) *Transnational Ties: Australian Lives in the World* (pp. 57–74). Canberra: Australian National University Press.
(16) Padman, E.L., compiler (1987) *The Story of Narrung, The Place of Large She-Oaks* (p. 25). E.L. Padman, self-published.
(17) Malcolm, D. (1992) *Neill Malcolm XIII Laird of Poltalloch*, booklet.
(18) 'Neill Malcolm 13th of Poltalloch', *Legacies of British Slavery database*. http://wwwdepts-live.ucl.ac.uk/lbs/person/view/2146637310 (accessed 19 March 2022).
(19) Ngarrindjeri Elders have never forgotten their old people's stories of the disappearance of folks made to work on Poltalloch Station, Narrung. Personal communications circa 2003.
(20) Ngarrindjeri Elders have never forgotten their old people's stories of the disappearance of folks made to work on Poltalloch Station, Narrung. Personal communications circa 1999.

(21) Berndt, R.M. and Berndt, C.H. (1993) *A World That Was: The Yaraldi of the Murray River and the Lakes, South Australia* (p. 296). Carlton, VIC: Melbourne University Press.
(22) Formed from the Greek οὐ, meaning not, plus τόπος, *topos*, a place. Hoad, T.F. (ed.) (1986) *The Concise Oxford Dictionary of English Etymology* (p. 518). Oxford: Oxford University Press.
(23) Formed from εὐ: good, plus τόπος: place. 'I was called Utopia in ancient times, because of its rarity … I am rightly called by the name of Eutopia' (1516) *De optimo reipublicae statu, deque nova insula Utopia*, quoted in 'eutopia, n.'. OED Online. March 2022. Oxford: Oxford University Press. See https://www.oed.com/view/Entry/65148 (accessed 21 March 2022).
(24) Kunoth-Monks, R. (2011) Foreword. In *The Land Holds Us: Aboriginal Peoples' Right to Homelands in the Northern Territory* (p. 5). Amnesty International. See https://www.amnesty.org/en/documents/asa12/002/2011/en/ (accessed 21 March 2022).
(25) 'In the north of England this period of bounds-beating was known as the Gang Days or Ganging Days'. Cooper, D. and Sullivan, P. (1994) *Maypoles, Martyrs & Mayhem: 366 Days of British Customs, Myths and Eccentricities* (p. 124). London: Bloomsbury.
(26) 'From the circumstance of the cross being carried in these processions, it is frequently, in old accounts, called Cross-week, and in the northern parts of England, Gang-week'. Gang Week, n. (1834) *British Magazine & Monthly Register, Dec.*, 615. *Oxford English Dictionary (Online)*. Oxford: Oxford University Press. See https://www.oed.com/view/Entry/76626?redirectedFrom=Gang+Week (accessed 21 March 2022).
(27) Harkin, N. (2015) Zero tolerance. In *Dirty Words* (p. 41). Carlton South, VIC: Cordite Books.
(28) Harkin, N. (2015) Zero tolerance. In *Dirty Words* (p. 43). Carlton South, VIC: Cordite Books.

Introduction
Embracing the Local Turn in Tourism to Empower Communities

Freya Higgins-Desbiolles and Bobbie Chew Bigby

Introduction

> '*Local* – From Middle English *local*, from Late Latin *locālis* ('*belonging to a place*'), possibly also via Old French *local*; ultimately from Latin *locus* ('*a place*')'. (Wiktionary, n.d.)

Arguably up until the COVID-19 global pandemic swept the world, the concept of 'localising tourism' would have seemed non-sensical. Afterall, tourism is defined by leaving one's home environment – one's locality – to travel a certain distance to somewhere else in order to enjoy touring and holidays. As a result, tourism has been defined as a system featuring the tourism industry's supply of tourism products and experiences to meet the demand of tourism consumers, the tourists. However, with the COVID-19 pandemic, when public health measures required borders to shut and travel curtailed, localising tourism suddenly became a necessary measure as people were compelled to enjoy leisure more nearby to their home communities. This has been entitled 'proximity tourism' and has been a noticeable outcome of the crisis (see Chapter 11). It was not only the reduction in travel circuits because of border closures that caused this but also a growing sense of solidarity as people with savings considered how their spending choices might support desperate hospitality and tourism businesses located nearby. Additionally, destination marketing organisations (DMOs) shifted from a predominant focus on marketing to international visitors to marketing to domestic tourism markets.

The pandemic not only changed tourists' travel patterns and tourism industry operations; it also changed the lives of local communities that had grown dependent on international and domestic tourists. For instance, Movono and Scheyvens (2021) reported that the pandemic caused Fijian people holding tourism jobs to return to their home villages and once again take up subsistence practices and community social networks. Using the lens of 'sustainable livelihoods', their work cast a spotlight on the vulnerabilities that tourism dependency can bring to community networks, lifeways and governance. COVID-19 also reinforced calls for a tourism reset in Key West, Florida where some locals formed the Key West Committee for Safer Cleaner Ships demanding the reduction of the footprint of cruise ship tourism arguing the economics did not stack up and the ecological costs were too much (Glenza, 2020). This suggests a need to rethink how tourism detracts from or enhances the lives of people, the uniqueness of place, the sustaining of ecologies and the interrelationships between these facets.

In this introduction, we engage with what has been called a 'local turn in tourism' (Higgins-Desbiolles & Bigby, 2022) and consider the multiple possibilities for localising tourism: defining tourism by the local community, local community empowerment, more localised geographies of travel and tourism, localising decision-making to the lowest level (subsidiarity) and the local interrelationships between people, place, ecology and all living things. In this introduction to the book, we plan to explain what localising tourism could mean, position it in relationship to phenomena such as community-based tourism and neolocalism, provide some examples that illustrate its possibilities and then overview the contents of the edited volume. The central thesis of this introduction is that local communities are not well served by their labelling as 'tourism destinations' and this localising agenda is essential to ensure long-term sustainability, equity and justice. The ethos for this was well established by Higgins-Desbiolles *et al.*: 'The place where tourism occurs is not a tourism destination; it is the local community's home, their standing place, a place of uncompromisable value' (2022a: 14).

With this edited volume, we hope to contribute to the establishment of a local turn in tourism studies with an agenda of redefining tourism by the local community, revamping our discourse and practices, redesigning our governance and reimagining the purposes and potentials of tourism. We are attentive to not fall into the 'local trap' of romanticising local communities (Born & Purcell, 2006) and failing to make visible the possibilities for conflicts, struggles for power and difficulties for inclusion (see Kenis & Mathijs, 2014). What we want to engage here are the multiple, diverse possibilities for radical reconfiguring in the tourism sphere, to place local communities in the centre and to place them in connection through 'local relatedness'.

Going Local

> 'Liveability over visitability' – Group Nao's DMOcracy campaign (2021, n.p.)

Rather than being simply about fun and holidays, tourism is 'a complex set of social discourses and practices' (Rojek & Urry, 1997: 1). One powerful framing of tourism is the concerted effort exerted to have it understood as a lucrative industry, that works to obscure its value as a powerful social force (Higgins-Desbiolles, 2006). It is under this ideological approach that we have seen an over-emphasis on the metrics of tourism – jobs, foreign exchange and growth – and a shift in power in favour of the industry and at the expense of communities and the environment. Under such a discourse, local communities are understood to be just one stakeholder in tourism, their home communities become converted into 'destinations' as they themselves become labelled as 'host communities' and they are admonished to allow the commodification of attractive aspects of their cultures and their ecologies in order to tap the benefits of tourism as a development strategy. For example, the World Tourism Organization (UNWTO), the most powerful global lobby group for the tourism industry, argues tourism is good for local communities:

> Tourism can be a powerful tool for community development and reducing inequalities if it engages local populations and all key stakeholders in its development. Tourism can contribute to urban renewal and rural development and reduce regional imbalances by giving communities the opportunity to prosper in their place of origin. Tourism is also an effective means for developing countries to take part in the global economy. (UNWTO, n.d.: n.p.)

The recent experience of overtourism before the pandemic brought greater attention to the concerns, rights and benefits of local communities. All over the world, from European cities such as Dubrovnik, Venice and Barcelona, to remote natural areas such as Indonesia's Komodo National Park and to small holiday towns such as Byron Bay, Australia (see Chapter 5), reports emerged outlining how local communities, and in some cases governing authorities, turned to various ad hoc responses to address overtourism and negative tourism impacts (see Figure 0.1). Analyses offered understanding of the meaning and import of these events both for tourism and for communities. In their study of Seville, Diaz-Parra and Jover explained the notion of the local community's 'right to the city':

> The Lefebvrian right to the city must be seen as a right of its residents to a non-alienated urban life and to collective appropriation by the locals that inevitably entails community and neighbourhood development, yielding authentic experiences for its inhabitants. (2021: 164)

Because of touristification, land speculation, gentrification and other sociospatial processes of tourism, communities around the world have become concerned and even active against tourism (see Chapter 13). Some have labelled this as 'anti-tourism' and 'tourismphobia' (Coldwell, 2017), thereby shortcutting needed analysis of the structural injustices that

Figure 0.1 (Left) Tourists read the sign asking visitors not to climb Uluru (Central Australia) and explaining the reasons the Anangu Traditional Owners ask this. (Right) Tourists climb Uluru, circa 2011. Credit: Co-author Freya Higgins-Desbiolles.

underpin these problems. In reference to Barcelona, Hughes stated 'it's not turismofobia, it's self-defence against barriocidio (death of the neighbourhood)' (Candidatura d'Unitat Popular, 2017 cited in Hughes, 2018: 475). Araya López (2021b) noted that instead of narrating protests by groups such as Arran through discourses of '*turismofobia*', these might be better understood as '*barriofilia*', or love of one's neighbourhood. Such power imbalances have been evident, however, well before the term 'overtourism' was coined in 2017 (see Figure 0.1). Tourism has long presented itself as a 'devil's bargain' for communities that have hoped they can tap its economic opportunities without diminishing their quality of life and enjoyment of place (Rothman, 1998).

It is in such contexts that localising tourism offers opportunities for addressing the power imbalances in tourism and rethinking its purpose and processes. There is an important precursor to localising tourism with 'buy local' initiatives. Michael Pollan helped inspire the 'buy local' movement with his call to 'think global, [but] buy local' (cited in Weinraub, 2006: n.p.). This was in reaction to a global free trade system that, while bringing the enjoyment of greater consumption to more people, did so unequally and at significant environmental, social and economic costs. More recently, neolocalism is an emerging phenomenon in tourism (Schnell, 2013). Neolocalism is defined as 'the conscious attempt of individuals and groups to establish, rebuild, and cultivate local ties, local identities, and increasingly, local economies' (Schnell, 2013: 56). In their study of an urban park and entertainment venue, Sipe identified neolocalism featuring these characteristics: 'being non-corporate, environmentally responsible, empowered and self-sufficient, authentic and community building' (2018: 36). This suggests evidence of a backlash against the commodification, homogenisation, alienation and ecocidal dimensions of modernising globalisation, characterised by a response to return to more local scale of living.

There is a truism that 'all culture is local' and we would suggest that similarly, all tourism is local; that is, tourism occurs in local places that are well loved by local people. However, since the 'tourism as industry' discourse took over under the era of neoliberalism, myopic focus has been on the tourists and the tourism industry. Local communities and local places have been ill-served by this system and that is in part the reason that it is opportune now to embark on a local turn in tourism. Peter Murphy foreshadowed this in his early advocacy for a 'community approach' to tourism claiming: 'It is the citizen who must live with the cumulative outcome of such developments and needs to have greater input into how the community is packaged and sold as a tourist product on the world market' (1985: 16). In this work, we want to move beyond this tendency to tourism-centricity evident in Murphy's and others' approaches and instead centre the local community, not as 'host' with 'greater input' at a 'tourism destination', but rather as custodians of a well-loved place with rights and obligations and who may or may not welcome tourists.

Community-based tourism (CBT) has been well examined in the tourism literature. CBT has been defined as: '[…] generally small scale and involves interactions between visitor and host community, particularly suited to rural and regional areas. CBT is commonly understood to be managed and owned by the community, for the community' (APEC, 2010). In her work on the role of tourism in community empowerment, Scheyvens defined CBT ventures as: 'those in which members of local communities have a high degree of control over the activities taking place, and a significant proportion of the economic benefits accrue to them' (2002: 10). CBT may also refer to the involvement of the 'host community' in the planning of tourism and some analysts argue it may result in more sustainable tourism (see Dangl & Jamal, 2016).

However, CBT was largely envisioned as a niche market and so its impacts have been narrowly felt and limited. Blackstock was critical of CBT stating: 'CBT can be perceived as an example of community development "imposter" driven by economic imperatives and a neo-liberal agenda, rather than values of empowerment and social justice' (2005: 40). In contrast, Blackstock argued that:

> Community development can be defined as 'building active and sustainable communities based on social justice and mutual respect' (Gilchrist, 2003: 22). Thus community development explicitly seeks to dismantle structural barriers to participation and develop emancipatory collective responses to local issues. (2005: 40)

This discussion of localising tourism is much more than advocating for community in tourism and as such it is attentive to these concerns with

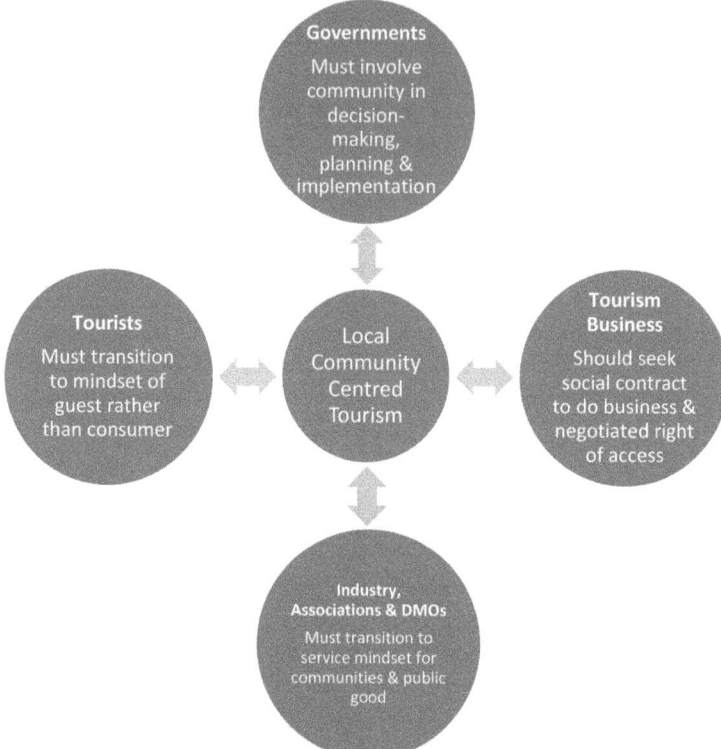

Figure 0.2 Community-centred tourism framework. Figure adapted with permission from Higgins-Desbiolles *et al.* (2019: 1937). Taylor and Francis Inc. http://tandfonline.com.

empowerment and social justice. The initial step is to shift to a new discourse on tourism starting with redefining tourism by placing local communities at the centre of the phenomenon (see Figure 0.2):

> Tourism for sustainability and degrowth must focus on the needs and interests of the local community; what tourism industry interests have usurped for themselves under the label of the 'host community'. A redefined tourism could be described as: the process of local communities inviting, receiving and hosting visitors in their local community, for limited time durations, with the intention of receiving benefits from such actions. Such forms of tourism may be facilitated by businesses operating to commercial imperatives or may be facilitated by non-profit organisations. But in this restructure of tourism, tourism operators would be allowed access to the local community's assets only under their authorisation and stewardship. (Higgins-Desbiolles *et al.*, 2019: 1936)

Such an approach would change the power relationships currently operative in tourism. This reorientation could work to transform all relationships in tourism including with tourists, with business, with governments and with the environment. This may be an answer to the tensions within local communities that has resulted from problems such as overtourism. It might also be an appropriate response to recent issues of undertourism that has resulted from the impacts of COVID-19 and exposed the vulnerabilities that accompany over-dependency on tourism.

It will be important to be fully inclusive of all aspects of the community if equity, justice and empowerment are to be achieved (see Chapter 4). The thinking of Boyle and Wyler (2021) on behalf of the Local Trust in the UK is essential reading for such efforts. Here they discuss 'left behind places' and an 'us versus them' mindset that features in communities where neoliberal hollowing out of community, connection and social solidarity sees gated communities of wealth and privilege next to ghettos and tower blocks, embodied in the tragedy of Grenfell Tower and the targeting of Windrush migrants by the UK's Home office. Boyle and Wyler's analysis of the 'top-down approach' to community development in the UK underscores the need for a decolonising and Freirian approach to localising tourism (see Higgins-Desbiolles *et al.*, 2022b). Boyle and Wyler explained:

> ...the persistent views that those in power seem to have held about local communities: that they are not capable; that they are not to be trusted with significant resources; and even that they do not know what is good for them. (2021: n.p.)

In addition to this issue of exclusion and the ongoingness of colonial mindsets, there is a need for a 'more than human' approach to communities when we think through the possibilities of localising tourism. Higgins-Desbiolles and Bigby provided their particular definition of community in their discussion of the 'local turn in tourism':

> In our articulation of the local community as the linchpin of the local turn, we mean more than just a certain group of people associated with a place. Instead, we are more broadly inclusive of the local community, the local ecology (living air, land and waterscapes and more-than-human beings) and all generations pertaining to that place (including future ones). (2022: 2)

Such a shift in understanding towards centring the community and viewing it from this more holistic perspective is already discernible. For instance, in Aotearoa/New Zealand, the Tourism Futures Taskforce Interim Report recognised:

> Communities sit at the centre of the visitor industry, and visitors sit at the heart of our communities. To thrive sustainably, the future visitor economy needs to be led by local communities. Our recommendations address the engagement, planning and funding mechanisms required to ensure

that communities are equipped with what they need to allow the visitor economy to strengthen their local identity, connect them better to their place and drive long-term well-being. This is where a regenerative tourism industry begins. By embracing the concept of Manaakitanga – care and consideration for those around us – we become a more connected, collaborative and richer society of hosts and visitors. (2020: 42)

We also see this shift in recent activities of leading DMOs. Perhaps most advanced is the 'DMOcracy' pilot project announced in Europe for implementation in 2022, involving some 22 destinations in efforts to build the democratic mandate for tourism and involve citizens in destination governance (see Chapter 8 and associated case study). In addition to this 'DMOcracy' project, there is the example of Agora Tourisme Bordeaux (Bordeaux Tourism, 2021). Julie Benisty Oviedo explained this strategy as a focus on 'empowering people', 'open talk on real impacts' and 'transparency' (Oviedo, 2021). Using design thinking methods and external facilitators to ensure integrity in the process, the aim is to have conversations and debates about the role of tourism so 'people can make up their own minds on tourism'. She explained that Bordeaux Tourism plans an annual forum for local peoples with a commitment to maintain this DMO's local engagement and accountability.

The principle of subsidiarity will be essential in better localising decision-making in tourism. The call for subsidiarity in tourism decision-making can be witnessed in the call for tourism decision-making for Waiheke Island, New Zealand to rest with the Local Board rather than the super-authority of Auckland Council (Project Forever Waiheke, 2021). Briassoulis explained this principle with regards to sustainably managing common pool resources:

Whenever possible, individuals and communities should be empowered to make and implement decisions locally. This is the subsidiarity principle calling for actions at the lowest possible level and for external intervention only when certain functions cannot be performed locally… Locals become 'stewards' of the commons and external appropriators accountable to them for the use of resources. Participatory and consensual approaches to policy design as well as development of horizontal, non-hierarchical relationships and networks among the appropriators and users of the commons should be encouraged. These might promote social equity and self-reliance and foster cooperation based on mutually agreed objectives and rules, open processes of entry and leave, and competition…. (2002: 1080)

Such a significant transformation will not be without its challenges and opponents. Hall offered this challenge:

… the local should not be romanticized, as so often seems to be the case in discussions of tourism planning. As Millar and Aiken (1995: 629) observed: 'Communities are not the embodiment of innocence; on the contrary, they are complex and self-serving …'. (2003: 99)

Relevant to illustrating this point is the research by Canan and Hennessy (1989) reporting on rural community dynamics in relation to tourism development on the island of Moloka'i, Hawai'i. They identified three divisions in this community: those supporting a tourism 'growth machine', those opposed to tourism developments largely for their negative cultural impacts (largely comprising Indigenous Hawai'ians) and those designated as 'in-between' who are local supporters of diverse development options (1989: 238). Importantly, one point marking these divisions was the perspective held on land, with the proponents of tourism development seeing land as a property and those opposed viewing land as 'the source of cultural identity, social life and spirituality' (1989: 232). Canan and Hennessy explained:

> One community, represented by the decisionmakers, orients itself by a class or socioeconomic basis of community identification, commonly referred to as gesellschaft (Tonnies, 1957). Another, represented by those who would limit development, places itself in the midst of a traditional, gemeinschaft or status-based community. The third, those we called the 'local supporters of diverse development,' places itself between these two communities in terms of the choices of the growth machine and might appropriately be called Zwischengruppe ('in between group'). (1989: 238)

Canan and Hennessy (1989) noted the paradox that the very thing that is attractive in the marketing of Moloka'i – undeveloped, laid-back and authentic – would be undermined by the growth machine proponents' agenda. But what they do not comment on is how embedded the growth machine proponents are in the local community or are they elite, mobile members of what Leslie Sklair would call the 'transnational capitalist class' (see Sklair, 2020). The members of the group that would limit such tourism development are comparatively more often native, poorer and possibly less mobile in this case of Molokai'i. This issue of mobility may be the key as it allows for the ability to 'bugger it up, and pass it down' (Wheeller, 1993: 125).

With this case in mind and in response to Hall, we would argue that while local communities may be diverse and fractious, we should understand how they have also been made that way by neoliberal market ideology with its totalising discourses of competition, selfish individualism and undermining of bonds of collective solidarity. Nonetheless, we also witness growing social solidarity in communities when tourism and hospitality providers work to build community and nurture the collective commons (e.g. Higgins-Desbiolles, 2012; Higgins-Desbiolles *et al.*, 2014; Williams *et al.*, 2012). This was evident during the recent pandemic when communities formed mutual aid groups often centred around a community café and other sources of social networks (Price, 2020). This suggests the need to continually promote dialogue between these segments of communities and to bridge their diverse interests in engaging with tourism.

Butcher offered a critique from a different angle, stating: 'development thinking, and the potential of tourism in development is ill served by the deference to local participation' (2012: 108). He argued this from his viewpoint that proponents of ecotourism and CBT support projects that limit the development and modernisation potential of tourism for local communities and thus are only empowered in '…participating in modifying the terms of [their] poverty' (Butcher, 2012: 108). This argument against community participation in tourism decision-making in the interests of unfettering tourism developments in the pursuit of greater income from tourism seems dubious and anti-democratic. If framing tourism as a power struggle over tourism's extractive capacities, this argument clearly sides with powerful tourism industry interests over those of impacted communities.

One final critique to mention is that localising tourism may see greater pressures on local communities to engage with tourism decision-making, tourism activities and tourism management. For instance, in a report on 'live like a local' tourism campaigns, Spinks noted the downside of these being a 'relatively new lack of separation between touristic and local life [with] knock-on effects…' (2018, n.p.). This is a legitimate concern, and it is for such reasons that communities must be empowered to set the pace, the scale and the conduct of tourism in their home communities. If tourism is designed to be community-enhancing and geared to local community well-being, such concerns should be greatly diminished. However, it is important to not fall into the 'local trap' identified by Born and Purcell in their discussions of food systems, referring to a tendency 'to assume something inherent about the local scale. The local is assumed to be desirable; it is preferred a priori to larger scales' (2006: 195).

Localising Tourism and a Degrowth Approach

There will be those that will respond to this localising agenda with the question: 'What about globalisation and all of its innumerable benefits'? We need to avoid such dualistic thinking that assumes that as we think through the possibilities of fully grounding tourism in the local community, we are by necessity rejecting the value of globalisation. The critical question is what form of globalisation do we refer to when we tout the benefits of globalisation? As Chomsky explained: 'The term "globalisation" has been appropriated by the powerful to refer to a specific form of international economic integration, one based on investor rights, with the interests of people incidental' (2002: n.p.). The concern here is with globalising capitalism and its practices of usurping the people, places and resources of others for elite capital accumulation. Rothman (1998: 339) noted how the growth in powerful multinational tourism organisations, enabled through the policies of capitalistic globalisation, resulted in the 'corporatization of place' on a massive scale.

Up until COVID-19, globalisation, global capitalism and global tourism seemed to be the unstoppable forces in shaping our world. As Appadurai (1996) explained transnational flows of people, media and commodities under global capitalism have resulted in culture and place become increasingly deterritorialised. It is such flows that have in fact enabled the expansion of the global tourism industry and together these forces have seen almost every local community and local ecology incorporated into the global system in particular ways that must be unpacked. As Fletcher *et al.* contended: 'Overtourism... must therefore provoke reconsideration of the political economy of tourism as a whole and not merely debate concerning the appropriate number of tourists ('carrying capacity') within a given location' (2019: 1747). Bianchi's work on the political economy of tourism is important in this regard, addressing forces of capitalist accumulation, state led development, labour exploitation and new practices of the gig economy:

> A 21st century political economy of tourism must address the systemic forces of accumulation, constellations of class power and models of innovation that will continue to radically restructure complex, multiscalar modes of industrial organisation and profit extraction in contemporary tourism. (2018: 100)

As Hickel unmasked in his discussion of growth-focused capitalism:

> In reality, what is going on is a process of elite accumulation, the commodification of commons, and the appropriation of human labour and natural resources – a process that is quite often colonial in character. This process, which is generally destructive to human communities and to ecology, is glossed as growth. (2021b: 1107)

From this perspective, there is a need to defend places from the assaults of external entities and economies which seek to appropriate and profit from tourism's commodification of the cultures, ecologies and resources of others. As Hickel stated: 'If capitalism depends on commodification, enclosure and accumulation, then decommodification, de-enclosure and deaccumulation spells its end' (2021a: n.p.). Degrowth is also essential to these processes. Latouche defined degrowth as:

> ...designed to make it perfectly clear that we must abandon exponential growth, as that goal is promoted by nothing other than a quest for profits on the part of the owners of capital and has disastrous implications for the environment, and therefore for humanity. It is not just that society is reduced to nothing more than an instrument or a means to be used by the productive mechanism; human beings themselves are becoming the waste products of a system that would like to make them useless and do without them. (2009: 8)

Defence from these forces can be mounted through cultivating long-term habitation and active commitment to place to counteract the rootlessness that globalisation has fostered for some of Earth's inhabitants. It necessitates a return to the 'real economy' that Norberg-Hodge refers to in her foreword. But it will take more than these things to counter the power arrays supporting such elite accumulation. This necessitates the formation of social movements for defence.

It is imperative that we also confront the growing support for exclusionary nationalism evident in Trumpism in the USA, Hindu nationalism in Modi's India and Orbán's policies of 'illiberal democracy' in Hungary, for instance. La Via Campesina observed that: 'Right wing populism and fascism consorts with global capital, increase[ing] their grip on governments on all continents' (2019: 1). Resistance to these trends might also be accomplished through embracing democratic localism and cooperation through genuine citizen engagement in the local economy, including the design, delivery and reception of public services that support all in the community. The roles of cooperatives, mutuals and social enterprises in these efforts is vital as they build connection, resilience and engagement (Higgins-Desbiolles, 2020: 619; Mtapuri & Giampiccoli, 2020).

As Yves Cochet has suggested, we should replace the World Trade Organization with a 'World Localization Organization' and its slogan should be 'Global protection for the local' (2005: 224). Works such as Serge Latouche's *Farewell to Growth* (2009) and Giorgos Kallis *et al*'.s 'Research on degrowth' (2018) are essential readings for those interested in the logics of degrowth, localisation and localising tourism. Post-development work by Arturo Escobar is also invaluable. He advocated for a 'defensive localization' (2001: 149). This emphasises autonomy from universalising knowledges dominated by the hegemonic imaginary of the North, which pushes forms of globalisation, development and modernisation beneficial to its own interests. Escobar asked: 'to what extent can we reinvent both thought and the world according to the logic of a multiplicity of place-based cultures?' (2001: 142).

Social Movements for Defence of the Local

> '...if everyone focused their love, care, and commitment to protecting and regenerating their local places, while respecting the local places of others, then a side effect would be the resolution of the climate crisis'.
> Charles Einstein (2018, n.p.)

In such contexts we find difficulties with politics and discover democracies are flawed by the growing sway of money, power and influence. Politics today emphasises competition and contest and is crippled by short-termism and self-interests. Resistance to the negative impacts of capitalist globalisation have taken the form of social movements including global responses found in the co-called 'anti-globalisation' movement and the World Social Forum, as well as in more local and regional activisms against *'les grands projets'*, such as the Zapatistas in Chiapas, Mexico and the Associations pour le Maintien d'une Agriculture Paysanne in France (Latouche, 2009: 6). Wiedmann *et al.* stated: 'It is important to recognise the pivotal role of social movements in this process [transformation for degrowth], which can bring forward social tipping points through complex, unpredictable and reinforcing feedbacks and create windows of opportunity from crises' (2020: 7).

It is possible to identify social movements operative in this space of concern with localisation and localising tourism. Arguably, most significant is the global advocate for the localisation movement called Local Futures (Local Futures, n.d.). It is in part based on the longitudinal learnings that Founder and Director Helena Norberg-Hodge made during her time living and working in Ladakh, India (see Foreword; Norberg-Hodge, 1992). This movement advocates:

> ... supporting a systemic shift towards localisation – working together to provide a framework that will allow the secure re-establishment of national, regional and local polities and economies that meet real human needs without compromising the natural world on which all life depends. (Norberg-Hodge & Read, 2016: 5)

La Via Campesina is also important in this space, who in 2018 was instrumental in getting the United Nations (UN) General Assembly to adopt the UN Declaration on the Rights of Peasants and Other People Working in Rural Areas. La Via Campesina described the import of this achievement to their movement:

> It strengthened our movement and organized our people. It is a symbol of pushback from us the people who are rooted to the land, against a global governance model that is heavily tilted and favourable to big business. It is a push back against the globalization of capital, from our people who are indeed determined to globalize our struggles and globalize our hope. (2019: 2)

Böhm *et al.* explained: 'La Via Campesina puts an alternative vision of autonomous development forward, highlighting local, community knowledges, self-determination and ecological sustainability based on long-term views, rather than short-term profits' (2010: 23). This is what Norberg-Hodge entitled 'positive counter-development' in the Foreword.

There is also the Transition Town movement also known as Transition. It described itself: 'Transition is a movement that has been growing since 2005. Community-led Transition groups are working for a low-carbon, socially just future with resilient communities, more active participation in society, and caring culture focused on supporting each other' (2021b). As Kenis and Mathijs (2014: 173) noted, it is not a small movement with groups and localities in multiple nations including the UK (where it began), Belgium, Japan, USA, Aotearoa/New Zealand and Australia among others.

The principles of the Transition movement include:

- respect for resource limits and creating resilience;
- promoting inclusivity and social justice;
- adopting subsidiarity (self-organisation and decision making at the appropriate level) ;
- paying attention to balance;
- being part of an experimental, learning network;
- freely sharing ideas and power;
- collaborating and looking for synergies;
- fostering positive visioning and creativity. (Transition, 2021a)

Quilley described Transition as a movement that is distinct from anti-globalisation:

> Transition politics is not anti-capitalist nor anti-globalisation precisely because the expectation is that an energy short fall will lead to a rapid relocalisation and simplification of economic life within a few decades. Transition speaks the language of localism and sufficiency and taps a deep frustration with consumer society. The expectation is that in the post-carbon, post-capitalist order, citizens will live more familial, authentic and creative 'hand made' lives, recovering a range of artisanal 'Transition skills'… and more rewarding gemeinschaftlich forms of community. (2013: 264–265)

This movement has been critiqued as mildly reformist (Trainer, 2009) and 'depoliticizing the local' (Kenis & Mathijs, 2014). Kenis and Mathijs (2014: 179) addressed the issue of a tendency to 'defensive localism' which they charge is in danger of being parochial, conservative, particular and even xenophobic, and when focused on local history and culture, even akin to chauvinistic patriotism. These authors make a valuable point that localisation efforts must not abandon the political in an effort to promote 'positive, open, constructive and collaborative' engagements for unity as they argue Transition does. Kenis and Mathjis explained:

> the crucial question concerning localisation initiatives is: 'are they significantly oppositional or primarily alternative?' In other words, do they recognise the existence of conflict, power and division or are they rather focussing on building small havens without [ant]agonising existing society? (2014: 182)

Political engagement must feature in social movements favouring localisation in tourism. Discussing the contestation for St Mark's Square in Venice, Araya López illustrates this point effectively:

> On the one hand, the square has been constructed as 'apolitical' ('no protest' zone) by local authorities and tourism stakeholders interested in using and maintaining this space as a centre for leisure and consumption. On the other hand, various collectives and citizens/inhabitants critical of

mass tourism aim at re-appropriating the square for political and social uses, while challenging powerful national and global forces that have appropriated this space. (2021a: 140)

In addition to these globally extensive examples, there are more local and regional social movements offering insights. One example is the Sydney Alliance in Australia, which brings together communities to work towards understanding, connections and joint action for the common good. In this case, the Alliance brings together faith organisations, trade unions and civil society groups in order to develop a more 'fair, just and sustainable city' (Sydney Alliance, n.d.).

There is also the slow movement which was inspired by Carlo Petrini's slow food movement in the 1980s. It is focused on slowing down the pace of life, enjoying local food and living and reducing global consumption patterns. The conceptualisation of slow tourism emerged from the broad sociocultural context of this slow movement (Fullagar *et al.*, 2012). Slow tourism '…encapsulates a range of spatio-temporal practices, immersive modes of travel and ethical relations that are premised on the desire to connect in particular ways and to disconnect in others' (Fullagar *et al.*, 2012: 3). It may be seen as a precursor to what we describe here as localising tourism.

More recently, the cooperative Fairbnb.coop promotes 'community powered tourism'. The movement began in 2016 from Venice, Amsterdam and Bologna 'seeking to create a just alternative to existing home-sharing platforms' (Fairbnb, n.d). Using principles of the circular economy, fair trade and the United Nations Sustainable Development Goals, this workers' cooperative works with hosts, guests and communities to create more regenerative tourism arrangements by ensuring sharing platform bookings contribute to community projects and do not cause negative impacts like other booking platforms such as Airbnb are alleged to cause. Specifically, Fairbnb claimed: 'We screen hosts according to destination specific rules. In specific areas, we promote the one host – one house rule: lawful hosts, preferably residents, with only one second home on the touristic market in their city' (Fairbnb, n.d.).

Social movements may be even more narrow in focus than these. There is a rich heritage of local communities taking ownership and control of declining key services such as hotels and shops in challenged urban and rural areas. One researched example is the Hotel Bauen in Buenos Aires, Argentina which became a workers-run cooperative after a workers' takeover in 2003 (Higgins-Desbiolles, 2012); it has unfortunately closed in 2020 due to COVID-19 impacts. In fact, in the wake of the pandemic it may become more imperative for community buy-outs and takeovers of such assets to maintain community thriving and amenity. These could be incentivised by urban and regional development bodies through collaboration with the co-operative and employee-owned business sector to provide communities with specialist advice, grants, loans, social investment and community funding platforms for viable businesses. These are business models that support livelihoods rather than growth and will become more important for building sustainable futures. The Albergo Diffusi model of 'dispersed hotels' offers another approach to localising tourism and hospitality. Instead of the standard hotel model, rooms are distributed in existing converted buildings in historic centres. As Italian founder Giancarlo dall'Ara explained:

> We want to create a community among locals and visitors. The hotel is an integral part of a virtuous network that promotes an economic and social revival outside of Italy's cosmopolitan centers. It is a driving force of sustainable development for a small town, for a region and for our country. (cited in Hengel, 2021)

These social movements, and alliances between them, will be essential in localisation and localising tourism (see Chapter 13). As Hickel advocated, post-growth, post-capitalist economic models will require alliances supporting the setting up of social systems and taking back the commons from the enclosures conducted for private elite accumulation:

> This is what the unions should demand – a social guarantee – and the environmentalist movement must unite with them to fight for it, because delinking livelihoods and well-being from capital would create political space for a just post-growth transition. And both movements must commit to supporting Southern struggles for self-determination, to permanently dismantle the imperial arrangement.... . (2021a: n.p.)

Latouche presented the key idea of 'related localisms':

> The key is the related localisms below of connecting together communities in resistance. From this point of view, the local is not a closed microcosm, but a knot in a network of virtuous and interdependent transversal relations, with a view to experimenter practices that can strengthen democracy (including participatory budgets) and make it possible to resist the dominance of neo-liberalism. (2009: 47)

This is an essential point: that localising tourism is not intended to be the narrow-minded, survivalist shutting off of communities as isolated microcosms but is rather communities in relatedness showing a way forward where tourism is redesigned to build quality of life for communities, ecologies and all the generations.

Key Terminologies for Localising Tourism

From this discussion, we can start to identify discourses and terminologies that are relevant to localising tourism and the local turn in tourism.

Local communities: A holistic inclusivity

Higgins-Desbiolles and Bigby described local communities broadly:

> In our articulation of the local community as the linchpin of the local turn, we mean more than just a certain group of people associated with a place. Instead, we are more broadly inclusive of the local community, the local ecology (living air, land and waterscapes and more-than-human beings) and all generations pertaining to that place (including future ones). (2022: 2)

Commoning

Petrescu *et al.* explained commoning:

> Commoning is the process by which commons, that is, cultural and natural resources that are held, governed, and produced collectively, are made... Commoning takes place when human and non-human agents come together to share access to, to take care of and responsibility for, and ultimately benefit from, a material or immaterial resource that supports livelihood and good living (as commons). The process of commoning creates a community and that becoming-community in turn creates a commons. (2021: 160)

Relatedly, there is a need to develop concerted strategies to maintain collective land governance where it continues to exist and to reinstate it in places where it has been abolished. This would assist in undoing the financialisation and privatisation of land that features in tourism and associated developments which has underpinned tourism-induced dispossession. It is an essential pillar in addressing crises arising from climate change.

Hickel's three Ds

These 'three Ds' – decommodification, de-enclosure and deaccumulation – are the pillars of degrowth and a transition away from a system geared to the accumulation of capital without end. In terms of tourism, such an approach would lead to a focus on supporting tourism for the public good, protecting the commons from privatisation and reducing the contribution of tourism to greenhouse gas emissions and other forms of waste, pollution and unsustainability.

Local currencies

Local currencies play a key role in delinking from the unequal free trade system and rebuilding a localised and more sustainable economy. Latouche explained: 'The role of local, social or complementary currencies has to be related to unsatisfied needs with regard to resources which would otherwise go unused' (2009: 49). See Chapter 6 in this volume.

The danger of 'localwashing'

As Signe Jungersted explains in Chapter 8, there is a danger that this language of localising will be co-opted and thereby undermined; she characterises this as 'localwashing'. This is an adaptation of the term 'greenwashing' which refers to tourism industry and DMO usage of terms such as 'eco', 'green' and 'sustainable' in deceptive and misleadingly ways to promote tourism products and experiences.

Local communities – not destinations

One of the foundational concepts in tourism management is the tourist destination, mostly commonly called a 'destination'. Importantly, this concept is defined by the tourists. For example, one publisher offers 11 definitions of a tourist destination, including these: 'geographic area or zone frequently visited by tourists' and 'A place for tourists to visit and stay, could be a country, state, region or city – usually due to its cultural or natural values' (IGI Global, n.d.).

> Perhaps the concept generates confusion rather than clarity, since there appears to be certain systematic self-contradictions in its use: the destination as a narrative, as an attraction, as a geographical unit, as an empirical relationship, as a marketing object, as a place where tourism happens ... and so on. It is interesting to investigate whether a destination is seen as a locality or as a production system, as an information system or as the production of services. Each approach tends to define the destination in a different way, emphasizing only one facet with no interest in certain others. (Framke, 2002: 92)

Critical consciousness for change

Paolo Freire offered an understanding of 'critical consciousness' (*conscientizacao*) in his discussion of *the Pedagogy of the Oppressed*: 'The pedagogy of the oppressed is an instrument for their critical discovery that both they and their oppressors are manifestations of dehumanization' (1970: 48). As Jackson explained, critical consciousness prepares people to change their world and challenge oppressions:

> This includes developing an understanding of how language frames current ways of thinking that represent humans as controlling the genetic basis of life. Additionally, we should remain critical of what 'universities have relegated to the category of low-status knowledge', knowledge that can contribute to the vitality of communities. The building of awareness and conscience around such issues is part of the project of critiquing globalization. (Jackson, 2007: 205)

Related localisms

As communities work to build community empowerment, economies for local thriving and forms of tourism, hospitality and leisure that strengthen their communities, cross-community learning and partnerships should be fostered that help build mutual support networks both nearby and far. This can be seen in the example of Fairbnb discussed earlier and others such as Worldwide Opportunities on Organic Farms (WWOOF) and Cittaslow (slow town movement) (see also Chapter 13).

Appreciative enquiry (AI) as a tool for community empowerment

Recent work discussing appreciative enquiry as a tool for stakeholder engagement opening up possibilities for 'dialogic democracy' (Guix & Font, 2022: 9) is promising. We view appreciative enquiry as tool communities can harness to shape their engagement with tourism in their own interests to build self-determining futures.

> The objective of AI is… to uncover and amplify existing strengths, hopes and dreams as a means of inspiring change from within. To achieve this, AI aims to be scientific (seeking socio-rational knowledge), metaphysical (seeking socio-rational knowledge), normative (seeking practical knowledge) and pragmatic (seeking knowledgeable action). These objectives have more recently been adapted to form 'the 4-D model', which essentially consists of the same process, but develops a label for each stage: Discovery, Dream, Design and Destiny. (Raymond & Hall, 2008: 284)

Citizen assemblies and participatory budgeting for participatory democracy

To promote citizen engagement and tap diverse insights, citizen assemblies can be used to consider challenges and crises in tourism.

> …These assemblies bring together 50 or more citizens over a number of days or weeks to learn about a particular policy challenge, deliberate together and recommend how to deal with it. Citizens are selected to reflect the demographic diversity of the population. The process is typically facilitated by an independent and apolitical organisation, which brings in experts across a wide range of disciplines, as well as competing interest groups and the voices of those personally affected by the issue in question. (Smith, 2019: n.p.)

Additionally, participatory budgeting practices are important to facilitating citizen engagement in determining community futures through the decisions made on managing public resources. First used from 1989 in Porto Alegre, Brazil, now more than 300 municipalities have adopted this approach (see UN-HABITAT, 2004). It has even been argued that the current war in Ukraine illuminates new forms of citizen activation significant to the future (Leighninger, 2022).

Inclusive tourism development approaches

Scheyvens and Biddulph explained inclusive tourism as:

> Transformative tourism in which marginalized groups are engaged in ethical production or consumption of tourism and the sharing of its benefits. This means something can only be considered inclusive tourism if marginalized groups are involved in ethical production of it, or they are

involved in ethical consumption of it, and in either case, marginalized groups share the benefits. Who is marginalized will vary from place to place but this could include the very poor, ethnic minorities, women and girls, differently abled people and other groups who lack power and/or voice... In terms of 'transformative', this could mean addressing inequality, overcoming the separation of different groups living in different places, challenging stereotypes or generalized histories, and opening people up to understanding the situation of minorities. (2018: 592)

Subsidiarity in policy and practice

As we consider localising tourism, subsidiarity is a key principle. Caritas Canada explained subsidiarity:

> The state is an instrument to promote human dignity, protect human rights, and develop the common good. Subsidiarity holds that such functions of government should be performed at the lowest level possible, as long as they can be performed adequately. When they cannot, higher levels of government must intervene. This principle goes hand-in-hand with Participation, the principle that all peoples have a right to participate in the economic, political and cultural life of society, and in the decisions that affect their community. (Caritas Canada, n.d.)

Subsidiarity in tourism and travel circuits

The concept of subsidiarity could also apply to the geographical reach of our travel circuits as well. It would create a framework for proximity travel (see Figure 0.3). It could be supported by policy and resourcing devised to encourage people to travel at the most local level possible and efforts could be made to induce a mindset of justifying tourism travels (Higgins-Desbiolles, 2010). This may be an essential component of a degrowth approach to tourism and an appropriate response to climate change. It also may be enforced by future pandemics and conflicts.

Participatory Action Research (PAR)

As we localise tourism, we will need to utilise research in both its formal and informal senses for communities to be guided and informed in their decisions. PAR is an important tool in the localising tourism toolkit. Freire provides an insight into PAR as praxis:

> To investigate the generative theme is to investigate people's thinking about reality and people's action upon reality, which is their praxis. For precisely this reason, the methodology proposed requires that the investigators and the people (who would normally be considered objects of that investigation) should act as co-investigators. The more active an attitude men and women take in regard to the exploration of their thematics, the more they deepen their critical awareness of reality and, in spelling out those thematics, take possession of that reality. (Freire, 1970: 149)

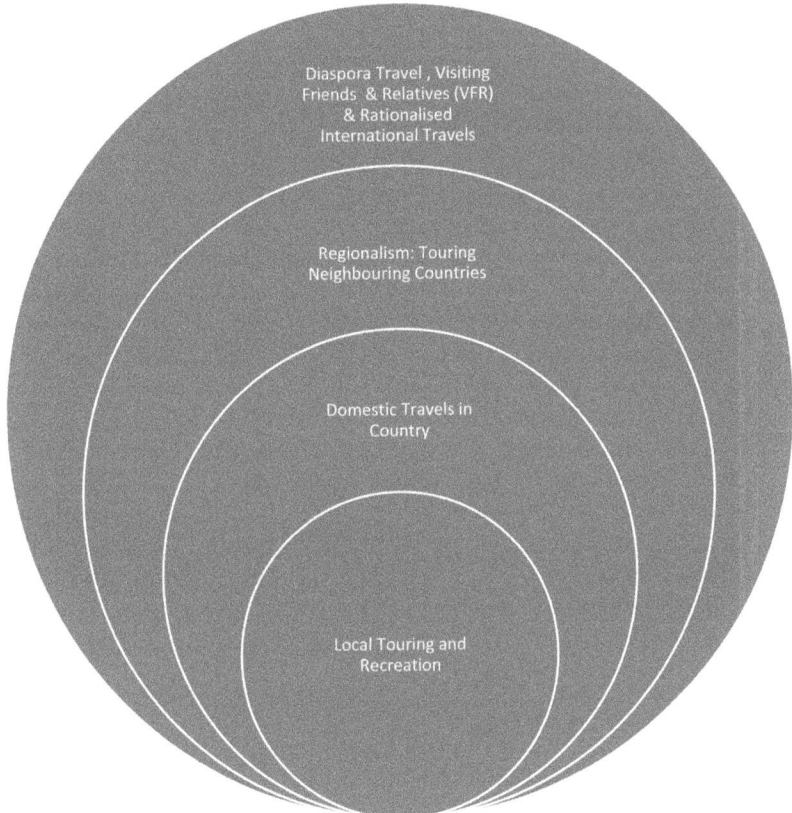

Figure 0.3 Applying subsidiarity to our travel circuits. A framework for proximity travel

Britton (2021) has explained a 'participatory city approach' used in her work with two UK boroughs as a pathway to building community so that ongoing participation and dialogue is possible.

Localising and Indigenising tourism research

In addition, localised protocols, epistemologies and ontologies must be the basis of research on tourism serving communities. An example can be found in the work of Movono and Scheyvens who referred to: 'The Fiji Vanua Research Framework ... as a central paradigm to guide the orientation and methodology of this long-term research project as it promotes the localisation of methods, decolonisation of research and focusses on the essence of Fijian society' (2021: 269). In Aotearoa/New Zealand, *Mātauranga Māori* describes Indigenous Māori knowledge, culture, values and world view and has recently been suggested as helpful in guiding community concern and engagement with conservation (Burnett *et al.*, 2021).

This Volume

With this opening, we can now surmise that Singh's observation that '…tourism [is] shaped to match the needs of the tourists' (2012: 22) is becoming outdated. As numerous scholars have noted (e.g. Moscardo, 2008; Scheyvens, 2002; Timothy, 2007), local communities have been recognised as the key to successful tourism. The more recent demand to define tourism by the local community is a response to newer pressures and problems, including overtourism (Higgins-Desbiolles *et al*., 2019). Localising the circuits of travel is likely to become increasingly necessary due to climate change, growing conflicts and instability. Proximity travel may have emerged from the constraints of the pandemic, but we propose localising tourism as a broader agenda to fully reorient tourism to the benefit of local communities and embedded within a love of place. Not only is all tourism local but also prosperous and thriving communities are the foundations for successful tourism futures. For those that may find these ideas impractical, the evidence for transitions is already apparent (NZ Government, 2021). The essential questions are: will these transitions be just transitions and how might we foster this? This edited volume is an initial attempt to answer these questions, surveying cases from around the world and offered by scholars and practitioners actively changing tourism practices on the ground.

This volume is divided into four sections.

- *Part 1 Theorising Local Communities in Tourism Anew*. In this section, Bigby, Edgars and Higgins-Desbiolles investigate aspects of place-based governance, arguing this may shape tourism to be better suited to people and place. Tomassini and Cavagnaro offer a framework for the local turn in tourism that is built on concepts of circular *oikonomia*, post-humanism and the sociology of space. Seyfi and Hall assume a critical stance and explore the example of boycotts, ethical consumption and destination communities to underscore that tourism embeds communities in capitalist structures and any discussion of alternatives requires grounding in such realities. Ooi calls for critical reflections on the moral limits of the market and how this might shape possibilities for a local turn in tourism.
- *Part 2 Case Studies of Local Community (Dis)/(Re)/Empowerment*. In this section, Canosa reports on engaged research with youth impacted by tourism in the tourist town of Byron Bay, Australia, and how they have used theatre performance, developed a tourist pledge and other interventions to shape tourism for the benefit of their community. Little reports on research from the tourist town of Monteverde, Costa Rica and how they have used community cryptocurrencies to support localised economies, build food security and support social and environmental goals as COVID-19 closed the taps on tourism income and employment. Jones and Zarb report on a 'Meet the Locals'

strategy used in communities of Malta which harnessed alternative tourism strategies to better shape tourism for community needs.
- *Part 3 Practitioners' Views and Insights.* Here, Jungersted presents a case study from her work developing the 'Localhood' strategy for Copenhagen, Denmark. Winkler narrates the story of the Cambodian Children's Trust (CCT), explaining how it evolved from an ethos of 'white saviourism' found in orphanage and volunteer tourism to a community-led strategy of development modelled in CCT's Village Hive approach. Bigby and Brown-Burdex report on efforts to use the Greenwood Cultural Centre and tours associated with it for truth-telling and community renewal in the aftermath of the well-known race massacre of 1921 in this community of Tulsa, Oklahoma, United States.
- *Part 4 Imagining New Futures.* In this final section, Müller, Fletcher and Blázquez-Salom take us to Mallorca, Spain to consider an example of the intersection of convivial conservation and proximity tourism at La Trapa, a public common that is used by and supported by the local community as a place to experience nature. Everingham and Francis-Coan use auto-ethnographic methods to consider a 'more-than-tourism' approach that may lead to local (re)enchantment with local places. Araya López considers the discourse of the 'right to the city' and the reasons communities in European cities such as Barcelona have formed or joined social movements in reaction to tourism. He also offers brief comparisons to other European cities, including Venice, Amsterdam and Lisbon.

This book offers only a beginning. It opens up many questions that are important to consider: how do we know the local, how do local places get made into usurped tourism destinations, how might we reclaim such places and for whom? The answers are many, varied and may differ based on one's positionality. With this edited volume, we hope to, at least partially, address these many important questions and thereby contribute to the furtherance of the local turn in tourism and tourism studies.

References

Appadurai, A. (1996) *Modernity at Large.* Minneapolis: University of Minnesota Press.
Araya López, A. (2021a) Saint Mark's Square as contested political space: Protesting cruise tourism in Venice. *Shima Journal* 15 (1), 137–165. https://doi.org/10.21463/shima.119.
Araya López, A. (2021b) A summer of phobias: Media discourses on 'radical' acts of dissent against 'mass tourism' in Barcelona [version 1]. Open Research Europe 1 (66).
APEC Tourism Working Group (2010) *Effective Community-based Tourism: A Best Practice Manual.* Sydney: Asia-Pacific Economic Cooperation.
Bianchi, R. (2018) The political economy of tourism development: A critical review. *Annals of Tourism Research* 70, 88–102. https://doi.org/10.1016/j.annals.2017.08.005.
Blackstock, K. (2005) A critical look at community based tourism. *Community Development Journal* 40 (1), 39–49. https://doi.org/10.1093/cdj/bsi005.

Böhm, S., Dinerstein, A.C. and Spicer, A. (2010) (Im)possibilities of autonomy: Social movements in and beyond capital, the state and development. *Social Movement Studies* 9 (1), 17–32. https://doi.org/10.1080/14742830903442485.

Bordeaux Tourism (2021) Agora Tourism Bordeaux. See https://agora-tourism-bordeaux.com/ (accessed 10 October 2021).

Born, B. and Purcell, M. (2006) Avoiding the local trap: Scale and food systems in planning research. *Journal of Planning Education and Research* 26, 195–207.

Boyle, D. and Wyler, S. (2021) 'Us and them': A mindset that has failed our communities. Essay for the Local Trust. See https://localtrust.org.uk/insights/essays/us-and-them-a-mindset-that-has-failed-our-communities/ (accessed 1 January 2022).

Briassoulis, H. (2002) Sustainable tourism and the question of the commons. *Annals of Tourism Research* 29 (4), 1065–1085.

Britton, T. (2021) Designing neighbourhoods for cohesion – You can't build cohesion the same way you create division and polarisation. See https://tessybritton.medium.com/designing-neighbourhoods-for-cohesion-you-cant-build-cohesion-the-same-way-you-create-division-99ada81753a2 (accessed 6 February 2022).

Burnett, S., Movono, A. and Scheyvens, R. (2021) How Māori knowledge could help New Zealanders turn their concern for the environment into action. *The Conversation* (Online). See https://theconversation.com/how-maori-knowledge-could-help-new-zealanders-turn-their-concern-for-the-environment-into-action-168831 (accessed 15 January 2022).

Butcher, J. (2012) The mantra of 'community participation' in context. In T.V. Singh (ed.) *Critical Debates in Tourism* (pp. 102–108). Bristol: Channel View Publications.

Canan, P. and Hennessy, M. (1989) The growth machine, tourism, and the selling of culture. *Sociological Perspectives* 32 (2), 227–243.

Candidatura d'Unitat Popular (2017) Els mites del turisme. See http://turisme.capgirembcn.cat/ (accessed 2 January 2022).

Caritas Canada (n.d.) Subsidiarity and the role of government. See https://www.devp.org/en/resource/catholic-social-teaching/subsidiarity-and-the-role-of-government/ (accessed 3 January 2022).

Chomsky, N. (2002) Noam Chomsky interviewed by Toni Gabric. See https://wayback.archive-it.org/all/20070807064137/https://www.zmag.org/content/TerrorWar/chomskygab.cfm (accessed 23 January 2022).

Cochet, Y. (2005) *Pétrole Apocalypse*. Paris: Fayard.

Coldwell, W. (2017) First Venice and Barcelona: Now anti-tourism marches spread across Europe. *The Guardian* (Online). See https://www.theguardian.com/travel/2017/aug/10/anti-tourism-marches-spread-across-europe-venice-barcelona (accessed 17 January 20220).

Dangl, T.B. and Jamal, T. (2016) An integrated approach to 'sustainable community-based tourism'. *Sustainability* 8 (5), 475. https://doi.org/10.3390/su8050475.

Diaz-Parra, I. and Jover, J. (2021) Overtourism, place alienation and the right to the city: insights from the historic centre of Seville, Spain. *Journal of Sustainable Tourism* 29 (2–3), 158–175. https://doi.org/10.1080/09669582.2020.1717504.

Einstein, C. (2018) *Climate: A New Story*. Berkeley: North Atlantic Books.

Escobar, A. (2001) Culture sits in places: reflections on globalism and subaltern strategies of globalization. *Political Geography* 20, 139–174.

Fairbnb (n.d.) About us. See https://fairbnb.coop/about-us-3/ (accessed 23 January 2022).

Fletcher, R., Murray Mas, I., Blanco-Romero, A. and Blázquez-Salom, M. (2019) Tourism and degrowth: An emerging agenda for research and praxis. *Journal of Sustainable Tourism* 27 (12), 1745–1763. https://doi.org/10.1080/09669582.2019.1679822.

Framke, W. (2002) The destination as a concept: A discussion of the business-related perspective versus the sociocultural approach in tourism. *The Scandinavian Journal of Hospitality and Tourism* 2 (2), 92–108. https://doi.org/10.1080/15022250216287.

Freire, P. (1970) *Pedagogy of the Oppressed*. New York: Continuum.

Fullagar, S., Markwell, K. and Wilson, E. (2012) *Slow Tourism: Experiences and Mobilities*. Bristol: Channel View Publications.

Gilchrist, A. (2003) Community development in the UK. *Community Development Journal* 38 (1), 16–25.

Glenza, J. (2020) 'We want our island back': The group taking on cruise ships in Florida Keys. The Guardian Online. See https://www.theguardian.com/us-news/2020/oct/01/florida-keys-cruise-ship-ban-covid-19 (accessed 21 December 2021).

Group Nao (2021) Time for DMOcracy. See https://groupnao.com/time-for-dmocracy/ (accessed 21 December 2021).

Guix, M. and Font, X. (2022) Consulting on the European Union's 2050 tourism policies: An appreciative inquiry materiality assessment. *Annals of Tourism Research* 93,103353. https://doi.org/10.1016/j.annals.2022.103353.

Hall, C.M. (2003) Politics and place: An analysis of power in tourism communities. In S. Singh, D.J. Timothy and R.K. Dowling (eds) *Tourism in Destination Communities* (pp. 99–115). Oxon: CABI.

Hengel, L. (2021) Alberghi Diffusi are the most sustainable hotels in Italy – Here's why. Forbes Online. See https://www.forbes.com/sites/liviahengel/2021/01/27/alberghi-diffusi-are-the-most-sustainable-hotels-in-italy—-heres-why/ (accessed 18 January 2022).

Hickel, J. (2021a) The age of imperialism is not over – but we can end it. *Current Affairs*. See https://www.currentaffairs.org/2021/12/the-age-of-imperialism-is-not-over-but-we-can-end-it (accessed 22 January 2022).

Hickel, J. (2021b) What does degrowth mean? A few points of clarification. *Globalizations* 18 (7), 1105–1111. https://doi.org/10.1080/14747731.2020.1812222.

Higgins-Desbiolles, F. (2006) More than an industry: Tourism as a social force. *Tourism Management* 27 (6), 1192–1208.

Higgins-Desbiolles, F. (2010) Justifying tourism: Justice through tourism. In S. Cole and N. Morgan (eds) *Tourism and Inequality: Problems and Prospects* (pp. 194–212). Oxon: CABI.

Higgins-Desbiolles, F. (2012) The Hotel Bauen's challenge to cannibalizing capitalism. *Annals of Tourism Research* 39 (2), 220–240.

Higgins-Desbiolles, F. (2020) Socialising tourism for social and ecological justice after COVID-19. *Tourism Geographies* 22 (3), 610–623. https://doi.org/10.1080/14616688.2020.1757748.

Higgins-Desbiolles, F. and Bigby, B.C. (2022) A local turn in tourism studies. *Annals of Tourism Research* 92, 103291. https://doi.org/10.1016/j.annals.2021.103291.

Higgins-Desbiolles, F., Carnicelli, S., Krolikowski, C., Wijesinghe, G. and Boluk, K. (2019) Degrowing tourism: rethinking tourism. *Journal of Sustainable Tourism* 27 (12), 1926–1944. https://doi.org/10.1080/09669582.2019.1601732.

Higgins-Desbiolles, F., Doering, A. and Bigby, B.C. (2022a) *Socialising Tourism: Rethinking Tourism for Social and Ecological Justice*. Abingdon: Routledge.

Higgins-Desbiolles, F., Moskwa, E. and Gifford, S. (2014) The restaurateur as a sustainability pedagogue: The case of Stuart Gifford and Sarah's Sister's Sustainable Café. *Annals of Leisure Research* 17 (3), 267–280.

Higgins-Desbiolles, F., Scheyvens, R. and Bhatia, B. (2022b) Decolonising tourism and development: From orphanage tourism to community empowerment in Cambodia. *Journal of Sustainable Tourism*. https://doi.org/10.1080/09669582.2022.2039678.

Hughes, N. (2018) 'Tourists go home': Anti-tourism industry protest in Barcelona. *Social Movement Studies* 17 (4), 471–477. https://doi.org/10.1080/14742837.2018.1468244.

IGI Global (n.d.) What is a tourist destination. *IGI Dictionary* (Online). See https://www.igi-global.com/dictionary/tourist-destination/39274 (accessed 13 May 2021).

Jackson, S. (2007) Freire reviewed. *Educational Theory* 57 (2), 199–213.

Kallis, G., Kostakis, V., Lange, S., Muraca, B., Paulson, S. and Schmelzer, M. (2018) Research on degrowth. *Annual Review of Environment and Resources* 43, 291–316.

Kenis A. and Mathijs E. (2014) (De)politicising the local: The case of the Transition Towns movement in Flanders (Belgium). *Journal of Rural Studies* 34, 172–183.

Latouche, S. (2009) *Farewell to Growth*. Cambridge: Polity.

Leighninger, M. (2022, 26 March) Ukraine is showing us how powerful modern civilians can be. *San Francisco Chronicle* (Online). See https://www.sfchronicle.com/opinion/openforum/article/Ukraine-is-showing-us-the-power-of-the-17028536.php (accessed 30 March 2022).

Local Futures (n.d.) About Us. See https://www.localfutures.org/about/ (accessed 23 January 2022).

Millar, C. and Aiken, D. (1995) Conflict resolution in aquaculture: A matter of trust. In A. Boghen (ed.) *Coldwater Aquaculture in Atlantic Canada* (2nd edn, pp. 617–645). Moncton: Canadian Institute for Research on Regional Development.

Moscardo, G. (ed.) (2008) *Building Community Capacity for Tourism Development*. Wallingford: CABI.

Movono, A. and Scheyvens, R. (2021) Tourism in a world of disorder: A return to the vanua and kinship with nature in Fiji. In Y. Campbell and J. Connell (eds) *COVID in the Islands: A Comparative Perspective on the Caribbean and the Pacific* (pp. 265–277). Singapore: Springer Nature. https://doi.org/10.1007/978-981-16-5285-1_15.

Mtapuri, O. and Giampiccoli, A. (2020) Toward a model of just tourism: A proposal. *Social Sciences* 9 (4), 34. MDPI AG. http://dx.doi.org/10.3390/socsci9040034 (accessed 23 January 20220).

Murphy, P. (1985) *Tourism: A Community Approach*. London: International Thomsen Business Press.

New Zealand Government (2021) 2021 support – Tourism Communities: Support, Recovery and Re-set Plan. See https://www.mbie.govt.nz/immigration-and-tourism/tourism/tourism-recovery/tourism-communities-support-recovery-and-re-set-plan/ (accessed 14 February 2022).

Norberg-Hodge, H. (1992) *Ancient Futures: Learning from Ladakh*. Oakland: Sierra Club Books.

Norberg-Hodge, H. and Read, R. (2016) *Post-growth Localisation*. Totnes: Local Futures. See https://www.localfutures.org/wp-content/uploads/Post-growth-Localisation.pdf (accessed 23 January 2022).

Oviedo, J.B. (2021) L'Agora du Tourisme à Bordeaux. European Cities Marketing Conference, 23 September (oral presentation).

Petrescu, D., Petcou, C., Safri, M. and Gibson, K. (2021) Calculating the value of the commons: Generating resilient urban futures. *Environmental Policy and Governance* 31, 159–174. https://doi.org/10.1002/eet.1890.

Price, C. (2020) How restaurants are supporting their communities during COVID-19. See https://pos.toasttab.com/blog/on-the-line/restaurants-supporting-communities-coronavirus (accessed 3 December 2021).

Project Forever Waiheke (2021) 'Waiheke is a community, not a commodity': Stakeholder perspectives on Waiheke tourism. Retrieved 20 February 2022, from https://static1.squarespace.com/static/5b1dd83a372b9624b25936a3/t/613ed1935f48cb52a413f469/1631506840241/Report+on+future+tourism+on+Waiheke+-+Project+Forever+Waiheke+140921+FINAL.pdf.

Quilley, S. (2013) De-growth is not a liberal agenda: Relocalisation and the limits to low energy cosmopolitanism. *Environmental Values* 22 (2), 261–285.

Raymond, E.M. and Hall, C.M. (2008) The potential for appreciative inquiry in tourism research. *Current Issues in Tourism* 11 (3), 281–292. https://doi.org/10.1080/13683500802140323.

Rojek, C. and Urry, K. (1997) *Touring Cultures: Transformations of Travel and Theory*. London: Routledge.

Rothman, H.K. (1998) *Devil's Bargains: Tourism in the Twentieth Century American West*. Lawrence, KS: University Press of Kansas.

Scheyvens, R. (2002) *Tourism for Development: Empowering Communities*. Harlow: Prentice-Hall.
Scheyvens, R. and Biddulph, R. (2018) Inclusive tourism development. *Tourism Geographies* 20(4), 589–609. https://doi.org/10.1080/14616688.2017.1381985.
Schnell, S.M. (2013) Deliberate identities: Becoming local in America in a global age. *Journal of Cultural Geography* 30 (1), 55–89. https://doi.org/10.1080/08873631.2012.745984.
Sklair, L. (2020) The transnational capitalist class. In J.G. Carrier and D. Miller (eds) *Virtualism* (pp. 135–159). London: Routledge.
Singh, T.V. (2012) *Critical Debates in Tourism*. Bristol: Channel View Publications.
Sipe, L.J. (2018) Leveraging neo-localism for experience innovation: A case study of an urban park and entertainment venue. *Journal of Themed Experience and Attractions Studies* 1 (1), Article 4. Available at: https://stars.library.ucf.edu/jteas/vol1/iss1/4.
Smith, G. (2019) Citizens' assemblies: How to bring the wisdom of the public to bear on the climate emergency. The Conversation (Online). See https://theconversation.com/citizens-assemblies-how-to-bring-the-wisdom-of-the-public-to-bear-on-the-climate-emergency-119117 (accessed 31 December 2021).
Spinks, R. (2018) The 'live like a local' travel ethos has failed – the question is what will replace it. Quartz. See https://qz.com/quartzy/1298546/the-live-like-a-local-travel-ethos-has-failed-locals-so-what-comes-next/ (accessed 21 December 2021).
Sydney Alliance (n.d.) Without community, there is no community recovery. See. https://www.sydneyalliance.org.au/ (accessed 24 January 2022).
Timothy, D.J. (2007) Empowerment and stakeholder participation in tourism destination communities. In A. Church and T. Coles (eds) *Tourism, Power and Space* (pp. 199–216). London: Routledge.
Tonnies, F. (1957) *Community & Society (Gemeinschaft und Gesellschaft)*. New York: Harper & Row.
Tourism Futures Taskforce (2020) Interim Report, December. See https://www.mbie.govt.nz/assets/the-tourism-futures-taskforce-interim-report-december-2020.pdf (accessed 23 December 2021).
Trainer, T. (2009) Strengthening the vital Transition Towns movement. *Pacific Ecologist* 18, pp. 11+. Gale Academic OneFile, link.gale.com/apps/doc/A205567727/AONE (accessed 22 January 2022).
Transition (2021a) Principles. See https://transitionnetwork.org/about-the-movement/what-is-transition/principles-2/ (accessed 14 January 2022).
Transition (2021b) What is Transition? See https://transitionnetwork.org/about-the-movement/what-is-transition/ (accessed 14 January 2022.
UN-HABITAT (2004) 72 frequently asked questions about participatory budgeting. Quito: UN-HABITAT.
UNWTO (n.d.) Tourism and the 2030 agenda. See https://www.unwto.org/tourism-in-2030-agenda (accessed 11 November 2021).
La Via Campesina (2019) 2018 Annual Report. See https://viacampesina.org/en/la-via-campesina-2018-annual-report/ (accessed 13 January 2022).
Weinraub, J. (2006) Q&A with Michael Pollan: Think global, eat local. *Washington Post*. See https://michaelpollan.com/interviews/qa-with-michael-pollan-think-global-eat-local/ (accessed 21 December 2021).
Wheeller, B. (1993) Sustaining the ego. *Journal of Sustainable Tourism* 1 (2), 121–9, https://doi.org/10.1080/09669589309450710.
Wiedmann, T., Lenzen, M., Keyßer, L.T. and Steinberger, J.K. (2020) Scientists' warning on affluence. *Nature Communication* 11, 3107. https://doi.org/10.1038/s41467-020-16941-y.
Williams, P.W., Gill, A.M., Marcoux, J. and Xu, N. (2012) Nurturing 'social license to operate' through corporate-civil society relationships in tourism destinations. In C.H.C. Tsu and W.C. Gartner (eds) *The Routledge Handbook of Tourism Research* (pp. 196–214). Abingdon: Routledge.
Wiktionary (n.d.) Local. See https://en.wiktionary.org/wiki/local (accessed 4 January 2022).

Part 1
Theorising Local Communities in Tourism Anew

1 Place-based Governance in Tourism: Placing Local Communities at the Centre of Tourism

Bobbie Chew Bigby, Joseph Edgar and Freya Higgins-Desbiolles

Introduction

'Much of what our nation has lost is that awareness that the earth can be for us a place of spiritual renewal, not just a place to stroll in a park, or hike in a forest, or find land to mine resources, but that it is a place where we can be transformed'. (hooks, 2009: 201)

With the globalisation of tourism and the mass tourism approach to development, many communities around the world find themselves in a state of vulnerable dependency on tourism (see Bianchi, 2018; Lacher & Nepal, 2010). This was true well before the COVID-19 global pandemic placed issues of dependency in stark relief. A potent mix of multinational corporations' investments, complex global supply chains and compliant governments that offer up local places to global corporations and global tourists results in local communities being pressed by tourism and even potentially being dispossessed (Higgins-Desbiolles, 2018). Signs of community discontent have long been apparent including expression of 'anti-tourism' sentiments, local activation to take back control of their local places and cooperation across jurisdictions to better empower local communities for self-determining futures (see Tomassini & Cavagnaro, 2020).

This chapter was inspired by a methodology of placed-based cooperative enquiry developed by Wooltorton *et al.* (2020), specifically in an Indigenous context. This is founded on Heron's (1996) work on cooperative inquiry and it offers a method by which local communities may be brought in better engagement and control of tourism development in their local communities. Wooltorton *et al.* argued that this approach has great value because it '...uses an extended epistemology inclusive of a relational ontology, in which knower and known are connected' (2020: 920).

In this chapter we explore issues of place and placelessness, the multitude of ways that local governance manifests in diverse locations, how decolonial and degrowth efforts are essential mechanisms and finally some tools that might serve local governance. We offer a detailed case study of Karajarri community engagement and governance of tourism to better understand how governance, relational ontologies and deep embedding in place may shape tourism for better futures for local communities, tourists and others. In light of the multiple crises we confront, place-based governance offers an approach through which just transitions can be negotiated.

Place and Placelessness

People are born in places and shaped by the communities that pertain to particular places. However, people are also mobile and so issues of attachment to place, return to place and care for place remain vital concerns. These connections have been made tangible in human cultures and ceremonies, for example in the widely used practise of placenta burial. Writing of this with regards to Eastern Polynesia, Saura *et al.* explained: '[This] custom is still widely observed and marks an essential connection between humans, the earth, plants and islands' (2002: 127).

Many cultures demonstrate the connection to place through their protocols, ceremonies and hosting. Thus, for instance, it is traditional in Māori public speaking for a speaker to place themselves with regard to their mountain, their river or sea, their locality, their tribal affiliation, their family and their own name (see Mason, 2021). This communicates that people and places are in relationship with one another and these ongoing relationships are practiced, nourished and respected. The stories of places may tell how the landscape, seascape and total environment came to be and the beings that inhabit these places were 'emplaced' in these places and set rules and protocols for good living. *Pakeha* (non-Indigenous) New Zealanders are beginning to observe these protocols as well, realising that deep connections to place can be nurtured and practiced. We see from human cultures that people and place are vitally interconnected.

There may be six ways of knowing place as explained in the work of Lukermann (1964).

(1) As a location that relates to other places.
(2) Place entails an integration of both cultural and natural elements.
(3) 'Although every place is unique, they are interconnected by a system of spatial interactions and transfers; they are part of a framework of circulation'.
(4) 'Places are localised – they are parts of larger areas and are focuses in a system of localisation'.

(5) 'Places are emerging or becoming…'.
(6) 'Places have meaning'. (citations from Relph, 1976: 3)

In geography, place refers to space to which people hold meaningful connection (Tuan, 2012). We might draw from the book *Detours*, the idea of physical spaces as 'storied places' (Aikau & Gonzalez, 2019: 16). Aikau and Gonzalez's book presents about knowing and caring for Hawai'i as a well-loved, storied place, explaining that the telling of Native Hawai'ian stories of place is a '…decolonial practice of restoring the relationship between people and places' (2019: 16). Wooltorton *et al.* emphasised the importance of Indigenous languages in knowing and taking care of places: 'This is because Indigenous languages function to enliven places through the naming and verbalising of animate life-giving energies and facilitate relational ways of understanding places, stories, narratives and verse' (2020: 919).

Places and being embedded in place are also sources of empowerment and well-being. For instance, in the Māori context in Aotearoa/New Zealand:

> Tūrangawaewae is one of the most well-known and powerful Māori concepts. Literally tūranga (standing place), waewae (feet), it is often translated as 'a place to stand'. Tūrangawaewae are places where we feel especially empowered and connected. They are our foundation, our place in the world, our home. (Te Ara, n.d.)

Place attachment can be defined as a positive, affective-emotional bond between people and places (Altman & Low, 1992). Such place attachment has meaningful outcomes: 'the main characteristic of [place attachment] is the tendency of the individual to maintain closeness to such a place' (Hidalgo & Hernández, 2001: 274).

However, both globalisation and tourism can weaken people's connections and attachments to place. Taking a decolonial perspective, Hirmer explained:

> Colonial relations persist undisturbed through the globalisation of markets, while capitalism (and its latest expression neoliberalism) remains largely unacknowledged as primary factor in persevering neo-colonial exploitations… fierce competition for the maximisation of production and the conquest of markets leads to the reification and universalisation of a linear time- and growth-scale that is in great part alien to peoples with cosmologies and metaphysics different from a western worldview entrenched in Enlightenment values. (2020: 125)

Receiving tourists into local communities can have an impact on people's sense of and enjoyment of place. Recent research undertaken in Amsterdam explained:

> The recent debate about the fact that some neighborhoods are drastically changing to primarily serve the high number of tourists who visit has led

to endangering the very unique character of cities, but, most importantly, residents don't feel at home any longer... As cities move to later stages of tourism development, where tourism starts to dominate the development of a city or even harm the quality of life in the city, residents' interpretations and evaluations of their cities are crucial. (Lalicic & Garaux, 2022: 202)

Placelessness is '...the weakening of distinct and diverse experiences and identities of place' (Relph, 1976: 6). Giridharadas argued our era may be characterised by the 'placeless' and the phenomenon of 'placelessness' as a result: 'What is arguably new is the influence of the placeless and the elevation of placelessness, in some quarters at least, to a virtue' (2010, n.p.). Tourism contributes to the creation of the placeless and placelessness, by supporting the disconnected, hypermobility of tourists and also through the displacement of local people from their home communities when these become commodified as tourism 'destinations'. This can be heard in the call from the small island of Waiheke (nearby to Aotearoa/New Zealand's largest city Auckland), 'Waiheke is a community, not a commodity' (Project Forever Waiheke, 2021).

The diverse economies field offers alternatives to the dominating and universalising forces of tourism and globalisation. For instance, Escobar mounted a defence of place in reaction to the 'de-localizing, disembedding and universalizing influence of modern economy, culture and thought' (2001: 141). Indigenous economies are examples of place-based economies built on relations, rights and obligations. Kuokkanen stated:

> The key principles of indigenous economies – sustainability and reciprocity – reflect land-based worldviews founded on active recognition of kinship relations that extend beyond the human domain. Sustainability is premised on an ethos of reciprocity in which people reciprocate not only with one another but also with the land and the spirit world. (2011: 219)

Kuokkanen recommended:

> ...reconceptualizing indigenous governance initiatives around the concept of the social economy. I suggest that situating the social economy at the center of indigenous governance enables the reinstatement of the vital social institutions that traditionally played a key role in the community governance. The concept of the social economy also allows us to see indigenous economic systems and subsistence activities as part and parcel of indigenous governance. (2011: 232)

In thinking through localising efforts, Latouche argued that it is not the size that is decisive but rather the identity of place: 'What matters is the existence of a collective project rooted in a territory, defined as a place for communal living that must be protected and cared for the good of all' (2009: 45). Parajuli has developed a valuable conceptualisation of place-based, grassroots forms of governance built on fostering ecological ethnicities and a simultaneous revitalisation of ecology and democracy; these stand in opposition to destructive totalising and globalising forces (Parajuli, 1996).

Essential to these diverse initiatives to push back against trans-local forces is a place-based leadership for place-based governance. Hambleton explained: 'It invites leaders to move outside their organisation [or self-interests] (be it a local authority, a business, a social enterprise, a university or whatever) to engage with the concerns facing the place' (2011: 15). Particularly as we face multiple, complex, compounding and cascading crises, such leadership is called on to defend place and peoples. However, this is not necessarily in prickly isolation but rather in related localisms, as explained in the Introduction. These practices must also engage with leading thinking on 'just transitions' which are emerging from dialogues on climate change, energy transitions, environmental justice and just sustainabilities (Kojola & Agyeman, 2021). Just transitions require addressing crises and challenges through approaches centring equity, inclusivity of diverse people and respect for human rights, with particular emphasis on real involvement of people in the development, implementation and enforcement of measures for transition (see Heffron & McCauley, 2018).

Understanding place-based governance in tourism can be served by learning from case study insights. Here we offer a brief case study from the Karajarri community and their Native Title Body, the Karajarri Traditional Lands Association (KTLA), who are located in the Kimberley region of Western Australia. Aboriginal communities such as the Karajarri name place as 'Country' and they have held roles of care and kinship for millennia, recognised through the use of the titles 'Traditional Custodians' and 'Traditional Owners' (TOs).

Case Study: Karajarri People, Country and Tourism

Behind every tourism idea, plan, programme and experience offered in Aboriginal tourism are individual people and families who are connected to their respective culture(s), communit(ies) and Countr(ies). In the case of Karajarri-led tourism, numerous individuals and families have been instrumental over the last decade in envisioning, building and delivering these tourism experiences. This section highlights the voices of some of the key tourism and cultural leaders in the Karajarri community and their reflections on the intersections of tourism, culture and place-based governance. Some of the research for this project occurred during the pandemic, so the impacts of COVID-19 are also part of this narrative. In many cases, these individuals are not only leaders of tourism development efforts among Karajarri, but also hold roles in the realms of culture, governance, business management and educational leadership that are vital to Karajarri people.

For well over the past decade, Karajarri people and KTLA leadership have been planning for increased tourism activities on their traditional Country, in particular at a coastal site near Port Smith. One of the key milestones in this tourism development journey was the launch of a Visitor Pass in 2016 that demarcates a 'Karajarri Tourism Zone' for certain areas of Country that are open for touring and approved recreational activities (KTLA, n.d.). The Pass also requires visitors and tour operators on Karajarri Country to pay for each day spent on Country, with generated revenues going directly to further tourism development. Following the launch of the Visitor Pass, KTLA embarked on its most ambitious tourism project – the launch of an official Karajarri Tourism Strategy (KTLA, 2016). Additionally, KTLA oversaw a Visioning report for the creation of a Karajarri Cultural and Tourism Hub, located on land acquired by KTLA at Port Smith. Plans for the Hub include: a Ranger's station, interpretive information, cultural walks, beach shelters and guided cultural tours (UDLA, 2018: 23–24).

In conversation with Karajarri tourism leaders on this tourism development journey, a top set of priorities articulated by most of the participants was to ensure that the tourism activities reflected a respect for Karajarri Country, Culture and the Traditional Custodians. According to Uncle Joe Edgar and Aunty Maria Morgan,[1] this respect for the Karajarri Country and its Karajarri TOs is reflected in having deep knowledge of the landscape where tours are conducted as well as ensuring that proper permissions are in place and the right people are informed about tours (see Figure 1.1). These permissions are viewed as essential to the tourism planning process on many different levels.

On one hand, the permissions are a critical tool in showing respect for the TOs that live in or near the areas being visited, or who maintain deep cultural, spiritual and stewardship connections to particular areas of land and water. Sam Bayley, who had worked as both Karajarri Indigenous Protected Area (IPA) Coordinator and CEO over several years and was deeply involved with Karajarri early tourism offerings, explained that if the TOs are not comfortable with visitors coming to their country, then tourism is simply not going to work. On another level, securing permissions also helps to ensure that Country is being looked after properly by knowing who is visiting certain areas of land and water, what they are doing there and how long they intend to visit. At the same time, these same permissions help to ensure the safety of visitors by informing TOs that tour activities might be happening on areas of Country that at times are inhospitable or dangerous due to excessive heat, cyclones, road conditions, bush fires or the presence of certain wildlife. Aunty Maria thus emphasised that being aware about the environment within which a tour operates is the key to not only looking after Country and being respectful to Traditional Owners, but also ensuring that the tour guests are comfortable and safe during the experience as well.

Figure 1.1 Uncle Joe Edgar shows visitors a *makapala* (bush banana) growing on a tree near the Karajarri Tourism and Cultural Hub. He explains the importance of these traditional bush foods and how many of them grow plentifully during the wet season. Credit: Co-author Bobbie Chew Bigby.

Emerging leader Wynston Shovellor pointed out three strategic priority areas that are driving and informing Karajarri-led tourism. The top priority is a focus on cultural maintenance and management, with tourism allowing Elders more opportunities to pass knowledge on to younger community members and inform outside visitors about Karajarri culture. He shared: 'This [culture and tourism] hub is to teach Karajarri TOs who don't have the cultural knowledge, but who want to learn about their culture'. The second key priority is generating the local economy for Karajarri people and creating a revenue source that is independent of grant funding from government and non-governmental partners. Wynston identified the third priority as proper land management. He explained how this land management is interconnected with other goals:

> In the past, there had been a lot of tourists that had been going to areas that are culturally significant and sensitive to TOs. We want to use best practices to properly manage the Country for ourselves and to make sure the public is safe as well. We have the authority to look after Country better and it is our responsibility to do so.

In conversation with Uncle Thomas 'Dooley' King, he immediately connected the ideas of priorities and values together with considerations of the direction of Karajarri tourism. From his perspective, the top priority of Karajarri-led tourism is to promote Karajarri values, which he articulated first and foremost as looking after Country and Culture, in line with the importance that Uncle Joe and Aunty Maria have put on the priority of necessary respect shown to Country, Culture and TOs. Uncle Dooley stated:

> From a tourism perspective, we need to keep our key values and have them as the underlying principle of how we intend to and what we want to promote in the tourism space... as opposed to promoting what the standard, mainstream values are... the priority is to reach a point where we've got balance between protecting and sticking to our values and principles and not sacrificing those for the sake of economic independence.

Uncle Dooley explained that this priority of upholding traditional Karajarri values is vital for Karajarri people themselves for remaining connected to their culture, Country, community and identity. Thus, these values should shape the tourism experiences offered by Karajarri people. But Uncle Dooley also emphasised the guiding importance of these values in relation to another top tourism priority, namely using the tours to impact and influence non-Indigenous visitors. Uncle Dooley terms this priority as an 'indoctrination of non-Indigenous people to an Indigenous way of thinking and looking at the world'. He explained:

> [The gap] between Western and Indigenous ways of thinking... that's been our biggest challenge and hurdle [for us as Aboriginal and Karajarri people], and it's been the main place where most of the misunderstandings between Indigenous people come up... If they [visitors] can see the world in the way that we see it as Karajarri people, then that will hopefully influence them to have more respect for Country and respect for what Indigenous people are saying and what we value, the world over. White people indoctrinated us into the Western system of thinking. Now it's our turn. Now we use cultural awareness and Indigenous tourism to do that.

In this vision, Uncle Dooley firmly connects the priority of sharing, educating and 'indoctrinating' visitors with the central values of caring for Country and Culture that Karajarri people hold as sacred. Ultimately this loyalty and commitment to traditional values through Indigenous-led tourism is understood not only to have an impact on the visitors, but also helps ensure that Karajarri continue to maintain their connection and stewardship over their own culture and Country. According to Uncle Dooley: 'If tourism is more Indigenous-controlled, then you're more in the driver's seat to determine to what extent you want to expose people to your culture or the impacts that you have on Country'.

For Petrine McCrohan, a non-Indigenous cultural enterprise facilitator who has worked with Karajarri people and numerous other Kimberley Aboriginal groups on tourism planning over decades, relationship-building and reciprocity are core values that are central to Indigenous-led tourism. These are then embodied in characteristics such as authenticity and integrity which underpin effective Indigenous tourism offers. Petrine stated:

> At the heart of tourism that embodies Aboriginal culture is the focus on reciprocity and relationships... relationship building is intrinsic to Indigenous people – relation to Country, to language, to each other... That sense of staying true and authentic, having integrity and respect for the ancestors and towards the culture and not moving away from that... to be honest, the only ones that are the most successful are the ones that do that – that come from a motivation of 'I'm not going to sell my soul to the broader system just because economic rationale says I should'.

Like Uncle Dooley, Petrine highlighted the fundamental importance of sticking with core values in Indigenous-led tourism, but also the inherent, deep tensions between implementing these traditional Indigenous values and Western economic systems and thinking. These values and approaches have been important pillars in the development and evolution of Karajarri's tourism model.

Reflections on the evolving Karajarri tourism model

The Karajarri tourism model has evolved over some 15 years. Indigenous tourism typically starts small to 'test the waters' and evolves in size and complexity based on learning through the evolution on how to manage and control engagement with tourism. Sam Bayley, a non-Indigenous professional who had worked with Karajarri over several years, gave the following summary of how Karajarri's tourism model started:

> It's really about starting slowly and getting something going... if you're waiting for the perfect product to happen, it's never going to happen, so you just need to start slowly with something people are comfortable with. It can be a big step up for some of the [remote] communities to deliver a [tourism] product. So simple things like self walks, self-drives, simple permit systems, bird viewing... you know, things people can do without having to rely too much on human capital to start with because a lot of tourism requires people to be there, present, all the time, and that's just not possible for a lot of Indigenous groups. For [Karajarri], it was a permit system where people could go to places under the right conditions. And then in time we're going to work with TOs to be tour leaders and tour guides and have accommodation... it's about good communication, going at the right pace and being open.

These points about transparent communication and securing the trust of the TOs are supported by Petrine's emphasis on the centrality of using a participatory planning model in Karajarri tourism. She noted that the fundamental strength of the participatory planning model is that community members are able to be in the driver's seat in building their own plan for tourism and choosing what levels of value and profitability are sought, understanding value in a plural sense including cultural, social, environmental and economic value. Petrine also explained the utility in engaging with the 'Stepping Stones' model as a participatory planning tool (Stepwise Heritage and Tourism, n.d.). She used this Stepping Stones approach with Karajarri Traditional Lands Association (KTLA) members in 2010 which enabled Karajarri members to learn from tourism practitioners and advisers such as herself in building their tourism product (see Figure 1.2). A participatory model also helps communities to explore their points of strength and difference so that tourism offerings can be tailored to these circumstances. Petrine particularly emphasised that for the participatory model to succeed in guiding tourism as it evolves, it should be used regularly rather than only once at the beginning of planning. This continuous check-in with community members and participatory planning helps to ensure that as many voices are included as possible, even while there are fluctuating movements of people in and out of the

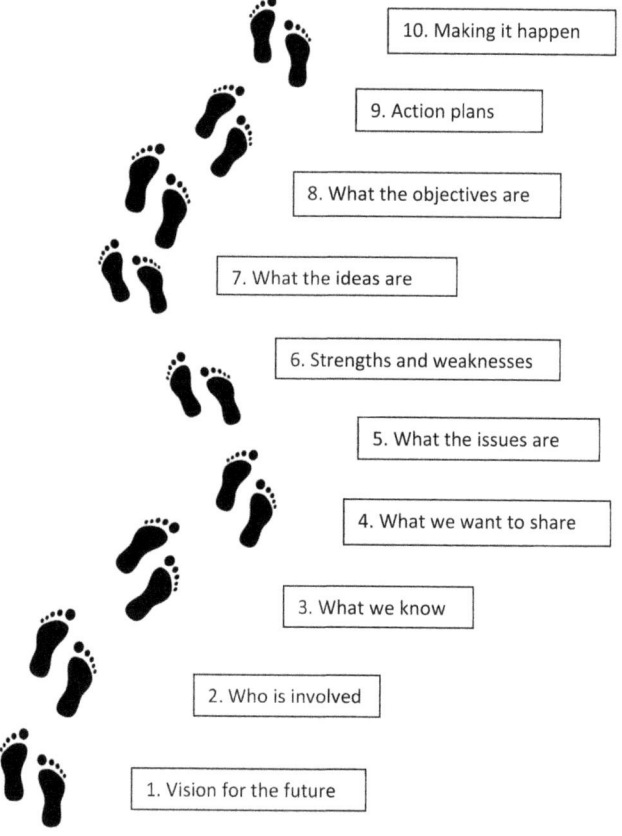

Figure 1.2 Stepping into tourism. Adapted from: Stepwise Heritage and Tourism (n.d.)

Figure 1.3 Karajarri tour leader Wynston Shovellor (centre) speaks to a group of students at a lagoon on Karajarri Country. He explains the significance of this waterscape to students, then demonstrates the Karajarri practice of *kuwiyinpijala*, where a person takes water in their hand and then blows it out, allowing the Country to know that person better. Credit: Co-author Bobbie Chew Bigby.

communities, so that communication remains strong, misunderstandings can be prevented and participation is ongoing.

Wynston spoke positively about the participatory and community-based nature of Karajarri's tourism model. He felt Karajarri tourism has developed at a steady pace that has been in line with what most Karajarri people are wanting (see Figure 1.3). He stated:

> Early on, there was a consultation with our Elders and the community about creating this strategic tourism plan. Tourism had started off with the rangers, how to manage their tourism area, and then with the strategic plan came the permit system up at Port Smith caravan park. Those were good first steps of trying to manage country better for us and for tourists. With the permit system, we just want to make sure everyone is comfortable and satisfied. What we generate from the permit fees, we want to give back to tourism and build shade shelters, picnic tables, creating better brochures, and employing more people on a casual basis… There are still gaps to be filled, but there's some promising outcomes in the near future.

This overview of the participatory and inclusive nature of the planning process presents an important reflection from a Karajarri man who is a former KTLA employee as well as a native of the Bidyadanga community. However, Uncle Joe offered a different perspective that lent vital insight into some of the difficulties and tensions regarding Karajarri governance and how this impacts tourism development both now and potentially in the future. According to Uncle Joe, many of these challenges arise from the fact that the Western governance models imposed through the required structure of Prescribed Bodies Corporate (PBCs) and Registered Native Title Bodies Corporate (RNTBCs) – are not appropriate or aligned with traditional Karajarri governance ideas and practices. Uncle Joe asserted:

> Our PBC is following rules and regulations that are quite foreign to us, to our Indigenous governance. They say that we are following the PBC's rules in that we have a sort of democracy, where majority rules. If you're polling on a particular issue, whoever raises the most hands passes the policy. But when it comes time to land management and issues that directly impact on our country, those protocols should not exist. Because in our culture, the custodians are the TOs, but they are also the custodians of a particular part of the country, the tribes and totemism.

This tension between Western and Karajarri forms of governance manifests not only in the protocols of PBC meetings or how the KTLA body and enterprises are structured, but as pointed out by Uncle Joe, also subsequently influences issues over which KTLA has authority, including aspects of land management. From Uncle Joe's perspective, this inappropriate imposition of Western governance also impacts how tourism is planned and implemented. Uncle Joe explained:

> You know with Karajarri we have three different dialects [or language-based groups] and they come from different parts of our homelands. Nadja is the coastal. To the northeast hinterland we've got the Naudu and to the southeast desert is the Nangu. And those people from those particular areas, they look after their own country and have particular say over what should happen – development, mining, tourism. They should be entitled to reject it or endorse it. But currently, we're working under a system of majority rules, so anyone can have a say over parts of other people's country, which is totally, totally foreign and inappropriate… all of a sudden we're having people establishing tourism ventures on somebody else's country without protocols being adhered to. I've been really unhappy with it, I suppose.

In a parallel discussion with Uncle Dooley on the extent to which Karajarri values are implemented in the planning and development of tourism, he shares Uncle Joe's observations on the differences in Western and Indigenous management. In terms of bridging Karajarri values with the Karajarri tourism model, Uncle Dooley stated: 'How do we do that? It's what we are all trying to rediscover. Because business from a Western perspective is totally different to running business from an Indigenous perspective'. Uncle Dooley's perspective underscores the cultural differences between governance and business management models and acknowledges that navigating this tension is a part of the journey for KTLA and the Karajarri community as whole in relation to tourism.

In his reflections on Karajarri's evolving tourism model, Uncle Dooley placed strong emphasis on the need for Karajarri tourism planners to try to understand their audiences and know who is being reached. This audience or market understanding is critical, in his eyes, not just in attempts to expand business to more targeted, interested visitors – a point shared by Petrine who advocates for finding and building niche markets that are receptive to Aboriginal cultural experiences. Moreover, Uncle Dooley believes that understanding audiences is critically important in order to know how to reach these people and open their eyes to what Karajarri tourism has to share. Uncle Dooley believes that speaking through the language of science is the key to this endeavour. He stated:

> I think that if we're going to promote our values in tourism ventures, we have to incorporate language, concepts and ideas from science where we can, to be able to explain. That's what I'm finding I'm doing now, having to explain it in a language that they [tourists] relate to. Because everyone is trying to explain it in an Indigenous way, and then they [tourists] scratch their heads. 'What are you blackfellas talking about?' ... Our relationship to country, why is it important, the importance of protecting spiritual sites related to water and across country... It's imperative on us, as keepers of the land, the keepers of country to say you have to listen to us. We are not talking bullshit. Your own science now has proven, substantiated what we have been saying all along... so I think science is one of our biggest tools as we use tourism as a vehicle to educate people and bring them into our domain.

This need for understanding audiences and finding the right science-based concepts and language to communicate with visitors is still part of an evolving process in developing the Karajarri tourism package. Uncle Dooley characterised this as part of the process of 'refining' the tourism offerings and ensuring that Karajarri 'get in full agreement amongst everybody about who is doing what... to work in unison so the [tourism] package appears as one, holistic package'. For Uncle Dooley, one part of what he refers to as 'refining the tourism package' implies ensuring that Karajarri people are informed, in agreement, clear and on board with tourism plans. But another critical piece of the refinement process in his eyes in geared towards the audience engagement. In this respect, Uncle Dooley believes that it is critical for Aboriginal tour leaders to communicate clearly with non-Indigenous visitors and explain why certain things are prohibited on Country or in relation to culture. He asserted:

> We have to teach visitors that when you come onto Country, there are these places you can and cannot go visit. And they have to understand why they can't go. For a lot of people who ask why they can't go there, Aboriginal people don't fully explain why... this is what I mean about refining [the tourism] more. You've got to explain to them why and how that relates to our beliefs, values, principles and spirituality that we must follow... I think we have to be able to accommodate peoples' inquisitiveness and thirst for better information... I think that's one of the key parts of more effective Indigenous tourism.

From Uncle Dooley's perspective, the task of planning and implementing Karajarri tourism that is in tune with traditional priorities and values is thus one that is seen to have important consequences not only for Karajarri people and Country, but for the visitors themselves as well. This Karajarri case study demonstrates one example of place-based governance which works at multiple levels and uses tourism as an implement of local community governance to achieve multiple goals. Country sits at the centre and Karajarri responsibilities as custodians shape the practices.

Multiple ways, multiple outcomes

This in-depth Karajarri case may lead readers to conclude that it is Indigenous peoples that are particularly guided to this place-based governance approach being considered in this chapter. In the Karajarri example, as one of many Aboriginal nations that hold pride as the oldest continuing human cultures on Earth, their place-based knowledge, custodianship and governance makes logical sense. But they are by no means the only communities with these forms of place-embedded governance.

There are many places that we could turn to in order to demonstrate the meaning, value, approaches and limits to place-based governance and place-based governance in tourism. Governance comes from the myriad local institutions, processes, cultures and value systems that people have co-constructed during their lifetimes, as well as during their ancestors' lifetimes. Numerous examples have been studied and shared, including:

- Ubuntu, described as:
  'A collection of values and practices that black people of Africa or of African origin view as making people authentic human beings. While the nuances of these values and practices vary across different ethnic groups, they all point to one thing – an authentic individual human being is part of a larger and more significant relational, communal, societal, environmental and spiritual world'. (Mugumbate & Chereni, 2020: vi)
- Buen Vivir, explained as:
  'The term Buen Vivir is best understood as an umbrella for a set of different positions... [Buen Vivir] are the Spanish words used in Latin America to describe alternatives to development focused on the good life in a broad sense. The term is actively used by social movements, and it has become a popular term in some government programmes and has even reached its way into two new Constitutions in Ecuador and Bolivia. It is a plural concept with two main entry points. On the one hand, it includes critical reactions to classical Western development theory. On the other hand, it refers to alternatives to development emerging from indigenous traditions, and in this sense the concept explores possibilities beyond the modern Eurocentric tradition'. (Gudynas, 2011: 441)

- Ol'lau in Palau:
  Palau has developed a Visitor Pledge and responsible tourism programme based on its culture of hospitality. 'Ol'au in Palauan means to invite someone into your space' (Galloway, 2022). Through this culturally infused approach, Palau shows a way that pledges can be made tools for community empowerment and expressions of real local hospitality as participating tourists '...can then redeem their points to unlock cultural and nature-based experiences that are normally reserved for Palauans and their close friends' (Galloway, 2022). This programme '...is offering a world-first initiative of 'gamifying' responsible tourism, whereby travellers will be offered exclusive experiences based on how they treat the environment and culture, not by how much they spend' (Galloway, 2022).
- Gross National Happiness (GNH) in Bhutan:
  'In his Coronation speech, the Fifth King, His Majesty Jigme Khesar Namgyel Wangchuck, said "I have been inspired in the way I look at things by Bhutan's development philosophy of Gross National Happiness ... to me it signifies simply 'Development with Values'". GNH at its core comprises a set of values that promote collective happiness as the end value of any development strategy. GNH might be described as:
    - Holistic: Recognising all the aspects of people's needs, be these spiritual or material, physical or social;
    - Balanced: Emphasising balanced progress towards the attributes of GNH;
    - Collective: Viewing happiness to be an all-encompassing collective phenomenon;
    - Sustainable: Pursuing well-being for both current and future generations;
    - Equitable: Achieving reasonable and equitable distributed level of well-being'. (Ura *et al.*, 2012: 6–7)

Each of these – Ubuntu, Buen Vivir, Ol'au and Gross National Happiness – has intersected and influenced forms of tourism developed in particular places. For instance, in the latter case of Bhutan, a low-volume, high-yield approach has been pursued in order to protect the Buddhist culture of the country and limit the negative impacts of tourism on community and ecology.

There are also the examples previously discussed in Higgins-Desbiolles *et al.* (2019: 1936–7): of the Guna of Panama and their Statute on Tourism (see also Pereiro *et al.*, 2012); the case of Lirrwi Tourism in Arnhem Land, Australia and their 'guiding principles' on tourism emphasising community-centric focus rather a tourism-centric one (see Lirrwi Tourism, n.d.); and the case of Kangaroo Island and its Tourism Optimisation Management Model initially prompted by community concern with an imposed day-tripper market (see Miller & Twining-Ward, 2005). It is also worthwhile revisiting Scheyvens (2006) detailed work on 'beach *fale*' tourism in Samoa which recounted how village control and benefit from tourism was secured through budget-friendly travel and following the principles of *fa'a Samoa*, the Samoan way of life (see Beautiful Samoa, n.d.). In numerous places around the world, including Ghana and India, the care for sacred groves and their responsible access by tourists presents another relevant case to consider (see Ormsby, 2012).

In drawing attention to these multiple examples, we highlight the fact that local embedding allows for pluralistic approaches for communities rather than imposed monocultures of integration into the global economy (see Figure 1.4). It is the monocultural approach to tourism that results in dangerous levels of dependency and sees power shift away from communities to remoter levels of governance, to multinational corporations and to possibly the 'placeless' international tourists. We therefore need to understand the structural contexts which enable such a power shift away from communities and how this might be countered.

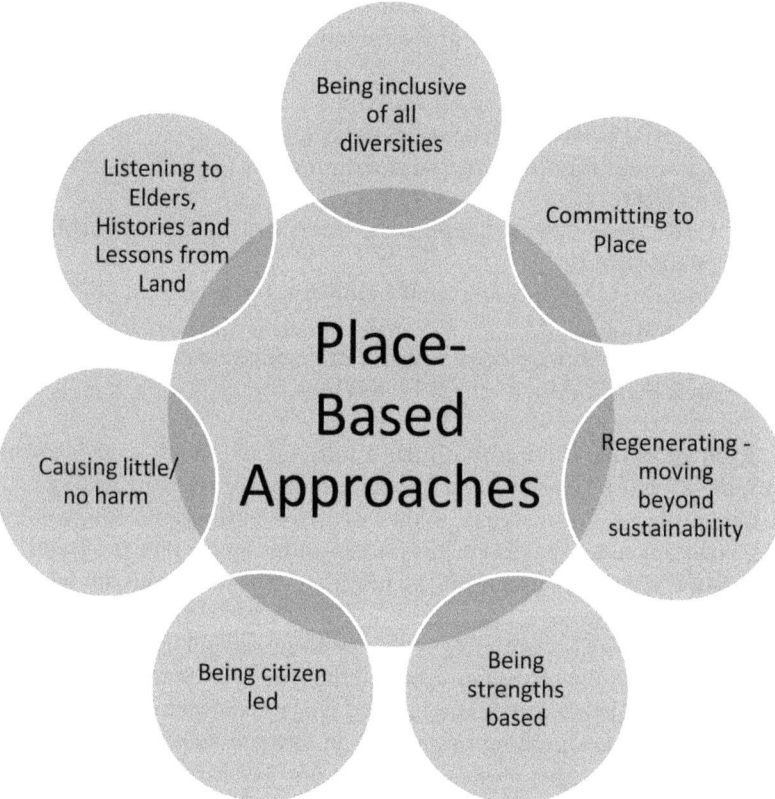

Figure 1.4 Principles of place-based approaches to tourism governance

Degrowth + Decolonising

> For some, placelessness has become the essential feature of the modern condition, and a very acute and painful one in many cases, such as those of exiles and refugees. (Escobar, 2001: 140)

Whether cast as the 'ongoingness of imperialism' (Higgins-Desbiolles, 2022), or globalising capitalism, modernising development (Escobar, 2001) or neoliberalism, the hegemonic and universalising practices of Western/Eurocentric pathways are contra to these ideas of place-based governance. It is important to understand that the wealth and power of these developed countries in many cases has been built on historic and ongoing colonial land and resource theft. More recently, this umbrella group of ideologies have in common a belief that everything in society as far as possible should be privatised and run for profit in the belief that this operates more efficiently. Increasingly fewer assets should be publicly owned and operated on a non-profit basis, but instead private ownership and profit-seeking are facilitated. Hospitals, universities, schools, social services, national parks, prisons and national security are no longer exempt. In this ideological worldview, employment should be kept low waged, non-unionised and 'flexible' with zero-hour contracts and precarity increasingly normalised. Most importantly, people are less enabled to see themselves as citizens with rights and responsibilities and instead encouraged to more strongly identify as consumers, in a system built on perpetual growth (see Sklair, 2016). Almost all governments, whether leftist, centrist or rightist, or designated as capitalist or socialist, are committed to perpetual growth, via market systems that are underpinned by a 'culture-ideology' of consumerism. A key feature of these systems is not the 'trickle down' of benefits to the poorest, but rather growing inequalities that are in fact a 'trickle up' to the wealthy elite. COVID has only compounded such dynamics; Chancel reported 'billionaires accumulated €3.6 trillion... of wealth during a crisis in which the World Bank estimates that some 100 million people have fallen into extreme poverty' (cited in Saraiva & Migliaccio, 2021).

Extractive forms of tourism and tourism monocultures have been facilitated in this context. Linehan *et al.* (2020: 9) noted: 'Ultimately, tourism dwells on the feelings of tourists rather than the toured objects where colonialism is viewed as symbol embodied with imagery, expectations and powers'. In this exploitative and extractive approach, local people become 'toured objects' and tourism industry interests prevail. As recounted in the introduction, this is why local communities sometimes oppose tourism development and form or join social movements to oppose the imposition of tourism. Additionally, these forms of extractive tourism overshadow, outnumber and out compete forms of tourism and leisure that are for the public good such as social tourism.

People are resisting these exploitative forces and continuing to challenge and respond to these modes of usurping space, place and community. Community struggles have adopted strategies that subvert these power plays and present alternative visions of community life, including delinking and degrowth (discussed in the Introduction). From the Zapatistas of Mexico with the practices of 'zapatismo', to the Kurds 'lab' in Rojava, to the cooperatives of Mondragón in Spain, communities are creating alternatives to the systems of domination described above. Importantly, these movements are connecting up and exchanging ideas through 'tourism'. For instance, the Zapatistas sent a delegation to Spain in 2021 marking the 500th anniversary of the Spanish 'conquest' of Mexico which was followed by tours throughout Europe to exchange ideas for social justice activism and solidarity (Vidal, 2021). These are one form of decolonial tourism, or 'Detours'. While tourism promotion and development are tied up with complex assertions of power to access, control and benefit from the resources held in communities around the globe, tourism is contested and a space for NGO advocacy, social movement formation and indeed these usurpations of tourism for social justice purposes such as Detours.

The idea of 'related localisms' introduced in the Introduction, offers some valuable solutions for furthering solidarity and these ideas have been explored from diverse corners of thought. For instance, sociologist Leslie Sklair proposed a similar concept:

> This is not the fantasy of cellular localism; my vision of an alternative, radical, progressive globalization envisages networks of small producer-consumer cooperatives (PCC) cooperating at a variety of levels, primarily to ensure a decent standard of living for everyone on the planet… all states end up being hierarchical, and that only in smallscale communities like PCCs, locally or globally linked via the Internet, can we avoid this inevitable slippery slope. (2016: 2–3)

In her 'Decolonial Manifesto', Hirmer envisioned:

> … a sustainable decolonised world-order is of a third kind: one where, instead of a reversal of centre-periphery relations or of an all-encompassing amalgamating centre, a multitude of interconnected units coexist, functioning at the same time as centres to themselves and peripheries to other centres. In this vision, power is diffused and boundaries between these centres may be porous, permitting non-hierarchical exchanges in multiple directions, as pares inter pares [equal among equals]. (2020: 124)

In such contexts, we could anticipate diverse forms and engagements with tourism as appropriate to particular places. Starting strategies might include practices of counter-mapping, a counter-cartographic practice that employs a diverse range of mapping methods, such as geospatial techniques, data visualisation, storytelling, art and performance that build on a body of collaborative work by critical cartographers, geographers, artists, educators and activists. Such counter-mapping practices have generated tools for collective mobilisation and produced alternative visions of contemporary space and their future possibilities (see for instance, Boukhris, 2017; Chapter 5 this volume).

Tools such as the model of placed-based cooperative enquiry offered by Wooltorton *et al.* (2020) could be adapted to offer communities another tool with which to build place-based governance approaches to tourism. Such approaches integrate experiential, creative, conceptual and post conceptual learning forms. Using multiples rounds of cooperative inquiry, human relationships are built and human–ecological relationships are nurtured. These may serve as foundations for co-construction of community and place-based governance. This process emphasises local attachments to place and appreciation of local ways of being, knowing and doing so that local places may be better sustained through tourism. These would be ongoing cycles of engagement in community building and tourism governance with iterations of these four stages at each cycle. These work to deepen experience and knowledge of people, place and tourism for building better places.

There are many more worthy of consideration. The use of experiential learning and place-based education would have relevant approaches to learn from and adapt to place-based governance in tourism and to integrate into tourism education. The recent popularity of regenerative approaches could be used for considerations of co-building placemaking and place renewal through tourism. Such models assist in conceptualising tourism as nested within community and ecological systems and help mitigate against the tourism-centric thinking found in some quarters.

We ourselves believe there is much to be learned from Indigenous scholars and knowledge-holders. Associate Professor Mary Graham (Kombumerri/Wakka Wakka) has explained an ethos of respecting a law of obligation rather than rights approach which is derived from thinking in relation and relatedness rather than the current predominate survivalist mode activated by the crises we face (Graham, 2020). Graham explained that caring for land teaches ethics and also that ethics is done in the doing. This points to the vital need for pluriversal thinking and research to guide future work in place-based governance. But it is a unity within this diversity that is also essential. This is some of the most vital work before us:

> There must be some convergence among nations on the idea of what the end objective of development and progress should be. There cannot be enduring peace, prosperity, equality and brotherhood in this world if our aims are so separate and divergent – if we do not accept that in the end we are people, all alike, sharing the earth among ourselves and also with other sentient beings, all of whom have an equal role and state of this planet and its players. (the Crown Prince of Bhutan cited in Ura & Galay, 2004: xii)

Conclusion

In 2020, it was reported that Royal Caribbean International (RCI) cruise company has an agreement with Vanuatu to develop a 'private cruise resort' at Lelepa (RCI, n.d.). Chua reported: 'Having a destination under cruise line control means arrivals and departures are guaranteed. The line may bypass strict customs and health screenings should a similar pandemic to COVID-19 reoccur' (Chua, 2020). Such developments are deeply problematic when we consider the need for greater place-based governance and care.

While we think and connect locally, we cannot be impervious to what is occurring globally as human impacts have wrought widescale change, including global climate change. Scientists have proposed a 'planetary boundary' (PB) framework with a view to finding a way to deal with complex Earth systems impacted by human demands and impacts. '…by identifying a safe operating space for humanity on Earth, the PB framework can make a valuable contribution to decisionmakers in charting desirable courses for societal development' (Steffen *et al.*, 2015: 736). This needs to be employed simultaneously to the myriad of place-based governance actions occurring in communities around the globe. It is both local and global approaches that are needed.

Values and relationships are essential pillars to these efforts. Engaging with the newest thinking on just transitions, we need all levels of governance empowering all diversity of beings to be recognised, valued and involved in devising pathways forward. The plural versions of these terms – 'communities', 'ecologies' and 'places' – guide us in respecting diversity but it is also diversity in relatedness not in disconnected, selfish isolation. The places where we live, work and play can teach us lessons for living well. Place-based governance for tourism can shape tourism to be suited to people and place. It presents a viable alternative to imposed and exploitative forms of tourism and helps us better live within place and planetary boundaries.

Note

(1) The titles of 'Uncle' and 'Auntie' are used in many Aboriginal communities as terms of respect for Elders. Elders are people who have gained recognition as custodians of knowledge and law and hold permission to share this with the right people, in the right contexts. Non-Aboriginal people should check the appropriateness of using these terms, which are often built on close relationships (http://www.indigenousteaching.com/glossary-terms).

References

Aikau, H.K. and Gonzalez, V.V. (2019) (eds) *Detours: A Decolonial Guide to Hawai'i*. Durham: Duke University Press.

Altman, I. and Low, S.M. (eds) (1992) *Place Attachment*. New York: Plenum Press.

Beautiful Samoa (n.d.) Samoan fales. See https://www.samoa.travel/stories/fales/ (accessed 11 February 2022).
Bianchi, R. (2018) The political economy of tourism development: A critical review. *Annals of Tourism Research* 70, 88–102. https://doi.org/10.1016/j.annals.2017.08.005.
Boukhris, L. (2017) The Black Paris project: The production and reception of a counter-hegemonic tourism narrative in postcolonial Paris. *Journal of Sustainable Tourism* 25 (5), 684–702. https://doi.org/10.1080/09669582.2017.1291651.
Chua, B. (2020) Confident Royal Caribbean charts Aussie expansion with new ships and private island. See https://cruisepassenger.com.au/news/quantum-of-the-seas-to-be-deployed-down-under/ (accessed 2 January 2022).
Escobar, A. (2001) Culture sits in places: Reflections on globalism and subaltern strategies of localization. *Political Geography* 20, 139–174.
Galloway, L. (2022) Palau's world-first 'good traveller' incentive. BBC Travel (Online). See https://www.bbc.com/travel/article/20220517-palaus-world-first-good-traveller-incentive (accessed 20 May 2022).
Giridharadas, A. (2010) The struggle of the global placeless. *New York Times* (Online). See https://www.nytimes.com/2010/03/27/us/27iht-currents.html (accessed 1 January 2022).
Graham, M. (2020) Connecting to place, caring for Country. Workshop 12 March 2020, Brisbane, Australia.
Gudynas, E. (2011) Buen Vivir: Today's tomorrow. *Development* 54 441–447. https://doi.org/10.1057/dev.2011.86.
Hambleton, R. (2011) Place-based leadership in a global era. *Commonwealth Journal of Local Governance* 8–9, 8–32. https://search.informit.org/doi/10.3316/informit.935224537151510.
Heffron, R.J. and McCauley, D. (2018) What is the 'Just Transition'? *Geoforum* 88, 74–77.
Heron, J. (1996) *Co-Operative Inquiry: Research into the Human Condition*. London: Sage.
Hidalgo, M.C. and Hernández, B. (2001) Place attachment: Conceptual and empirical questions. *Journal of Environmental Psychology* 21, 273–281.
Higgins-Desbiolles, F. (2018) Sustainable tourism: Sustaining tourism or something more? *Tourism Management Perspectives* 25, 157–160.
Higgins-Desbiolles, F., Carnicelli, S., Krolikowski, C., Wijesinghe, G. and Boluk, K. (2019) Degrowing tourism: Rethinking tourism *Journal of Sustainable Tourism* 27 (12), 1926–1944. https://doi.org/10.1080/09669582.2019.1601732.
Higgins-Desbiolles, F. (2022) The ongoingness of imperialism: The problem of tourism dependency and the promise of radical equality. *Annals of Tourism Research* 94, 103382. https://doi.org/10.1016/j.annals.2022.103382.
Hirmer, M. (2020) A manifesto for decolonial subversions. *Decolonial Subversions*, 120–130. See http://decolonialsubversions.org/manifesto.html (accessed 11 February 2022).
hooks, b. (2009) *Belonging: A Culture of Place*. New York: Routledge.
KTLA (n.d.) Visitor pass. See https://www.ktla.org.au/visitor-pass (accessed 23 February 2022).
KTLA (2016) Karajarri Tourism Strategy 2016–2021. See https://www.ktla.org.au/publications (accessed 23 February 2022).
Kojola, E. and Agyeman, J. (2021) Just transitions and labor. In B. Schaefer Caniglia, A. Jorgenson, S.A. Malin, L. Peek, D.N. Pellow and X. Huang (eds) *Handbook of Environmental Sociology. Handbooks of Sociology and Social Research* (pp. 115–138). Cham: Springer. https://doi.org/10.1007/978-3-030-77712-8_7.
Kuokkanen, R. (2011) Indigenous economies, theories of Subsistence, and women: Exploring the social economy model for Indigenous governance. *American Indian Quarterly* 35 (2), 215–240. https://doi.org/10.5250/amerindiquar.35.2.0215.
Lacher, R.G. and Nepal, S.K. (2010) Dependency and development in Northern Thailand. *Annals of Tourism Research* 37 (4), 947–968.

Lalicic, L. and Garaux, M. (2022) Tourism-induced place change: The role of place attachment, emotions, and tourism concern in predicting supportive or oppositional behavioral responses. *Journal of Travel Research* 61 (1), 202–213. https://doi.org/10.1177/004728752096.

Latouche, S. (2009) *Farewell to Growth*. Cambridge: Polity.

Linehan, D., Clark, I.D. and Xie, P.F. (2020) Introduction. In D. Linehan, I.D. Clark and P.F. Xie (eds) *Global Transformations in Tourist Destinations* (pp. 1–11). Cheltenham: Edward Elgar.

Lirrwi Tourism (n.d.) Lirrwi Tourism guiding principles. See https://www.lirrwitourism.com.au/guiding-principles (accessed 11 February 2022).

Lukermann, F. (1964) Geography as a formal intellectual discipline and the way in which it contributes to human knowledge. *Canadian Geographer / Le Géographe Canadien* 8, 167–172. https://doi.org/10.1111/j.1541-0064.1964.tb00605.x.

Mason, C. (2021) Virtues, vices and place attachment. Keynote address to the Undergraduate Philosophy Journal's Virtual Conference, 19–20 June 2021. See https://ir.canterbury.ac.nz/handle/10092/102389 (accessed 7 February 2022).

Miller, G. and Twining-Ward, L. (2005) Tourism optimization management model. In G. Miller and L. Twining-Ward (eds) *Monitoring for a Sustainable Tourism Transition: The Challenge of Developing and Using Indicators* (pp. 201–232). Oxon: CABI.

Mugumbate, J.R. and Chereni, A. (2020) Editorial: Now, the theory of Ubuntu has its space in social work. *African Journal of Social Work* 10 (1), v–xv.

Ormsby, A. (2012) Perceptions of tourism at sacred groves in Ghana and India. *Recreation in Society in Africa, Asia and Latin America* 3 (1), 1–18.

Parajuli, P. (1996) Ecological ethnicity in the making: Developmentalist hegemonies and emergent identities in India. *Identities* 3 (1–2), 15–59.

Pereiro, X., de Leon, C., Martinez Mauri, M., Ventocilla, J. and del Valle, Y. (2012) Los Turistores Kunas Anthropologia del Turismo Etnico en Panama. Palma: Universitat de les Illes Balears.

Project Forever Waiheke (2021) 'Waiheke is a community, not a commodity': Stakeholder perspectives on future Waiheke tourism. See https://tinyurl.com/2ewbptey (accessed 5 February 2022).

Relph, E. (1976) *Place and Placelessness*. London: Pion.

RCI (n.d.) Perfect day at Lelepa. See https://www.royalcaribbean.com/aus/en/lelepa-cruises (accessed 5 February 2022).

Saraiva, A. and Migliaccio, A. (2021, 7 December) Pandemic has been great for billionaires, says Piketty Lab. *Financial Review*. See https://www.afr.com/wealth/investing/pandemic-has-been-great-for-billionaires-piketty-lab-says-20211207-p59fnm (accessed 11 February 2022).

Saura, B., Capestro, M. and Bova, H. (2002) Continuity of bodies: The infant's placenta and the island's navel in Eastern Polynesia. *The Journal of the Polynesian Society* 111 (2), 127–145. http://www.jstor.org/stable/20707058.

Scheyvens, R. (2006) Sun, sand, and beach fale: Benefiting from backpackers – the Samoan way. *Tourism Recreation Research* 31 (3), 75–86. https://doi.org/10.1080/02508281.2006.11081507.

Sklair, L. (2016) The end of the world or the end of Capitalism? *Global Dialogue: Newsletter for the International Sociological Association* March, 22–23.

Steffen, W., Richardson, K., Rockström, J., Cornell, S.E., Fetzer, I., Bennett, E.M., Biggs, R., Carpenter, S.R., De Vries, W., De Wit, C.A., Folke, C., Gerten, D., Heinke, J., Mace, G.M., Persson, L.M., Ramanathan, V., Reyers, B. and Sörlin, S. (2015) Planetary boundaries: Guiding human development on a changing planet. *Science* 347 (6223), 1259855. https://doi.org/10.1126/science.1259855.

Stepwise Heritage and Tourism (n.d.) Stepping Stones for Tourism. See http://www.stepwise.net.au/planning/steppingstones_tourism.php (accessed 3 January 2022).

Te Ara (n.d.) Tūrangawaewae – a place to stand. *The Encyclopedia of New Zealand* (Online). See https://teara.govt.nz/en/papatuanuku-the-land/page-5 (accessed 1 January 2022).

Tomassini, L. and Cavagnaro, E. (2020) The novel spaces and power-geometries in tourism and hospitality after 2020 will belong to the 'local'. *Tourism Geographies* 22 (3), 713–719. https://doi.org/10.1080/14616688.2020.1757747.

Tuan, Y.-F. (2012) Space and place: Humanistic perspective. In S. Gale and G. Olsson (eds) *Philosophy in Geography* (pp. 387–422). Amsterdam: Springer.

UDLA (2018) Karajarri Cultural & Tourism Hub: A Visioning Report. Prepared for KTLA, October 2018.

Ura, K., Alkire, S., Zangmo, T. and Wagdi, K. (2012) An extensive analysis of GNH Index. See https://ophi.org.uk/wp-content/uploads/Ura_et_al_Extensive_analysis_of_GNH_index_2012.pdf (accessed 11 February 2022).

Ura, K. and Galay, K. (2004) Gross National Happiness and development. Proceedings of the First International Seminar on Operationalization of Gross National Happiness. See https://www.bhutanstudies.org.bt/publicationFiles/ConferenceProceedings/GNHandDevelopment/1GNH%20Conference.pdf (accessed 3 January 2022).

Vidal, M. (2021) Zapatistas 'invade' Madrid to mark Spanish conquest anniversary. See https://www.aljazeera.com/news/2021/8/13/zapatistas-invade-madrid-to-mark-spanish-conquest-anniversary (accessed 13 February 2022).

Wooltorton, S., Collard, L., Horwitz, P., Poelina, A. and Palmer, D. (2020) Sharing a place-based Indigenous methodology and learnings. *Environmental Education Research* 26 (7), 917-934. https://doi.org/10.1080/13504622.2020.1773407.

2 Circular *Oikonomia*, Posthumanism and Local Space to Socialise Tourism

Lucia Tomassini and Elena Cavagnaro

Introduction

The rapid growth of tourism and global mobility before the COVID-19 outbreak and the global crisis experienced afterwards have been challenging the perception of the global and local context within which we live and travel. In *Down to Earth: Politics in the New Climate Regime* (2018) and *After Lockdown: A Metamorphosis* (2021a), Bruno Latour stresses how this historical time urges us to adopt a novel perspective on the enmeshed environmental and socioeconomic crisis. For him, this means adopting the 'down to earth' perspective of terrestrials entangled with other terrestrials, all belonging to a flattened topography in which the dimensions of global and local are merged. Such an approach contrasts with the 'out of this world' perspective that has been allowing exploitative approaches both to nature and to human and non-human beings (Latour, 2018). Moreover, Latour argues that, after the COVID-19 outbreak:

> It is very difficult for most people used to the industrialised way of life, with its dream of infinite space and its insistence on emancipation and relentless growth and development, to suddenly sense that it is instead enveloped, confined, tucked inside a closed space where their concerns have to be shared with new entities: other people of course, but also viruses, soils, coal, oil, water, and, worst of all, this damned, constantly shifting climate. (2021b: n.p.)

The pandemic crisis therefore appears as a warning and, as such, also as an opportunity for a change of perspective over the space within which we live and travel, since 'all the resources of sciences, humanities, and arts, will have to be mobilised once again to shift attention to our shared terrestrial condition' (Latour, 2021b: n.p.).

Within this context, our study explores how the COVID-19 global crisis – together with its uneven social justice, unbalanced power relations and global-local (im)mobilities – prompts us to rethink the space inside and outside tourism. It does so by investigating the active role that the local space can play to 'socialise tourism', by re-centring it within the society and re-orienting it towards the environmental and social needs and well-being of the local dimension and its dwellers (Higgins-Desbiolles, 2020; Higgins-Desbiolles *et al.*, 2022; Higgins-Desbiolles & Bigby, 2022; Tomassini & Cavagnaro, 2020). To clarify the terminology, we concur with Higgins-Desbiolles and Bigby (2022) in referring to 'local' not merely as a group of people pertaining to a place, but also encompassing the local (human) community and the local ecology of human and non-human beings, spanning both present and future generations. Moreover, drawing on Latour's (2007) Actor Network Theory, we use the notion of 'local' not in juxtaposition to the global, but as part of a flattened topography where the dimensions of the local and the global are entangled and merged while hosting interdependent terrestrials. Such human and non-human terrestrials share what the cultural geographer Doreen Massey (2005) identifies as 'throwntogetherness'. James Oliver (2020: n.p.) stresses that Massey (2005) used this term 'for thinking about (and with) place: as a "time-space" of relational encounters, open and progressive, full of potential, and of the eventfulness of place'. Hence the notion of throwntogetherness conveys an engagement understood as an emplacement through time (Oliver, 2020).

Building on the above, we further explore the pivotal role that the local space can play to socialise tourism by: (1) re-focusing it within the society and places in which it takes place (Higgins-Desbiolles *et al.*, 2022), (2) reconnecting its sociological space with the one of citizenship (Tomassini *et al.*, 2021) and (3) critically socialising animal-based tourism (Kline, 2022). In so doing, we use the lens of posthumanism and we refer to 'social' not as a solely human domain (Braidotti, 2011, 2013; Ferrando, 2019). Drawing on theorisations of the circular economy (Geissdoerfer *et al.*, 2017; McDonough & Braungart, 2003; Stahel, 2019) and the circularity archetype (Bradley, 2012) – here reformulated as a 'circular *oikonomia*' (Leshem, 2013, 2016) – this study interweaves features of the sociology of space (Massey, 1994, 2005) and posthumanism (Braidotti, 2011, 2013) to offer a novel theoretical framework to explore the local turn in tourism studies (Higgins-Desbiolles & Bigby, 2022). These theoretical lenses are combined in our analysis because they all share an advocacy of the creation of multiple ethical relations with 'multiple' entities within an active space. Thereby, this novel theoretical ground allows opening the way to imagine a future tourism rooted in the 'eventfulness' of the local space – as space is understood as an 'event' made of relational encounters (Massey, 2005) – together with the possibilities of affirmative ethics in such local space. Moreover, this theoretical framework allows delineating new lines of research to envision tourism as interdependent upon the well-being of the local space in which it takes place.

The Circular Economy as a Circular Oikonomia

The notion of a circular economy was conceived in the 1970s and is rooted in studies of non-linear systems mimicking living systems (Stahel, 2019). In its main theorisations and practical implications, the circular economy is understood as the regeneration and upcycling of products and (raw) materials in product-oriented industries. Murray *et al.* (2017: 369) define a circular economy as: 'the redesign of processes and cycling of materials'. McDonough and Braungart (2003) consider it as an eco-effective, 'cradle-to-cradle' practice where products and (raw) materials are designed to be continuously regenerated through loops of 'make–use–upcycle'. The Ellen MacArthur Foundation states that: 'a circular economy decouples economic activity from the consumption of finite resources… It is a resilient system that is good for business, people, and the environment' (n.d., para. 2). In 2015, the European Commission adopted the *Circular Economy Action Plan* as a strategy to accelerate Europe's transition towards a circular economy (European Commission, 2015), with the plan being updated and renewed in 2020 (European Commission, 2020). Moreover, the United Nations Industrial Development Organisation and the Department of Economic and Social Affairs acknowledged this European Action Plan as a best practice to stimulate the implementation of the UN Sustainable Development Goals (SDGs) and the 2030 Agenda. Despite this increasing interest, however, the circular economy paradigm is still largely under-investigated and under-theorised in tourism studies (Boluk *et al.*, 2019; Murray *et al.*, 2017; Sørensen & Bærenholdt, 2020).

The idea of circularity is an ancient archetype that has been used through the centuries to comprehend the biological processes of the Earth, its ecosystems, the cycle of the seasons, the planets' orbits, the carbon cycle, spiritual life and religious rituals (Bradley, 2012). Hence the circular economy – together with the idea of circularity – appears as a promising basis from which to critically rethink the future of tourism, its sustainability and its relationship with the local space in which it takes place (Boluk *et al.*, 2019; Geissdoerfer *et al.*, 2017; Stahel, 2019). Hitherto, the circular economy has mainly been investigated and theorised with regards to a product-oriented vision focusing on the recycling, upcycling and redesign of products and materials (McDonough & Braungart, 2003). In contrast, the potentialities of the circular economy and circular regenerative processes to socialise tourism through the creation of social value in the local space in which such processes take place has not yet been explored. Thus, in this study we seek to begin to address that omission by arguing that the circular economy can contribute in a broad sense to the task of socialising tourism: by articulating the multiplicity of novel relations, connections and networks among (human and non-human) stakeholders, i.e. terrestrials (Latour, 2007, 2018); by reshaping the existing power relations among them; by re-orienting the tourism experience towards a geographical proximity; and by re-centring tourism on the well-being of both the local space and its terrestrial dwellers.

By exploring the potential of a circular economy for the local turn in tourism, we adopt a concept of 'economy' that is derived from the ancient Greek notion of *oikonomia* (from the ancient Greek: *Oikos* – household – and *Nomos* – rule). Oikonomia diverges from contemporary economics in its deeper relationship to ethics and worthy goals that are not purely economic in the management of the resources of the 'oikos' (Leshem, 2013, 2016). Latouche (2004) argues that the notion of oikonomia – with a reference to Aristotle – is historically juxtaposed to the idea of an economics purely aimed at increasing profit. As such, Latouche defines oikonomia as intersecting with Ivan Illich's (1973, 1979) reflections on vernacular and convivial economics able to re-incorporate economics into the social dimension and society. This means embracing a holistic approach and a stronger relationship with ethics by integrating the idea of circular processes into an economics model that seeks the achievement of just and stable equilibria for a society made of interdependent terrestrials. In this study, therefore, the notion of a circular economy moulds the circularity archetype into an oikonomia, prompting an ethical engagement and entanglement with our 'household' – i.e. the Earth – as encompassing all the human and non-human beings that inhabit it as 'terrestrials' thrown together in one shared place. As Massey highlights:

> What is special about place is precisely that throwntogetherness, the unavoidable challenge of negotiating a here-and-now (itself drawing on a history and a geography of thens and theres); and a negotiation which must take place within and between both human and nonhuman. This is no way denies a sense of wonder: what could be more stirring than walking the high fells in the knowledge of the history and the geography that has made them here today. This is the event of place. (2005: 140)

This notion of 'throwntogetherness' (Massey, 2005) resonates with Italo Calvino's reflection in *Invisible Cities* (2003) of possible, multiple, imaginary, eventful and metaphoric cities as fictionally narrated by Marco Polo to the Grand Kahn. Marco Polo concludes his recount of *Invisible Cities* by saying:

> The hell of the living is not something that will be. If there is one, it is what is already here, the hell we live in every day, that we make by being together. There are two ways to escape suffering it. The first is easy for many: accept the hell, and become such a part of it that you can no longer see it. The second is risky and demands constant vigilance and apprehension: seek and learn to recognize who and what, in the midst of hell, are not hell, then make them endure, give them space. (Calvino, 2003: 166)

This therefore re-elaborates the notion of throwntogetherness (Massey, 2005) focusing on the ethical engagement required in an open, fluid 'eventfulness' space in which multiple relations and connections can happen (Oliver, 2020).

Hence, we offer an understanding of a circular economy as an oikonomia ethically driven towards praiseworthy goals beyond the pursuit of merely economic outcomes (Leshem, 2013, 2016); an oikonomia grounded and entangled in the local space where it takes place. By integrating into the conceptualisation of tourism the notion of circular regenerative processes of products, services, mobilities and nature as a way of enacting sustainability (Geissdoerfer *et al.*, 2017), we open up a route to rethink tourism and its practices, as an agent of positive transformation and hope (Ateljevic, 2020), as well as a force for social and ecological justice (Higgins-Desbiolles, 2006, 2020). We argue that a circular oikonomia in the sociorelational space of tourism can activate deeper regenerative processes for places, nature and living creatures by enacting a multiplicity of novel relations, connections and networks among a plurality of stakeholders understood as interdependent and entangled terrestrials (Latour, 2007; 2018; Massey, 1994, 2005). This means prompting an ethical engagement as emplacement (Oliver, 2020) because the 'where' in which these relations take place plays a pivotal role (Barad, 2007; Bright *et al.*, 2013). Using the words of Tuck and McKenzie:

> We urge readers and colleagues to reconsider place and its implications, not because it offers a generalizable theory or universal interpretation, but because generalizability and universality are impossibilities anyway, in no small part because place matters and place is always specific. (2015: 637)

By drawing on the idea of an oikonomia (Leshem, 2013, 2016) grounded in a local time-space and made of thrown together terrestrials (Latour, 2007, 2018; Massey, 1994, 2005), we understand the circular economy as enduring, cyclic and regenerative processes that are building a persistent equilibrium in a flattened interconnected topography made of entangled and merged global and local dimensions. This understanding is compatible with the ancient Greek notion of *Aion* (αἰών), identifying a vital force and never-ending lifetime and, as such, a cyclic time casting across eternity (Braidotti, 2013; Parker, 2016). It also conforms with the concept of *Zoe* (ωη): the non-human, productive, enduring, and vital force that Rosi Braidotti (2011, 2013) identifies in her Posthuman Critical Theory. Envisioning the future of tourism and its 'local turn' through the lens of a circular oikonomia therefore means envisioning an enduring equilibrium made of regenerative, cyclical processes that socialise tourism, via multiple ethical relations, connections and links of affectivity, responsibility and interdependence with multiple others, in an open space permeated by an eternal vital force – i.e. *Aion* (αἰών) and *Zoe* (ωη). In this way, a novel tourism space can arise grounded in the well-being of both the human and non-human local community and the global 'household' we inhabit, belong to and share (Tomassini & Cavagnaro, 2020). For this scope, our reflection also inevitably embraces a posthuman approach challenging anthropocentric ontology and epistemology (Braidotti, 2011, 2013; Ferrando, 2019).

Posthumanism

Posthumanism is an anti-individualistic philosophical approach and an ethical position extending moral concerns to non-human animals, beings and entities that inhabit the world (Braidotti, 2006; Haraway, 2003, 2006). It does so by questioning humanism and the classical ideal of man as the measure of all things that has served to assign to the human being the position of a distinctive 'terrestrial', superior to other non-human animals and entities (Braidotti, 2013). Such humanist conceptualisations have widely permeated Western philosophy, consolidating a pervasive dualism between nature and culture (Braidotti, 2013). In contrast, posthumanism denies that dualism by comprehending the human as entangled with the environment and nature (Latour, 2007, 2018; Braidotti, 2013). Moreover, it takes a critical stance on the existence of ideal (Western/Eurocentric) civilisation models, that have been affirming a sexualised, racialised, naturalised difference between subjects embodying such ideal models and the 'others'.

In our study we argue that posthumanism offers a novel lens to investigate the local turn in tourism studies and the socialisation of tourism (Higgins-Desbiolles *et al.*, 2022; Kline, 2022). It does so by postulating a networked sociological space in which all 'terrestrials' are interconnected and interdependent in a flat ontology (Latour, 2007, 2018). The posthuman approach is largely grounded in the rhizomatic subjectivity and epistemology elaborated by Deleuze and Guattari (1987) as anti-hierarchical, without centre or periphery, without a beginning, middle, end or a privileged point of view. The posthuman condition understands the human subjectivity as co-created with non-human entities (Hayles, 1999), as Braidotti argues:

> I define the critical posthuman subject within an eco-philosophy of multiple belongings, as a relational subject constituted in and by multiplicity, that is to say a subject that works across differences and is also internally differentiated, but still grounded and accountable. (2013: 49)

Posthumanism challenges the idea of 'otherness' and its implications in terms of cultural dominance, power relationships and anthropocentric stances (Braidotti, 2011, 2013). In other words, it takes a critical stance on the idea of difference and otherness as a negative counterpart where the others are usually sexualised, racialised and naturalised with a sense of exclusion or disqualification. Such an approach is still novel and barely explored in tourism studies (Cohen, 2019; Guia, 2021; Guia & Jamal,

2020). Nevertheless, for Guia and Jamal (2020), applying posthumanism to tourism studies can offer a different theoretical ground on which to reconsider research about neo-colonialism and neoliberalism, while offering a post-anthropocentric understanding of tourism. As they explain:

> Posthumanist methodologies are thus needed in tourism research if we are to challenge the habitual anthropocentric gaze taken by tourism researchers [...] to rethink our conceptions of tourists' experiences [...] to challenge tourists' visual imagery (as well as that of hosts), which tends to reproduce 'everyday banalities' [...] to identify and avoid indefensible binaries of 'either-or' commonly used in research projects. (Guia & Jamal, 2020: 2–3)

Since the Enlightenment period, tourism has been historically linked to the idea of exploring an 'otherness' made of distant, exotic places and different cultures, often based on unbalanced power relationships and unjust practices (Bianchi, 2018; Higgins-Desbiolles, 2018; Jamal, 2019; Jamal & Higham, 2021; Tomassini *et al.*, 2021). Therefore, envisioning a tourism future beyond the dualistic, divisive notion of 'otherness' and rooted instead in the needs and well-being of the local space – with its relational encounters within the human and non-human local community – is an exercise of imagination for new – and fairer – tourism practices (Cohen, 2019; Guia, 2021; Guia & Jamal, 2020) grounded in the interplay between socialism (Higgins-Desbiolles, 2020; Higgins-Desbiolles *et al.*, 2022) and posthumanism (Braidotti, 2011, 2013; Kline, 2022).

We argue that posthumanism can contribute deeply to socialising tourism in the local space by enacting an ethics that is affirmative. The interplay between the notion of throwntogetherness (Massey, 2005) and affirmative ethics (Braidotti, 2011, 2013) can in turn contribute to the re-centring of tourism practices on the safety and well-being of the local community – both human and non-human – because affirmative ethics 'rests on an enlarged sense of inter-connection between self and others, including the non-human or "earth" others, by removing the obstacles of self-centred individualism on the one hand and the barriers of negativity on the other' (Braidotti, 2013: 190). This means enacting constructive and positive collective bonds of ethical engagement to construct a global affective community – emplaced locally – which is consolidated and strengthened by the positive ground of ethical joint projects and activities (Braidotti, 2013). 'To be posthuman [...] implies a new way of combining ethical values with the well-being of an enlarged sense of community, which includes one's territorial or environmental interconnections' (Braidotti, 2013: 190). Thus, 'the pursuit of collective projects aimed at the affirmation of hope, rooted in the ordinary micro-practices of everyday life, is a strategy to set up, sustain and map out sustainable transformations' (Braidotti, 2013: 192).

Moulding a posthumanist, affirmative ethics into the circular *oikonomia* of the local space can open up tourism practices rooted in the affirmation of hope and belonging to a global community which is locally emplaced and affectively connected (Barad, 2007; Bright *et al.*, 2013; Thrift, 2008). This means being engaged with the positive affirmation of an ethical, just tourism, as Guia argues:

> Taking Posthumanism seriously means actively resisting the co-optation of tourism by the market, that is, learning to contest neo-liberalism with others […] it paves the way for re-introducing political responsibility, solidarity and advocacy as positive world-making practices with which to subvert the current commodification and de-politicization of all forms of tourism. (2021: 517)

A Theoretical Framework of Circular Oikonomia and Posthumanism

This study offers a critical reflection on the socialisation of tourism (Higgins-Desbiolles *et al.*, 2022) by focusing on the local space of tourism practices (Tomassini & Cavagnaro, 2020) and the local turn of tourism (Higgins-Desbiolles & Bigby, 2022). It does so by proposing a theoretical framework grounded in a conceptualisation of the circular economy (Bradley, 2012; Stahel, 2019) as a circular oikonomia (Leshem, 2013, 2016) able to manage the well-being of our 'household', and the dwellers that inhabit it, by re-incorporating economics into the broad concept of society in the form of a convivial economics imbued with the local dimension and its vernacular values (Latour, 2004; Illich, 1973, 1979). Such an oikonomia is conceived here as 'circular' since the circularity archetype envisions and enacts regenerative processes and cyclical patterns, not just of products and (raw) materials (McDonough & Braungart, 2003), but also of mobility practices, relations, connections and natural and social assets, ultimately creating a stable and enduring equilibrium for social and ecological justice (Higgins-Desbiolles, 2020). The proposed theoretical framework (see Figure 2.1) interweaves this conceptualisation with features of the sociology of space (Massey, 2005) and posthumanism (Braidotti, 2006, 2011, 2013).

Massey's (2005) ruminations on space cast light on an open space that is multiple, always under construction and never given. Within such space, all terrestrials (Latour, 2007, 2018) share the condition of throwntogetherness (Massey, 2005; Oliver, 2020) which requires an engagement meant as an emplacement through time (Oliver, 2020). Because as Barad stresses, justice is about the connections and responsibilities to each other and, as such, it is about entanglements:

> [justice] is not a state that can be achieved once and for all. There are no solutions: There is only the ongoing practice of being open and alive, each intra-action, so that we might use our ability to respond, our responsibility, to help awaken, to breathe life into ever new possibilities for living justly. (2007: x)

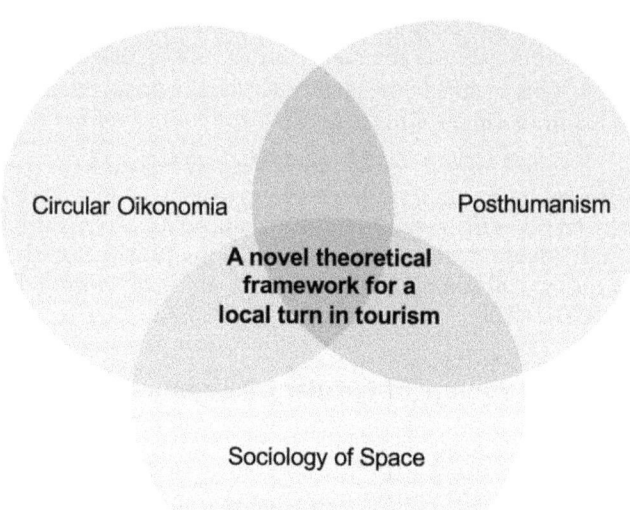

Figure 2.1 Visual model of the proposed theoretical framework for a local turn in tourism

These entanglements occur in a 'space-time' that is constantly productive and open (Massey, 2005) and in a place where 'porous networks of social relations' (Massey, 1994: 121) take shape eventually through local and context-related power geometries (Massey, 1994). Moreover, such entanglements are affective; they move through what Thrift (2008) has identified as spatialities of feelings. This understanding of a circular oikonomia locally emplaced in an open space always under construction and made of multiple relations, prompting ethical engagement via affective entanglements among terrestrials (Latour 2007, 2018), is interlaced with posthumanism (Braidotti, 2006, 2011, 2013). By rejecting the dualist nature–culture divide, the anthropocentric ontology and epistemology according to which man is the measure of all things, as well as the cultural dominance and unbalanced power relations resulting from the idea of Eurocentric civilisation models, posthumanism offers a novel ground to rethink the local turn in tourism via circular oikonomia (Braidotti, 2011, 2013). While the Enlightenment's humanistic tradition set the basis for and legitimised travel and tourism as an interest in an otherness different from the self, posthumanism challenges the idea of advantaged, elite tourists travelling to see a sexualised, racialised and naturalised other (Cohen, 2019). In doing so it offers an affirmative ethics grounded in encounters and relations advocating political responsibility for the vulnerable and disempowered. As such, posthuman ethics is an ethics of relational virtuosity; it vigorously and positively affirms a self that endures in a complex and deep relationship with the other (Guia, 2021).

We argue that a theoretical framework of circular oikonomia and posthumanism offers an innovative theoretical lens to examine the local turn in tourism, as well as a novel perspective on the current environmental and social crisis (Latour, 2018, 2021). It does so by challenging the disentanglement of tourists with regards to the hosting local space and hosting local community (Higgins-Desbiolles & Bigby, 2022; Tomassini *et al.*, 2021). Enacting circular and cyclic regenerative processes for resources, natural assets, practices and relations via affective entanglements that aim to reconnect the space of tourism and citizenship (Tomassini *et al.*, 2021) where this latter is here extended to both human and non-human terrestrials (Latour, 2007, 2018) makes it possible to envision a regenerative future tourism rooted in actions of posthuman affirmative ethics (Braidotti, 2011, 2013; Guia, 2021). This means conceiving a local turn in tourism as a constructive, positive planning ability entailing ethical commitment and political responsibility for a sustainable and just tourism within an open space which is locally situated and re-centred on the safety and well-being of the locality in which it takes place (Higgins-Desbiolles, 2020; Tomassini & Cavagnaro, 2020). The enacting of a circular oikonomia in tourism prompts a rethinking of the whole structure of tourism services and practices and opens the way to a socialising of tourism via multiple regenerative cyclical processes of political responsibility and affirmative ethics. Furthermore, the theoretical framework of circularity and posthumanism makes it possible to envision multifunctional tourism spaces (destinations) and multifunctional hospitality facilities that open up their spaces to the local community of both human and non-human dwellers.

There are numerous examples of such approaches emerging around the world. For instance, projects like Migrantour (see Migrantour, n.d.) transcend the boundaries between visitors and hosts via intercultural guided walking tours focused on migrant heritage (Ormond & Vietti, 2022). There is also the example of 'Roots Guide' which invites visitors 'to deeply connect with people and places' (Roots Guide, n.d.; Ormond & Vietti, 2022). Both of these initiatives, Migrantour and Roots Guide, can be interpreted as expressions of posthuman affirmative ethics to socialise tourism in the local space. Similarly, initiatives like 'Marry an Amsterdammer for a day' (Nicholls-Lee, 2019) seek to overcome the dualism between 'tourists' and 'locals' by proposing temporary affective relationships which are locally emplaced. In 2016, the official tourism organisation of Copenhagen, Wonderful Copenhagen, launched the 'localhood' long-term vision (Wonderful Copenhagen, n.d.) according to which 'locals and visitors not only co-exist but interact around shared experiences of localhood' (para. 2) in a co-creational perspective (Phi & Dredge, 2019; see also Chapter 8 this volume). With regards to animal-based tourism and wildlife tourism, the project *How is the water* (Cyan Planet, n.d.) connects people with the ocean via emotional experiences created with immersive media, using empathy to spark action for marine environment conservation. This is a project rooted into a posthuman affirmative ethics with the idea of prompting meaningful encounters to socialise wildlife tourism and rethink its future via a posthuman approach (Bertella, 2022; Bertella *et al.*, 2019; Kline, 2022). Similarly, Airbnb's 'animals on Airbnb experiences' (Airbnb, n.d.) – designed in accordance with the World Animal Protection's policy (World Animal Protection, n.d.) on animal welfare and well-being – seeks to prompt an ethical focus on animal-based tourism together with a deeper reflection on ethical regenerative encounters between human and non-human animals in tourism (Glusac, 2019; Kline, 2022).

Additionally, the hospitality industry demonstrated willingness to socialise tourism in the local space via a multifunctional approach during the COVID-19 outbreak. For instance, several hospitality businesses worldwide opened up their facilities to serve the needs and well-being of their local communities by hosting medical workers, as recounted by Hardingham-Gill (2020) in the USA and by Vimercati *et al.* (2020) in Italy. Hospitality's interest in activating cyclical, enduring, regenerative processes in the local oikonomia is also clearly evident. For example, locally emplaced projects such as *Circular Hospitality in Friesland and De Friese Doorlopers* in the Northern Netherlands (Stenden AIHR, 2021) and the Circular Hotels in Amsterdam (*Koplopergroup Circulaire Hotels*) (Gemeente Amsterdam, n. d.) gather small hospitality firms (the former) and hotels (the latter) willing to commit to circular economy principles and practices. These act at environmental and social levels involving a multiplicity of local stakeholders as well as the local community in initiatives for waste collection and waste upcycling, social inclusiveness projects and forms of green mobilities to help realise the co-development of circular and regenerative destinations. These demonstrate important achievements in transforming tourism and hospitality for more just and regenerative futures.

Conclusion

This study has explored a theoretical framework of circular oikonomia and posthumanism for a local turn in tourism. The premise is that the COVID-19 global crisis needs to be understood both as a warning and as an opportunity for a change of perspective over the space within which we live and travel. By conceiving a circular oikonomia as a vernacular economy prompting conviviality in the local space where it is contextualised, we built a theoretical ground to socialise tourism by recentring it on the well-being of the local space and its dwellers. For us, this means making sense of the condition of throwntogetherness that human and non-human terrestrials experience in an open space – locally emplaced – and imbued with a posthuman affirmative ethics.

This study therefore contributes to the limited but emerging academic research focused on both circular economy and posthumanism in tourism studies, as well as to their theoretical implications for tourism. The novelty of our study lies in trying to offer a new theoretical ground for a local turn in tourism grounded in the multiple, cyclical and regenerative processes of an oikonomia that allows 'the reincorporation of the social into the economic, or rather into sociality' (Pieroni, 2004: 12) and values the construction of an enduring equilibrium, as well as the well-being of the local space and its dwellers. Such a theoretical ground has practical implications for envisioning, designing and implementing novel policies for tourism practices that address unjust and unbalanced power relations. Moreover, on a practical level, our study stresses the potential for investigating regenerative, cyclical patterns of production and consumption within a tourism space permeated by multiple and heterogeneous relationships underpinned by actions of affirmative ethics and political responsibility. The aforementioned projects and initiatives – Migrantour, the Roots Guide, Marry an Amsterdammer for a day, Wonderful Copenhagen's Localhood, What is Water, Airbnb Animal Experiences, Circular Hospitality in Friesland and De Friese Doorlopers and Circular Hotels in Amsterdam – exemplify this potential as well as the practical implications of a local turn in tourism via a posthuman circular approach in an 'eventful' space. In consideration of this, we recommend future research on the interplay between the circular economy and posthumanism in tourism studies via qualitative and quantitative approaches, since this will allow further exploration and investigation of the circularity paradigm and its practical implications for a local turn in tourism.

References

Airbnb (n.d.) Animals on Airbnb experiences. See https://www.airbnb.com.au/s/experiences/animal (accessed 31 January 2022).

Ateljevic, I. (2020) Transforming the (tourism) world for good and (re)generating the potential new normal. *Tourism Geographies* 22 (3), 467–475. https://doi.org/10.1080/14616688.2020.1759134.

Barad, K. (2007) *Meeting the Universe Halfway.* Durham, NC: Duke University Press.

Bertella, G. (2022) Wildlife tourism in 2150: Uplifted animals virtual and augmented reality and everything in-between. In I. Yeoman, U. McMahon-Beattie and M. Sigala (eds) *Science Fiction, Disruption and Tourism* (pp. 97–108). Bristol: Channel View Publications.

Bertella, G., Fumagalli, M. and Williams-Grey, V. (2019) Wildlife tourism through the co-creation lens. *Tourism Recreation Research* 44 (3), 300–310. https://doi.org/10.1080/02508281.2019.1606977.

Bianchi, R. (2018) The political economy of tourism development: A critical review. *Annals of Tourism Research* 70, 88–102. https://doi.org/10.1016/j.annals.2017.08.005.

Boluk, K.A., Cavaliere, C.T. and Higgins-Desbiolles, F. (2019) A critical framework for interrogating the United Nations Sustainable Development Goals 2030 Agenda in Tourism. *Journal of Sustainable Tourism* 27 (7), 847–864. https://doi.org/10.1080/09669582.2019.1619748.

Bradley, R. (2012) *The Idea of Order: The Circular Archetype in Prehistoric Europe*. Oxford: Oxford University Press.

Braidotti, R. (2006) Posthuman, all too human: Towards a new process ontology. *Theory, Culture & Society* 23 (7–8), 197–208. https://doi.org/10.1177/0263276406069232.

Braidotti, R. (2011) *Nomadic Theory: The Portable Rosi Braidotti*. New York: Columbia University Press.

Braidotti, R. (2013) *The Posthuman*. Cambridge: John Wiley & Sons.

Bright, N.G., Manchester, H. and Allendyke, S. (2013) Space, place, and social justice in education growing a bigger entanglement: Editors' introduction. *Qualitative Inquiry* 19, 747–755. https://doi.org/10.1177/1077800413503794.

Calvino, I. (2003) *Invisible Cities*. New York: Mariner Books.

Cohen, E. (2019) Posthumanism and tourism. *Tourism Review* 74 (3), 416–427. https://doi.org/10.1108/TR-06-2018-0089.

Cyan Planet (n.d.) Discover the power of virtual reality and experience the ocean. See https://www.cyanplanet.org/ (accessed 31 January 2022).

Deleuze, G. and Guattari, F. (1987) *A Thousand Plateaus: Capitalism and Schizophrenia*. Minneapolis: University of Minnesota Press.

Ellen MacArthur Foundation (n.d.) What is a circular economy? See https://ellenmacarthurfoundation.org/topics/circular-economy-introduction/overview (accessed 18 January 2022).

European Commission (2015) Closing the loop – An EU action plan for the circular economy. COM/2015/0614 final. See https://eur-lex.europa.eu/legal-content/EN/TXT/ (accessed 18 January 2022).

European Commission (2020) A new circular economy action plan for a cleaner and more competitive Europe. COM/2020/98. See https://eur-lex.europa.eu/legal-content/EN/TXT/ (accessed 18 January 20220).

Ferrando, F. (2019) *Philosophical Posthumanism*. London: Bloomsbury.

Geissdoerfer, M., Savaget, P., Bocken, N.M.P. and Hultink, E.J. (2017) The circular economy – A new sustainability paradigm? *Journal of Cleaner Production* 143, 757–768. https://doi.org/10.1016/j.jclepro.2016.12.048.

Gemeente Amsterdam (n. d.) Duurzam ondernemen. See https://www.amsterdam.nl/wonen-leefomgeving/duurzaam-amsterdam/duurzaam-ondernemen/hotels-amsterdam/ (accessed 31 January 2022).

Glusac, E. (2019) New Airbnb Excursions to Focus on Animals. See https://www.nytimes.com/2019/10/03/travel/airbnb-animal-experiences.html (accessed 28 January 2022).

Guia, J. (2021) Conceptualizing justice tourism and the promise of posthumanism. *Journal of Sustainable Tourism* 29 (2–3), 502–519. https://doi.org/10.1080/09669582.2020.1771347.

Guia, J. and Jamal, T. (2020) A (Deleuzian) posthumanist paradigm for tourism research. *Annals of Tourism Research* 84, 102982. https://doi.org/10.1016/j.annals.2020.102982.

Hayles, N.K. (1999) *How We Became Posthuman: Virtual Bodies in Cybernetics, Literature, and Informatics*. Chicago: University of Chicago Press.

Haraway, D. (2003) *The Companion Species Manifesto: Dogs, People, and Significant Otherness*. Chicago: Prickly Paradigm.

Haraway, D. (2006) A cyborg manifesto: Science, technology, and socialist-feminism in the late 20th century. In J. Weiss, J. Nolan, J. Hunsinger and P. Trifonas (eds) *The International Handbook of Virtual Learning Environments* (pp. 117–158). Berlin: Springer.

Hardingham-Gill, T. (2020) What happened when luxury hotels swapped tourists for medical workers. See https://edition.cnn.com/travel/article/hotels-swap-tourists-for-medical-workers/index.html (accessed 28 January 2022).

Higgins-Desbiolles, F. (2006) More than an 'industry': The forgotten power of tourism as a social force. *Tourism Management* 27 (6), 1192–1208. https://doi.org/10.1016/j.tourman.2005.05.020.

Higgins-Desbiolles, F. (2018) Sustainable tourism: Sustaining tourism or something more? *Tourism Management Perspectives* 25, 157–160. https://doi.org/10.1016/j.tmp.2017.11.017

Higgins-Desbiolles, F. (2020) Socialising tourism for social and ecological justice after COVID-19. *Tourism Geographies* 22 (3), 610–623. https://doi.org/10.1080/14616688.2020.1757748.

Higgins-Desbiolles, F. and Bigby, B.C. (2022) A local turn in tourism studies. *Annals of Tourism Research* 92, 103291. https://doi.org/10.1016/j.annals.2021.103291.

Higgins-Desbiolles, F., Doering, A. and Bigby, B.C. (2022) *Socialising Tourism: Rethinking Tourism for Social and Ecological Justice*. New York: Routledge.

Illich, I. (1973) *Tools for Conviviality*. New York: Harper & Row.

Illich, I. (1979) Vernacular values and education. *Teachers College Record* 81 (1), 31–76.

Jamal, T. (2019) *Justice and Ethics in Tourism*. New York: Routledge.

Jamal, T. and Higham, J. (2021) Justice and ethics: Towards a new platform for tourism and sustainability. *Journal of Sustainable Tourism* 29 (2–3), 143–157. https://doi.org/10.1080/09669582.2020.1835933.

Kline, C. (2022) Socialising animal-based tourism. In F. Higgins-Desbiolles, A. Doering and B.C. Bigby (eds) *Socialising Tourism: Rethinking Tourism for Social and Ecological Justice* (pp. 195–213). New York: Routledge.

Latouche, S. (2004) *Altri Mondi, Altre Menti, Altrimenti: Oikonomia Vernacolare e Società Conviviale*. Catanzaro: Rubettino Editore.

Latour, B. (2007) *Reassembling the Social: An Introduction to Actor-Network-Theory*. Oxford: Oxford University Press.

Latour, B. (2018) *Down to Earth: Politics in the New Climatic Regime*. Cambridge: Polity Press.

Latour, B. (2021a) *After Lockdown: A Metamorphosis*. Cambridge: Polity Press.

Latour, B. (2021b) The pandemic is a warning: We must take care of the earth, our only home. The Guardian. See https://www.theguardian.com/commentisfree/2021/dec/24/pandemic-earth-lockdowns-climate-crisis-environment (accessed 18 January 2022).

Leshem, D. (2013) Oikonomia redefined. *Journal of the History of Economic Thought Perspectives* 35 (1), 43–61. https://doi.org/10.1017/S1053837212000624.

Leshem, D. (2016) What did ancient Greeks mean by *Oikonomia*? *Journal of Economic Perspectives* 30 (1), 225–231. DOI: 10.1257/jep.30.1.225.

Massey, D. (1994) *Space, Place and Gender*. Cambridge: Polity Press.

Massey, D. (2005) *For Space*. London: Sage.

McDonough, W. and Braungart, M. (2003) *Cradle to Cradle: Remaking the Way We Make Things*. New York: North Point Press.

Migrantour (n.d.) Migrantour sustainable routes. See http://www.mygrantour.org/en/ (accessed 31 January 2022).

Murray, A., Skene, K. and Haynes, K. (2017) The circular economy: An interdisciplinary exploration of the concept and application in a global context. *Journal of Business Ethics* 140 (3), 369–380. https://doi.org/10.1007/s10551-015-2693-2.

Nicholls-Lee, D. (2019) Amsterdam invites tourists to 'marry' a local for a day. See https://www.theguardian.com/travel/2019/jun/05/amsterdam-fake-wedding-tourists-local-untouristy (accessed 28 January 2022).

Oliver, J. (2020) Throwntogetherness: Engagement as emplacement, past present, and future. See https://nitro.edu.au/articles/2020/12/4/throwntogetherness-engagement-as-emplacement-past-present-and-future (accessed 18 January 2022).

Ormond, M. and Vietti, F. (2022) Beyond multicultural 'tolerance': Guided tours and guidebooks as transformative tools for civic learning. *Journal of Sustainable Tourism* 30 (2–3), 533–549. https://doi.org/10.1080/09669582.2021.1901908.

Parker, R. (2016) Aion. *Oxford Classical Dictionary*. Oxford: Oxford University Press. https://doi.org/10.1093/acrefore/9780199381135.013.227.

Phi, G.T. and Dredge, D. (2019) Collaborative tourism-making: An interdisciplinary review of co-creation and a future research agenda. *Tourism Recreation Research* 44 (3), 284–299. https://doi.org/10.1080/02508281.2019.1640491.

Pieroni, O. (2004) Introduzione 'Le marteau dans la tête'. In S. Latouche (ed.) *Altri Mondi, Altre Menti, Altrimenti: Oikonomia Vernacolare e Società Conviviale* (pp. 5–21). Catanzaro: Rubettino Editore.

Roots Guide (n.d.) Deeply connect with people and places. See https://rootsguide.org/ (accesed 31 January 2022).

Sørensen, F. and Bærenholdt, J. (2020) Tourism practices in the circular economy. *Annals of Tourism Research* 85, 103027. https://doi.org/10.1016/j.annals.2020.103027.

Stahel, W.R. (2019) *The Circular Economy: A User's Guide*. New York: Routledge.

Stenden AIHR (2021) News. See https://www.stendenaihr.com/news/circular-hospitality-in-friesland-de-friese-doorlopers-community-outcomes-of-the-first-year (accessed 31 January 2022).

Tomassini, L. and Cavagnaro, E. (2020) The novel spaces and power-geometries in tourism and hospitality after 2020 will belong to the 'local'. *Tourism Geographies* 22 (3), 713–719. https://doi.org/10.1080/14616688.2020.1757747.

Tomassini, L., Schreurs, L. and Cavagnaro, E. (2021) Reconnecting the space of tourism and citizenship: the case of tourists' hubris. *Journal of Tourism Futures* 7 (3), 337–349. https://doi.org/10.1108/JTF-10-2020-0176.

Thrift, N. (2008) Spatialities of feeling. In N. Thrift (ed.) *Non-representational Theory: Space, Politics, Affect* (pp. 171–197). New York: Routledge. https://doi.org/10.4324/9780203946565.

Tuck, E. and McKenzie, M. (2015) Relational validity and the 'where' of inquiry: Place and land in qualitative research. *Qualitative Inquiry* 21 (7), 633–638. https://doi.org/10.1177/1077800414563809.

Vimercati, L., Tafuri, S., Chironna, M., Loconsole, D., Fucilli, F.I.M., Migliore, G. and Gesualdo, L. (2020) The COVID-19 hotel for healthcare workers: An Italian best practice. *Journal of Hospital Infection* 105 (3), 387–388. https://doi.org/10.1016/j.jhin.2020.05.018.

Wonderful Copenhagen (n.d.) Localhood. See https://localhood.wonderfulcopenhagen.dk/ (accessed 31 January 2022).

World Animal Protection (n.d.) Our work. See https://www.worldanimalprotection.org/our-work (accessed 31 January 2022).

3 Travel Boycotts, Ethical Consumption and Destination Communities: Expanding the Morality of Neighbourliness

Siamak Seyfi and C. Michael Hall

Introduction

> *'I still think that people should not come to Burma (Myanmar) because the bulk of the money from tourism goes straight into the pockets of the generals. And not only that, it's a form of moral support for them because it makes the military authorities think that the international community is not opposed to the human rights violations'.* Aung San Suu Kyi, Leader of Myanmar's National League for Democracy (cited in Pears, 2019: n.p.)

With greater public attention to corporate social responsibility and increased vulnerability of brands and corporate reputations, boycotts have increasingly become a consumer tool to coerce corporate or political change (Friedman, 1991; Klein *et al.*, 2004; Seyfi & Hall, 2020). Many contemporary consumers choose brands, products and services according to how ethical, environmentally friendly and socially and politically responsible they view them to be (Seyfi & Hall, 2020; Stolle & Micheletti, 2013). Tourism is argued as becoming 'commodified', or at least intimately tied in with processes of commodification, and has therefore become a ready arena for 'political consumer actions' and consumer activism (Boström *et al.*, 2019; Gretzel, 2017). As a result of such processes, and the various meanings that are associated with travel, for many, tourism has become recognised as a global phenomenon with substantial justice implications (e.g. Higgins-Desbiolles, 2020; Mowforth & Munt, 2015). Some argue that as tourists develop a greater sense of civic responsibility or have become more politically conscious, so the number of boycotts increases (Shaheer *et al.*, 2018, 2019; Shepherd, 2021). However, while from a social justice perspective such a situation would be laudable, it must also be treated with caution as the increase in the number of boycotts may be the result of social media growth rather than a real increase in the number of people that care enough about an issue to make it a boycott action (Seyfi *et al.*, 2021). Nevertheless, regardless of the context, travel and business boycotts clearly are of importance to destinations, industry and tourism dependent communities alike, and their tourism-related futures.

Ethics now occupies an important place for many people in the choice of destination and tourism product. How tourists practice political and ethical consumerism in their wider travel context is of growing interest (Castañeda & Burtner, 2010; Lovelock & Lovelock, 2013; Weeden & Boluk, 2014). The ethics of tourism has long been concerned with what tourists purchase and the effect of tourism on communities and the environment (Duffy & Smith, 2004; Lovelock & Lovelock, 2013). Although it should be noted that such concerns have often been expressed as more generic ethical questions rather than dealing with the specific ethical concerns of consumers and businesses, with some even questioning the extent to which leisure tourists actually care about the ethics of their tourism decisions so long as they have a cheap holiday (Hall, 2014a; Tomazos & Butler, 2012). Nevertheless, the ethics of tourism is addressed by: institutions, i.e. religious bodies; organisations, i.e. non-governmental organisations (NGOs); and individuals, all of which address in various ways what is and what is not appropriate to consume and supply and, beyond that, what should be done to influence or prevent such production and consumption. Such is the interest in conscious consumption in some quarters that tourist and hospitality products are increasingly focusing on branding, communicating and promoting the supposed environmental and other benefits of purchase (Hall, 2014b). But what then is ethical in tourism consumption and production? A number of different dimensions can be suggested:

- appropriate and inappropriate purchasing and shopping behaviours;
- consumer and destination resistance and activism;
- consumption and producer morality in relation to concerns over sustainability;
- corporate and social entrepreneurial attempts to promote ethical consumption and production opportunities; what is sometimes referred to as corporate social responsibility and/or societal marketing;
- ethical consumption and production as a conscious political project of individuals and small groups as part of a broader politics of consumption (Hall, 2014a, 2014b).

Importantly, the above elements are not mutually exclusive, but should be seen as deeply entwined with each other. The consumption and production of tourism are interrelated and ethical decisions made by consumers have flow on effects in destinations and vice versa. Such relationships also highlight the way in which the idea of ethical consumption has expanded from the immediate point of consumption throughout the entire system through which the tourist experience is produced. This shift in understanding of the morality of consumption therefore reinforces the significance of a growing trend to understand ethical consumption not just in terms of the act or experience of consumption and the cost and utility of what they consume, but also in terms of the ways in which they are produced, processed and transported (Trentmann, 2007). As a result of such expanded notions of ethical consumption, individuals may be able to do two things: 'Firstly, at a personal level, they can lead lives that are more moral. Secondly, at a public level, they can use their purchases to affect the larger world, by putting pressure on firms in a competitive market to change the ways that they do things' (Carrier, 2010: 672). One of the main ways that ethical consumption has grown at the public level has been with respect to boycotts.

Over the years, many destinations have been boycotted by tourists for a range of reasons including human rights violations, treatment of LGBTQ-people and other minorities, violations of animal welfare and rights and inappropriate environmental actions (Shaheer *et al.*, 2018, 2019). Boycotts usually mean fewer tourists, less business and a dent in the public image of a destination, but what of the communities affected by boycotts? The idea of community in tourism is often used in a far too monolithic manner and in a way that serves to hide the different competing interests and values that emerge with respect to the economic and political uses of tourism (Hall, 2008). Moreover, as suggested above, ethical contestation and decision-making is not just the domain of the consumer but also resides with the producers of the tourism experience which, at a community level includes not just businesses but also the residents themselves.

To address this question and drawing on different cases, this chapter discusses such discourses as a way of problematising the notions of community which are often central to tourism planning discourse. These have already been the focus of growing research (Higgins-Desbiolles & Bigby, 2022; Higgins-Desbiolles *et al.*, 2021, 2022) and proposed as a 'local turn in tourism' (Higgins-Desbiolles & Bigby, 2022). In response, we argue that not only does there need to be greater recognition of the lack of homogeneity in communities, but that the very embeddedness of destination communities in the capitalist production of tourism makes it extremely hard for them to develop truly inclusive forms of tourism.

Boycotts in Tourism

Friedman (1999: 4) defined boycott as 'the attempt by one or more parties to achieve certain objectives by urging individual consumers to refrain from making selected purchases'. The boycott phenomenon has been widely studied in the social sciences but has only recently become a subject of inquiry in tourism (Seyfi & Hall, 2020; Seyfi et al., 2021; Shaheer et al., 2018, 2019; Shepherd, 2021), although it had previously been discussed in relation to work on the politics of tourism (Richter, 1989).

Individuals, civil society organisations and NGOs are increasingly involved in supporting boycotts in order to change institutions, organisations or societal situations that are seen as ethically, ecologically or politically disagreeable (Rössel & Schenk, 2018). The growth in boycotting can arguably be interpreted as the 'flip side' to neoliberal globalisation. This is because the increased global stretch of capital, communication and mobility has made businesses and places more vulnerable to crises or problems in the tourism system, including opposition to tourism industry and destination actions, decisions or policies.

Significantly, boycotters strive to get the media's attention in order to harm the reputation of a specific business, destination or a country (Hawkins, 2010; Yousaf et al., 2021). The growth of social media has only enhanced this capacity (Gretzel, 2017; Seyfi et al., 2021). For instance, the study of Yousaf et al. (2021) on the #BoycottMurree campaign showed how campaign momentum on social media affected Murree, which is a popular tourist destination in Pakistan. This is also especially evident in the case of Boycott, Divestment and Sanctions (BDS) campaigners pushing for a boycott of Israel's hosting of the Eurovision Song Contest 2019 and accusing Israel of exploiting the artistic event to whitewash its human rights violations against Palestinians (Morrison, 2022; Shepherd, 2021). Nonomura (2017) argued that, owing to increased political engagement thanks to social networks, many individuals have developed a greater awareness of their agency and active position as consumer citizens in society. Lamers et al. (2019) identified specific practices of political and ethical consumption within a tourism context including sustainable consumption, climate change, slow travel, conservation tourism and voluntourism that have strong ethical and political consumerism dimensions grounded on individualised responsibility-taking. Furthermore, there appears to be a surge in the boycotting of destinations and tourist firms for sustainability-related reasons (Shaheer et al., 2019) and 90% of boycotts in the last 70 years have occurred within the last decade (Seyfi & Hall, 2020; Shaheer et al., 2018). For many, such an increase in tourism-related boycotts is related to technological innovations which have facilitated boycott organisation and communications (e.g. Gretzel, 2017; Mkono, 2018; Shaheer et al., 2018; Yousaf et al., 2021), while Castañeda and Burtner (2010) attributed such a surge in boycotting to the growing use of tourism as a vehicle for social and political change. However, it is likely that the reality for this increase lies at the intersection of all these factors given that technologies should be best understood as something embedded within socioeconomic structures, rather than as having separate agency (Seyfi et al., 2021).

Boycott and Social Justice

Boycotts have long been heralded as exemplary nonviolent movements working for social justice and have been used as a 'tried-and-true tactic of social justice movements' to respond to injustice or wrongdoing, and as a means of dismantling systems of injustice (Chaitin *et al.*, 2017; Maira, 2018; Radzik, 2017). Boycott has become a widespread and established part of social and environmental justice activism in tourism in recent years (Mkono, 2018; Seyfi *et al.*, 2021). This is evidenced by the growth in tourists' boycotting of products, brands, destinations and services on the basis of their political or ethical characteristics (e.g. sustainability, social justice, fair-trade, corporate responsibility or animal welfare) (Boström *et al.*, 2019; Stolle & Micheletti, 2013). These are believed to be increasingly important in global struggles for social justice and human rights (Seyfi *et al.*, 2021; Stolle & Micheletti, 2013) and serve to connect tourism consumption with broader political and ethical issues. This is something the industry has often tried to position itself as being outside of, together with many consumers. These issues have long been demonstrated in the politics of tourism in south-east Asia, for example. Countries such as Myanmar, Laos and Cambodia have long had goals with respect to tourism that are similar to many others in terms of the attraction of foreign exchange, encouragement of economic development and generating employment opportunities. However, since the 1990s the creation of a positive image via tourism promotion and having tourists come visit is also important in terms of not only attracting much-needed foreign investment, but also providing political legitimacy internationally, if not domestically. No more has this been clearer than in the case of Myanmar. However, the political dimensions of tourism, including such things as the abuse of human rights in the name of tourism development, challenges not only the tourism industry but also the student of tourism with respect to the ethics of tourism. Is money 'more important than human rights'? (Oo & Perez, 1996: n.p.). Or is the most appropriate approach that of TravelAsia (1996: n.p.) when trying to deal with issues of human rights abuse in Myanmar and the encouragement of tourism via Visit Myanmar Year:

> As travel business professionals, we have to keep a clear head and not take sides. Indeed, the only side we should be on is our customers. The only questions we should ask are, does Myanmar have the products, the infrastructure and the will to become one of Asia's great destinations?

In their study, Stephan and Chenoweth (2008) reported that boycotts were becoming more successful in achieving their aims in a number of cases. They further argue that such success could be because such activism enhances legitimacy and encourages broad participation. Traditionally, the boycott has been viewed as social responses to injustice and wrongdoing. In his interpretation of the term 'social', Radzik (2017) argued that boycotters lack legal power and that their own activities are often legitimate. Boycotting is a moral instrument in the toolkit of everyday people. For Hallward (2013), boycotts have been used by civil society organisations as a tool for exerting political pressure in a non-violent manner in the international realm. Nonetheless, as Sharp (2013) pointed out, these tactics continue to be problematic since they seek to bring social, political and economic change. The South African boycott movement as well as Palestinian BDS movement are two examples of boycotts as a way for fighting for social justice. For many commentators, when tourists travel to countries with poor human rights record, they might unintentionally support businesses linked to war crimes and human rights violations. From a tourism perspective, such ethical debates have been applied to sanctions on tourism to a destination (Seyfi & Hall, 2019; 2020). For example, as noted above, in the longstanding debate over sanctions on Myanmar tourism, some proponents of sanctions have suggested boycotting Myanmar travel and tourism to effect 'regime change' and argue that it is unethical for individuals to support the military dictatorship, which controls most of the country's productive capacity including tourism, which attracts considerable foreign exchange (Hall & O'Sullivan, 1996). In contrast, those who argue against sanctions suggest that visiting such destinations might help to reduce the isolation of the wider population as well as generating needed employment; additionally, visitors can advocate for justice (e.g. Seyfi *et al.*, 2021). However, what is important to stress here is that these competing arguments do not just occur in the abstract, but they reflect the real-life tensions and strains that exist within communities.

Boycott or Not: Is Boycotting Always the Best Practice?

It has become an increasingly common debate to consider the pros and cons of boycotting attractions, brands and destinations. When deciding whether or not to visit a specific destination, especially those with a poor human rights record, travellers may sometimes face a significant ethical dilemma. Many have questioned who does such a kind of action really hurt. Do such measures harm the people they are intended to protect more than the ruling classes they are targeting (e.g. Delacote, 2009), given that such policies may reduce the population's well-being more than the governing class? There are two schools of thought. One view believes that boycotting a destination is a way to demonstrate and to take a stand against injustice, cruelty and/or discrimination. The supporters of this view argue that tourists should avoid at all costs going to a destination, for example, where human or animal rights are violated in order not to condone abuse. Perhaps, one longstanding case subjected to such actions is Myanmar, but it is significant to note that although tourism has fluctuated in terms of arrivals and economic impact it has never stopped. For example, the legitimately elected leader of Burma, Aung San Suu Kyi declared in 1996:

> ...They (tourists) should stay away until we are a democracy. Look at the forced labour that is going on all over the country. A lot of it is aimed at the tourist trade. It's very painful. Roads and bridges are built at the expense of the people. If you cannot provide one labourer you are fined. If you cannot afford the fine, the children are forced to labour. (cited in Pilger, 1996)

Even though such remarks were widely reported, it is significant that although tourism slowed, the tourists never stopped coming and investment in tourism development also continued.

More recently, some tourists and travel companies alike boycotted Brunei following the introduction of the death penalty for homosexuality. The second view contends that it is the very people who are being oppressed who are most likely to suffer as a result of travel boycotts and going to such destinations might help to get the population out of their isolation by avoiding government-run businesses as much as possible. The supporters of this view believe that travel and human interaction can bring about more and deeper change than a reflective boycott.

It should also be acknowledged that there is also a third view, which is essentially one that is not bothered by the issue either way, and in fact does not even engage with it, because it focuses completely on personal or business satisfaction, i.e. it is based on self-interest rather than notions of community interest. In fact, in one of the seminal works on the politics of tourism, Hall (1994) argued that this was more likely the dominant perspective among consumers. He noted for example how the growth of mass tourism to Portugal and Spain in the late 1960s and 1970s was the result of the actions of fascist dictatorships to generate foreign income and that the vast number of visitors did not care because it represented a cheap holiday package. While on the community side he noted that even though many community members recognised their powerless in front of change and the loss of traditional agriculture and fishing there was still often support for the employment, even though of poor quality, that was created. And for many in such communities, employment is the major issue if they wish to stay in place, even though place and communities change as a result.

In his study on the ethical arguments for and against tourism in Myanmar, Hudson (2007) reported that Myanmar visitors were generally in favour of tourism in Myanmar but were uncomfortable with the ethical implications of their visit. He further continued that visitors who disregarded a suggested boycott of Myanmar emphasised the advantages of their tourism to Burmese residents, the transforming nature of their interactions with locals and the chance tourism afforded them to demonstrate their sympathy with the Burmese people. Tourism guidebooks such as

Lonely Planet and websites such as Responsible Travel argue that it is possible to travel very responsibly in destinations with poor ethical records and argue that visitors to Myanmar may visit and not support the regime as long as they travel 'responsibly' (Shepherd, 2021). Radzik (2017) argued that boycotting implies 'damage to innocents', in which third parties are economically or socially penalised for government policies over which they have little influence.

The Debates over Boycotts and the Problem of Community

The debates over sanctions are therefore just one case of the problem of community writ large. One of the problems in discussing community issues in tourism is the extent to which they have often become implicitly homogenised in notions of community which have emphasised commonality of interest rather than difference. This is of especial importance in tourism given that tourism development can serve to dramatically alter the web of relations that residents have with place and further contribute to different perspectives on the benefits of tourism. As Millar and Aitken (1995: 620) commented in a resource management context:

> Conflict is a normal consequence of human interaction in periods of change, the product of a situation where the gain or a new use by one party is felt to involve a sacrifice or changes by others. It can be an opportunity for creative problem solving, but if it is not managed properly conflict can divide a community and throw it into turmoil.

Communities exist within a web of kinship, physical interdependency and social obligation, and because of this, tourism cannot be separated from the social issues of property and morality (Hall, 2008; Millar & Aitken, 1995). But it is important to recognise that such morality is not necessarily the ethical position of the tourism consumer.

Different places and positions in the tourism system will have different ethical positions on the nature and value of the economic, social and environmental dimensions of tourism. In the absence of a universal ethical position on the relative worth of the collective versus the individual in notions of rights, or even different notions of rights, it seems readily apparent that difference within communities is normal. From a consumer perspective, for example, Zapata Campos et al. (2018) suggested that Swedish consumers of package travel were far more concerned with animal rights than labour rights in the destinations they travelled to. While from a destination community perspective, Hall (2008) drawing on Millar and Aitken (1995) argued that there is a two-part morality of neighbourliness in which, while there is a recognition that everyone has the right to make a living, there is also a belief that everyone who is affected by tourism development should have the right to be consulted. Promotion of boycotts therefore will almost invariably lead to conflict and debate within destination communities as it inherently contributes to the embodied difference of the two-part morality of neighbourliness. In societies where there are strong, impartial, legal and regulatory institutions, the moral divisiveness of boycotts may not necessarily lead to long-term splits within communities. However, the majority of boycotted (and at the extreme, sanctioned) countries are those in which governments tend to operate, almost by definition, against international legal norms and have legal systems that only have limited independence from government (Seyfi & Hall, 2019, 2020; Seyfi et al., 2020). Therefore, destination communities affected by boycotts and sanctions not only face undoubted difference with respect to the morality of neighbourliness but the economic difficulties that sanctions create may even favour economic arguments for tourism and employment regardless of the ethical cost.

If seeking to reorient tourism communities then the experience of the role of sanctions suggests that alternative community futures lie in consideration of alternative economic frameworks, as the central role of economic exchange means that communities are bound to the morality of the 'right' to make a living which is interpreted from within a capitalist context. The result of which is that rather than ethical consumption, procotting, and boycotting creating an alternative economic space, new spaces of capitalism have instead been opened in existing systems of provision (Hall, 2014a). Ethical tourism consumption, including switching destinations in the case of tourism sanctions, is still consumption. Similarly, acknowledgement of the morality of the right to make a living as the dominant morality in many communities means that the right to production cannot be ignored even if, from other notions and framing of rights, other human rights are being substantially damaged.

Questions of what precisely makes such projects 'alternative' are central to issues of sustainable tourism and rethinking community futures, as are a set of questions about the relationship between the alternative and its constituent other, the mainstream or conventional capitalist economy (Whatmore et al., 2003). Few alternative tourism products and the communities within which they are embedded are so alternative that they eschew the circulation of capital in commodity form altogether; rather, they attempt to harness intrinsic dynamics of capitalism to progress destination projects. By definition, a destination community cannot exist unless it is a part of the economic system of tourism (Hall, 2022). Tourism is therefore a driver of capitalism and extends and deepens neoliberalism within destination and other tourism communities (Brockington & Duffy, 2010). If capitalism is identified as part of the problem for alternative community futures, it is therefore difficult, if not impossible, for it to offer the solution. As witnessed in responses to boycotts in countries like Israel, neoliberalism does not necessarily displace or obliterate existing ways of valuing place and community relations. Instead, it mixes with the local context to create new dynamics (Brockington & Duffy, 2010) and thereby reflects the fundamentally uneven nature of the neoliberal capitalist project (Hall, 2014a).

Conclusion

The way that tourism acts to provide an economic value for communities is integral to the way that neoliberalism is fundamentally concerned with 'the financialisation of everything' (Harvey, 2005: 33). The problem with the rhetoric of ethical consumption, including enacting and support of boycotts, is that tourism's contribution to often laudable goals as being such a positive form of ethical consumption, 'presents us only with market solutions' (Brockington & Duffy, 2010: 481) which only further embed the consumer *and* the destination community in processes of contemporary capitalism. Undoubtedly, boycotts do open a contested moral terrain with respect to their real impacts. However, as we have suggested, the intrinsic embeddedness of destination communities in the capitalist tourism system means that the economic aspect of the morality of neighbourliness dominates community structures and thinking. While the ethical consumer selects another destination to travel to within the tourism system, the boycotted destination often only further deepens its dependency on tourism and creates further fractures in entwined community social and economic bonds in the desire to make a living (Seyfi & Hall, 2020).

There are certainly attempts by members of destination communities and tourist consumers to escape the 'totalizing logic of the market' by attempting to construct localised 'emancipated spaces' that are constructed by 'engaging in improbable behaviors, contingencies, and discontinuities' (Firat & Venkatesh, 1995: 255). Such improbable behaviours include acts that appear to be outside the logic of commercialisation, such as voluntary simplicity, anti-consumption and the ethics of the local and cooperativism (Gibson-Graham, 2003a, 2003b). Nevertheless, focusing on lifestyle alone, without challenging the role of structure and the cultural and political forces of production, may well mean that alternative consumption and production paths remain cut off over the longer term (Chassagne & Everingham, 2019; Hall, 2014a, 2022; Hall *et al.*, 2021).

There is therefore a need in tourism for a clearer theorisation of what are alternative community futures or not (Whatmore *et al.*, 2003). Although this is a long-held point of debate in tourism studies, given the long-standing search for alternative tourisms, such a discussion may need to connect more to some of the broader debates on alterity and alternative economic and political spaces. Problematising boycotts as a form of ethical consumption and the ethics and morality of its effects on communities is a way of opening thinking and reflexivity regarding future tourism communities. As we have argued, the two-part morality of neighbourliness that is central to communities and considerations of their decision-making and future is typically dominated by economic reasoning, in great part because being a destination community inherently means that you are embedded in the capitalist logic of the tourism system. To think other, therefore means to engage with non-capitalist alternatives as to how tourism can be reimagined, along with the communities within it. Such theorisations need to engage much more substantially with the way tourism embeds communities in contemporary neoliberal capitalism and its implications for the way members of communities live their lives and how power is structured in communities. And ask why?

References

Boström, M., Micheletti, M. and Oosterveer, P. (2019) *The Oxford Handbook of Political Consumerism*. Oxford: Oxford University Press.
Brockington, D. and Duffy, R. (2010) Capitalism and conservation: The production and reproduction of biodiversity conservation. *Antipode* 42, 469–484.
Carrier, J.G. (2010) Protecting the environment the natural way: Ethical consumption and commodity fetishism. *Antipode* 42, 672–689.
Castañeda, Q. and Burtner, J. (2010) Tourism as 'a force for world peace': The politics of tourism, tourism as governmentality and the tourism boycott of Guatemala. *The Journal of Tourism and Peace Research* 1 (2), 1–21.
Chaitin, J., Steinberg, S. and Steinberg, S. (2017) Polarized words: Discourse on the boycott of Israel, social justice and conflict resolution. *International Journal of Conflict Management* 28 (3), 270–294.
Chassagne, N. and Everingham, P. (2019) Buen Vivir: Degrowing extractivism and growing wellbeing through tourism. *Journal of Sustainable Tourism* 27 (12), 1909–1925.
Delacote, P. (2009) Boycotting a dictatorship: Who does it really hurt? *Economics Bulletin* 29 (3), 1859–1867.
Duffy, R. and Smith, M. (2004) *The Ethics of Tourism Development*. London: Routledge.
Firat, A.F. and Venkatesh, A. (1995) Liberatory postmodernism and the reenchantment of consumption. *Journal of Consumer Research* 22, 239–267.
Friedman, M. (1991) Consumer boycotts: A conceptual framework and research agenda. *Journal of Social Issues* 47 (1), 149–168.
Friedman, M. (1999) *Consumer Boycotts*. London: Routledge.
Gibson-Graham, J.K. (2003a) An ethics of the local. *Rethinking Marxism* 15 (1), 49–74.
Gibson-Graham, J.K. (2003b) Enabling ethical economies: Cooperativism and class. *Critical Sociology* 29, 123–161.
Gretzel, U. (2017) Social media activism in tourism. *Journal of Hospitality and Tourism* 15 (2), 1–14.
Hall, C.M. (1994) *Tourism and Politics*. Chichester: John Wiley & Sons.
Hall, C.M. (2008) *Tourism Planning* (2nd edn). Harlow: Pearson.
Hall, C.M. (2014a) You can check out anytime you like but you can never leave. In C. Weeden and K. Boluk (eds) *Managing Ethical Consumption in Tourism* (pp. 32–55). London: Routledge.
Hall, C.M. (2014b) *Tourism and Social Marketing*. London: Routledge.
Hall, C.M. (2022) Tourism and the Capitalocene: From green growth to ecocide. *Tourism Planning & Development* 19 (1), 61–74.
Hall, C.M. and O'Sullivan, V. (1996) Tourism, political stability and violence. In A. Pizam and N. Mansfield (eds) *Tourism, Crime and International Security Issues* (pp. 105–121). New York: John Wiley & Sons.
Hall, C.M., Lundmark, L. and Zhang, J.J. (eds) (2020) *Degrowth and Tourism: New Perspectives on Tourism Entrepreneurship, Destinations and Policy*. London: Routledge.

Hallward, M.C. (2013) *Transnational Activism and the Israeli-Palestinian Conflict.* New York: Palgrave Macmillan.
Harvey, D. (2005) *A Brief History of Neoliberalism.* Oxford: Oxford University Press.
Hawkins, R.A. (2010) Boycotts, buycotts and consumer activism in a global context: An overview. *Management & Organizational History* 5 (2), 123–143.
Higgins-Desbiolles, F. (2020) Socialising tourism for social and ecological justice after COVID-19. *Tourism Geographies* 22 (3), 610–623.
Higgins-Desbiolles, F. and Bigby, B.C. (2022) A local turn in tourism studies. *Annals of Tourism Research* 92, 103291.
Higgins-Desbiolles, F., Bigby, B.C. and Doering, A. (2021) Socialising tourism after COVID-19: Reclaiming tourism as a social force? *Journal of Tourism Futures* 8 (2), 208–219. https://doi.org/10.1108/JTF-03-2021-0058.
Higgins-Desbiolles, F., Doering, A. and Bigby, B.C. (eds) (2022) *Socialising Tourism: Rethinking Tourism for Social and Ecological Justice.* London: Routledge.
Hudson, S. (2007) To go or not to go? Ethical perspectives on tourism in an 'outpost of tyranny'. *Journal of Business Ethics* 76 (4), 385–396.
Klein, J.G., Smith, N.C. and John, A. (2004) Why we boycott: Consumer motivations for boycott participation. *Journal of Marketing* 68 (3), 92–109.
Lamers, M., Nawijn, J. and Eijgelaar, E. (2019) Political consumerism for sustainable tourism: A review. In M. Boström, M. Micheletti and P. Oosterveer (eds) *The Oxford Handbook of Political Consumerism* (pp. 349–365). Oxford: Oxford University Press.
Lovelock, B. and Lovelock, K. (2013) *The Ethics of Tourism: Critical and Applied Perspectives.* London: Routledge.
Maira, S. (2018) *Boycott!: The Academy and Justice for Palestine.* Oakland, CA: University of California Press.
Millar, C. and Aiken, D. (1995) Conflict resolution in aquaculture: a matter of trust. In A. Boghen (ed.) *Coldwater Aquaculture in Atlantic Canada* (2nd edn) (pp. 617–645). Moncton, New Brunswick: Canadian Institute for Research on Regional Development.
Mkono, M. (2018) The age of digital activism in tourism: Evaluating the legacy and limitations of the Cecil anti-trophy hunting movement. *Journal of Sustainable Tourism* 26 (9), 1608–1624.
Morrison, S. (2022) Border-crossing repertoires of contention: Palestine activism in a global justice context. *Globalizations* 17 (1), 17–33.
Mowforth, M. and Munt, I. (2015) *Tourism and Sustainability: Development, Globalisation and New Tourism in the Third World.* London: Routledge.
Nonomura, R. (2017) Political consumerism and the participation gap: Are boycotting and 'buycotting' youth-based activities? *Journal of Youth Studies* 20 (2), 234–251.
Oo, A.N. and Perez, M. (1996) Behind the smiling faces. *Newsletter – The International Communication Project*, 28. See http:/www.comlink.apc.org/fic/newslett/eng/28/page_36.htm (accessed 14 February 2022).
Pears, L. (2019) Travel boycotts: should we boycott countries on political grounds? See https://www.theplanetedit.com/should-we-boycott-countries/ (accessed 14 February 2022).
Pilger, J. (1996, 1 June) The land of fear. *The Sydney Morning Herald, Spectrum.*
Radzik, L. (2017) Boycotts and the social enforcement of justice. *Social Philosophy and Policy* 34 (1), 102–122.
Richter, L.K. (1989) *The Politics of Tourism in Asia.* Honolulu: University of Hawaii Press.
Rössel, J. and Schenk, P.H. (2018) How political is political consumption? The case of activism for the global south and fair trade. *Social Problems* 65 (2), 266–284.
Seyfi, S. and Hall, C.M. (2019) International sanctions, tourism destinations and resistive economy. *Journal of Policy Research in Tourism, Leisure and Events* 11 (1), 159–169.

Seyfi, S. and Hall, C.M. (2020) *Sanctions, Boycotts and Tourism*. London: Routledge.

Seyfi, S., Hall, C.M. and Vo-Thanh, T. (2020) Tourism, peace and sustainability in sanctions-ridden destinations. *Journal of Sustainable Tourism* 30 (2–3), 372–391. https://doi.org/10.1080/09669582.2020.1818764.

Seyfi, S., Hall, C.M., Saarinen, J. and Vo-Thanh, T. (2021) Understanding drivers and barriers affecting tourists' engagement in digitally mediated pro-sustainability boycotts. *Journal of Sustainable Tourism*. https://doi.org/10.1080/09669582.2021.2013489.

Shaheer, I., Carr, N. and Insch, A. (2019) What are the reasons behind tourism boycotts. *Anatolia* 30 (2), 294–296.

Shaheer, I., Insch, A. and Carr, N. (2018) Tourism destination boycotts–are they becoming a standard practise? *Tourism Recreation Research* 43 (1), 129–132.

Sharp, G. (2013) *How Nonviolent Struggle Works*. East Boston, MA: Albert Einstein Institution.

Shepherd, J. (2021) 'I'm not your toy': Rejecting a tourism boycott. *Tourism Recreation Research*. https://doi.org/10.1080/02508281.2021.1998874.

Stephan, M.J. and Chenoweth, E. (2008) Why civil resistance works: The strategic logic of nonviolent conflict. *International Security* 33 (1), 7–44.

Stolle, D. and Micheletti, M. (2013) *Political Consumerism: Global Responsibility in Action*. Cambridge: Cambridge University Press.

Tomazos, K. and Butler, R. (2012) Volunteer tourists in the field: A question of balance? *Tourism Management* 33 (1), 177–187.

TravelAsia (1996) Our say: Balancing politics and tourism. *TravelAsia*, 26 July.

Trentmann, F. (2007) Citizenship and consumption. *Journal of Consumer Culture* 7, 147–58.

Weeden, C. and Boluk, K. (eds) (2014) *Managing Ethical Consumption in Tourism*. London: Routledge.

Whatmore, S., Stassart, P. and Renting, H. (2003) What's alternative about alternative food networks? *Environment and Planning A* 35, 389–91.

Yousaf, S., Razzaq, A. and Fan, X. (2021) Understanding tourists' motivations to launch a boycott on social media: A case study of the #BoycottMurree campaign in Pakistan. *Journal of Vacation Marketing* 1356766721993861.

Zapata Campos, M.J., Hall, C.M. and Backlund, S. (2018) Can MNCs promote more inclusive tourism? Apollo tour operator's sustainability work. *Tourism Geographies* 20 (4), 630–652.

4 The Local Turn in Tourism: Place-based Realities, Dangers and Opportunities

Can-Seng Ooi

Introduction

International tourism is sometimes seen as the continuation of colonisation and domination of developing countries by wealthy Western countries (Chambers & Buzinde, 2015; Hales *et al.*, 2018; Mietzner & Storch, 2019). Serious concerns about justice, fairness and equity are raised in the tourism industry as issues of climate change, worker enslavement and community exploitation are still prevalent (Fennell, 2018; Jamal, 2019). The current domination of multinational companies and rich developed countries has led to the concentration of wealth in the hands of a small number of people and countries, while wanton tourism consumption and growth have polluted the world and depleted natural resources (Higgins-Desbiolles, 2020). The expansive influence of international tourism and hospitality companies, the increased economic dependency on the visitor economy and the touristification of society in many destinations have led to demands for a more responsible form of tourism development. Tourism development must be respectful of the destination and the community, be of the right size and must benefit the host society; these also mean taking the 'local turn' and being place-based.

The local turn here follows Higgins-Desbiolles and Bigby's (2022) formulation. The local turn refers to a form of tourism that is respectful and more inclusive of the local community, ecology and to its past and future generations. The other chapters in this book have discussed the importance and the necessity of taking this approach. This chapter's position is totally aligned with them. Many chapters in this book have also discussed alternative views of making tourism development more independent and self-sustaining for the community. This chapter complements these perspectives by advocating for approaches that will also challenge and eradicate locally established social inequality, exploitation and injustices. As it will be elaborated later, all communities are socially stratified, consist of different interest groups and are diverse. Conflicts and change in society are inevitable. Sociopolitical negotiations and contestations are inadvertent. When taking the local turn, the question remains on which and whose local perspectives to accept and respect. The local turn in tourism should not merely be 'local washing' and 'community washing' (see Chapter 8). Local washing will happen when the champions for tourism development appropriate the discourse of 'community focus' but they remain focused on the views and interests of local elites and the privileged few, rather than on the wider diverse community. Consequently, any future tourism development will perpetuate entrenched local systems of exploitation, abuses, nepotism and cronyism.

In accepting that tourism is primarily a set of economic activities but with social, cultural and political implications for society, many of the social injustices and inequities are embedded in the local tourism market mechanisms. The market is a set of mechanisms to distribute resources and benefits in society, even though distribution occurs unevenly and imperfectly (North, 1991). Tourism is only possible because of the market, through which visitors lacking in local knowledge and local competence can buy air tickets, eat local produce, meet residents and experience local heritage. The market is not immoral, but it has two moral limits. As will be discussed soon, there are consequences when market processes determine and shape aspects of community life. The consequences have partly led to the call for more respect for local approaches to tourism development.

In the next section, I discuss the moral limits of the market. Subsequently, I will demonstrate that many common tourism development methods not only have local elements, they are also addressing the two moral limits of the market. The discussion however will also show that a local community and society is inevitably complex, heterogeneous and evolving. In taking the local turn in tourism development, there is a need to address place-based challenges that are anti-social, discriminatory and unfair. While there are many challenges, there are also opportunities, as we have to decide which and whose local perspectives and causes we want to advocate, support and champion.

Two Moral Limits of the Market

There are many perspectives on various tourism challenges such as authenticity, social impacts of tourism, sustainable and regenerative practices, social justice and multinational exploitation. Many of these perspectives tacitly acknowledge the moral limits of the market. The market is a human-made socioeconomic institution (North, 1991). It serves important functions in modern society; it distributes and allows for the convenient exchange of goods and services. The market brings great benefits and welfare to society. All modern societies have place-based variations of the market, differentiated by regulations, norms and informal practices. For all of them, when compared to bartering, modern market transactions are straightforward and efficient. The market we know today facilitates exchange between individuals who may not necessarily know each other (Fligstein, 2002; North, 1991; Roth, 2015). Money is the common denominator that enables us to trade and to acquire products and services seamlessly. However, markets that use money as a medium of exchange have at least two moral limits (Ooi, 2022; Sidelsky & Skidelsky, 2015; Simmel, 1978).

Moral limit one: Repugnant transactions

Money as a means for universal exchange has consequences (Simmel, 1978). This does not deny the many advantages including the liberation of individuals to have access to countless products and services. It also enhances personal liberty and freedom. Monetary exchanges are efficient and effective (Fligstein, 2002; North, 1991; Williamson, 1998). Instead of having to build trust and closer relations between persons to facilitate bartering, money is used. Tourism today is made possible because people can travel to places without knowing their hosts personally. Visitors have access to experiences and attractions in exotic places.

But money cannot buy everything. Some things are not supposed to be priced and sold because they are sacred, revered or supposedly priceless. If they are exchanged monetarily, they become repugnant transactions (Roth, 2015). For example, how much should a Christian family charge tourists to join in their traditional Christmas family dinner? Is it appropriate to let paying visitors witness private weddings and funerals? Moral limit one of the market points to how economic exchange transforms products, services and/or experiences in ways that denigrate and even destroy the intrinsic values of what is being bought. The Christmas family dinner, for instance, would lose its traditional familial values if paying tourists join the family. Issues of authenticity often tacitly assume this moral limit of the tourism market. Repugnant transactions may be too lucrative and thus still occur even though many people may feel that they should not. This is one of the two moral limits of the market.

However, what are considered appropriate and what are repugnant transactions are locally determined. For instance, many Chinese funerals in Southeast Asian countries are loud public events and it is normal for crowds – including tourists – to gather to witness the band and the procession. In contrast, funerals in many Western societies are private and sombre where tourist curiosity is definitely not welcomed. For tourism, it is thus important to rely on locals to decide what can be experienced and offered to tourists. As markets are locally embedded, place-based norms, values and practices determine what are acceptable and what are repugnant transactions.

Moral limit two: The unfair market

Volunteer tourism is a popular alternative form of tourism that attempts to satisfy the altruistic and touristic desires of tourists (Sin, 2009). It is an attempt to convert tourism into local community development resources. Without doubting the positive impact of volunteer tourism, such activities however still cater more to the needs of visitors than to that of the community, resulting in possibly slower local projects, poor quality workmanship, decreased demand for local workers and creation of dependency (Guttentag, 2009; Higgins-Desbiolles *et al.*, 2022; Ong *et al.*, 2013). In this context, the volunteer tourists do not only have significant say on how they want to help (since they are paying), but additionally the benefits of such activities do not necessarily go to the community but to the agents and operators that manage such schemes. The volunteer tourism market does not distribute profits to the local people who need the benefits most. Instead, the market enriches a small group of businesses while the wider society suffers. Even in a developed country like Australia, data from its island-state Tasmania indicate that one-third of the tourism and hospitality workforce do not earn enough and live below the official poverty standard even though the industry has become an economic driver of the state's economy (Denny *et al.*, 2019). The tourism industry is lucrative, but not everyone working in it benefits adequately from it.

Related to how benefits and costs are distributed through the market, accessibility to goods and services is based largely on a person's ability to pay in the market, rather than based on a person's needs. For instance, tourists visit places like South Korea and Thailand for cosmetic surgery and other medical services. The medical tourism industry leads to an expanded private, technology-intensive health care which unfortunately may remain out of reach to poorer local residents (Chen & Flood, 2013). Local doctors are also tempted to specialise and cater to the lucrative visitor market instead. Sick locals find that they could not compete with wealthier visitors when seeking medical care. Moral limit two of the market points to how the market fails to distribute the benefits of market exchange equitably and to those who need them most.

To ensure that the benefits from market exchanges are more equitably distributed, local circumstances, contexts and perspectives must be incorporated. A tourism market in one system cannot be transposed into another destination but the benefits of the market must be distributed fairly and more equitably. Like with moral limit one, this is more easily said than done.

Localisation in Popular Approaches of Making Tourism More Sustainable

To reiterate, moral limit one points to how economic exchanges transform many products, services and/or experiences in ways that denigrate and even destroy their intrinsic values. The market allows for repugnant transactions. Moral limit two refers to how the market distributes the benefits of market exchange in unequitable ways. Accessibility to goods and services is based on people's ability to pay rather than their needs, and thus the benefits of the market do not always go to more deserving parties. In both set of challenges, local perspectives must be part of the solution to addressing any moral challenges in a market. In fact, many of the common models and approaches to making tourism activities and development better address the moral limits of the market, albeit implicitly and tacitly. These complementing models and approaches include the triple bottom line, public–private partnerships (PPPs), redesigning and regulating the market and community-led initiatives.

Stakeholder theory and the triple bottom line

Freeman's (2010) stakeholder framework assumes the importance of the local community and industry (Budeanu et al., 2016). It advocates that different stakeholders' – industry, workers, residents, civil society, the environment – diverse needs and interests must be served, albeit through negotiation and collaboration (Angelo & Maria, 2010; Currie et al., 2009; Nilsson, 2007). One popular way of operationalising stakeholder theory is the triple bottom line (TBL) accounting framework, addressing profits, people and planet (Ringham & Miles, 2018).

In a balance sheet, the economic bottom line is the easiest to quantify as that is the original purpose of the accounting framework. TBL also measures the firm's contribution to the community and to the environment. Many tourism establishments support local community projects and causes, such as sponsoring local charities, community events, forest regeneration projects and waste reduction schemes. Each community has its own set of contexts and circumstances that TBL can be adapted to address. The social and environmental causes that firms adopt in their three bottom lines should reflect the local situation and those relevant to the community. TBL encourages businesses to find the most relevant and their favourite local community and environmental causes to support. In this way, repugnant transactions can be avoided, and the benefits from tourism can be distributed according to the desires of the society and reflected in the TBL approach.

Localising the TBL is more easily said than done. Many tourism companies promote a more sustainable form of tourism, but commerce, environment and community interests do not necessarily overlap. The balance between the different stakeholders is often influenced by those with more resources to push for their agendas (Liu, 2003; Ooi, 2013). A more localised practice of TBL has the danger of perpetuating existing local social abuses, thus aggravating the entrenched moral challenges in the tourism market. And within a community, there are differences, with groups and individuals hustling for influence, resources and power. It is easy to say that being respectful of the local situation and circumstance is central but tourism businesses respecting the local must also decide which groups and causes they want to support. They may be influenced and blocked by local gatekeepers and power brokers or find it more convenient to work with established non-governmental organisations and community parties which have their own interests and agendas.

Public–private partnerships

The state has the responsibility of ensuring the well-being of the population and the environment. There are many public goods, such as infrastructure and national parks that are important to both residents and visitors. Developing and upholding these public goods is expensive, and the public service may not have the expertise to build, operate or maintain them (Ruhanen, 2013). Partnering with the private sector may help. Ideally, by cooperating and collaborating, joint benefits for industry, local community and the environment can be realised. Many tourist attractions are public goods (e.g. national parks, heritage sites, cultural institutions). With the commercial expertise of businesses and the competences of the public sector in managing the population, local solutions can be found to provide public tourism services that are effective, efficient and even profitable. Progress can be measured and managed, such as through the TBL framework (Andersson & Getz, 2009; Castellani & Sala, 2010; Zapata & Hall, 2012). While keeping the local context central, the PPP approach breaks sectoral boundaries, promotes a whole-of-destination approach to tourism development that removes red-tape and ensures supporting regulations on new projects, shares business risks among parties and jointly aims for social, environmental and economic viability.

Unfortunately, challenges found in many public projects are also found in PPP projects, such as lengthy procurement processes, long-term inflexible contracts, delays in project implementation and incorrect estimations of project costs (Geoffrey Deladem *et al.*, 2021). The desire to get the best from both the public and private approaches may end up with a PPP project that also has the worst from both approaches. Relations embedded in PPP may also be unequal. In many places, influential businesses and/or authoritarian political partners may dictate PPP projects. Local social and environmental causes are appropriated by businesses and politicians to further their own selfish goals (Iossa & Martimort, 2016; Lai & Ooi, 2015; Zapata & Hall, 2012). Nepotism, corruption and dictatorships in many countries make a mockery of such partnerships, as social and environmental interests are acknowledged only in name (Papathanassis *et al.*, 2017). This may aggravate the second moral limit of the market – only a small group of elites control and benefit from the PPP. These considerations however should not detract us from the principles of engaging stakeholders with complementary skills.

Redesigning and regulating the market

Markets are human constructs and are central to capitalism. Markets are also designed for specific purposes. All markets are regulated even though it is popularly assumed that market forces are beyond human control; they can be manipulated. Thus, during the COVID-19 pandemic, the tourism market collapsed as public health regulations were imposed, resulting in the decimation of the global visitor economy. Many governments provided economic support for businesses and workers in tourism and hospitality. The state continues to play an important role in the market and the economy, even when it chooses not to regulate. In the age of neoliberal governance – giving primacy to individual freedom, market mechanisms, personal responsibility and small government – market outcomes are seen as the collective result of individuals in society (Chandler, 2019; Higgins & Larner, 2017; Joseph, 2013). But market mechanisms are designed, and the outcomes can be changed. Therefore, following moral limit one, there are things that cannot be priced but can still be bartered through a set of well-designed market mechanisms. For instance, the kidney exchange is the classic example of how an exchange market can function. Considered as repugnant transactions, many countries disallow human organs to be commercially traded, but they can be exchanged (Roth, 2015). Giving away a kidney is framed as a donation even though economic resources are needed for an exchange to take place. A tourism example is couchsurfing.org; through this website, a host offers a 'couch' to guests through a hospitality exchange network to which that they all belong (Decrop *et al.*, 2018; Germann Molz, 2013). Similarly, Worldwide Opportunities on Organic Farms (WWOOF) uses a bartering system to attract holiday-workers into the organic farming movement (Deville *et al.*, 2016). WWOOF is a work exchange network and participants work on organic farms while on holiday in exchange for food and lodging.

Market mechanisms are shaped by regulations. Taxing profits, for instance, is a common strategy to address moral limit two of the market. Taxes can be levied and increased for highly profitable businesses and then spent on community initiatives (Arguea & Hawkins, 2015; Burns, 2010; Nepal & Nepal, 2019). Ride-hailing app Uber and short-term accommodation app Airbnb have transformed the local mobility and short-term accommodation markets respectively. Un-utilised assets are activated and accessed through their digital marketplaces, bypassing traditional industry gatekeepers. Authorities around the world have started regulating such digital platform markets. For instance, cities from San Francisco to Singapore have started regulating the short-term rental service to limit the transformation of neighbourhoods and disrupting established landlord–tenant relations (Tun, 2020). These controls include banning such rentals in some neighbourhoods, limiting the number of days allowed for short-term rental and taxing rental income.

Market regulations can be changed to allow or disallow certain transactions, define how and who can access the market and also how profits are (re)distributed. Markets work within the local social, political, cultural and economic frameworks. The role of the state in the market and how welfare is distributed are determined by local practices, circumstances and situations. Unfortunately, a redesigned market does not ensure a more transparent and more equitable manner for the benefits to be distributed. Redesigned markets often face the same moral limits as the more laissez faire ones. They are subjected to corporate lobbyism, selective and weak enforcement and illegal practices.

Community-focused tourism: For the local, by the local

In giving primacy to local stakeholders, many researchers and practitioners propose tourism development strategies that are ground-up and community-led and focused (Asker *et al.*, 2010; Higgins-Desbiolles *et al.*, 2022; Muganda *et al.*, 2013; Sofield, 1993). Local communities know their culture, heritage and environment better than anyone else. Over the generations, they have created opportunities for themselves to thrive and have found solutions to the challenges they face. They should thus be consulted extensively and also allowed to lead any development projects (Okazaki, 2008; Tolkach & King, 2015). A more ground-up approach offers bespoke ways of conducting place-based tourism development. Lessons can be learned from the Global South and from Indigenous communities (Cave & Dredge, 2020; Higgins-Desbiolles *et al.*, 2022). Residents find their own local ways to do tourism while protecting their own welfare. There will then be less economic leakage and more local control (Nyaupane *et al.*, 2006). In the manual published by the Asia Pacific Economic Cooperation (APEC), there are many positive examples of community-based tourism projects (Asker *et al.*, 2010). For instance, in the Canadian town of Chemainus, with a population of less than 5000, members of the town participated in the planning and in workshops that led to a widely accepted community plan to manage the growth, development and tourism there. They developed a local vision on how they want to reenergise and revitalise their place. Local member, Karl Schultz, played the pivotal initiator role, as he drove the idea of using public wall murals to enliven the town. The public art works sparked interest in the community, and they were integrated into the larger plan to increase visitor numbers; in turn, this resulted in the mushrooming of new small businesses (Asker *et al.*, 2010: 15).

Along this line of argument, more and more researchers and practitioners are championing regenerative tourism (Ateljevic, 2020; Pollock, 2019). This view advocates that tourism should be first a resource for community development and environmental protection (Pollock, 2019). In the context of orphanage tourism in Cambodia, for instance,

Higgins-Desbiolles *et al.* (2022) proposed a Freirian method that aims to empower the community. That method involves building a dialogue of critical consciousness, co-development towards transformative outcomes, capacity sharing and trust in the capabilities of the locals (see Chapter 9 and associated case study). This contrasts (but not necessarily contradicts) with many development project experiences in Cambodia. It is common for social enterprises there to benefit from an increased awareness of economic sustainability in various non-profit sector organisations, which also leads to transformative effects on their goals, motives, methods, income distribution and governance components. However, many of these social enterprises have found it more difficult to strike a balance between their social mission and the demands of the market (Khieng & Dahles, 2015). A substantial portion of the revenue earned has been used for business operational purposes instead of supporting social and community aims. This demonstrates that local ways still face the costs of operating in the market.

The Local Turn and Dealing with Local Realities

The above discussion highlights that some form of local turn is already assumed, if not embedded, in common sustainable tourism development practices. At their own level of practice, each approach taps into local concerns, issues, values, resources and perspectives. The TBL is a business-level response to include selected local community and environmental concerns into measuring the success of the business. PPP taps into private business and management expertise for furthering public good; public resources will be more efficiently utilised, and the benefits distributed more equitably to the local community. As for redesigning the market, the market can be regulated to ensure that the local community will benefit from tourism by both avoiding repugnant transactions and enhancing a more equitable distribution of benefits. A community-focused approach, by its own terms, concentrates on local values, practices and agendas; as a result, tourism activities, practices and values will be determined by locals.

The above discussion also points to the challenges of deploying these interrelated tourism development approaches. Going local is a laudable move but local societies are also imbued with social inequality, inequity, exploitation and illiberal values. For instance, in Indonesia, Lasso and Dahles (2021) evaluated the introduction of ecotourism in Komodo and Labuan Bajo. There was extensive local participation. Local residents were even leading and self-advocating tourist development. Those parties were essential players in the industry as service providers, entrepreneurs, tourism producers and community leaders. Regrettably, much of their involvement and initiatives were not recognised, if not thwarted by local authorities. While many locals continued to push in the ecotourism direction, the tourism-derived benefits remained concentrated in the hands of a handful of people who were better connected and were receiving formal institutional support. In such a context, officials as agents for public partners might aim to serve only their cronies. Surely cronyism, nepotism and corruption should not be part of the local turn. They aggravate the moral limits of the market by allowing the vast majority of benefits to go to only the local elites.

Reflecting the challenges imposed on the local community by less responsible tourism practices, a local-driven tourism development must not harm the community. This inevitably also means a new way of doing tourism that dismantles the status quo that entrenches local inequalities, exploitation and abuses (Nguyen *et al.*, 2021; Ooi, 2019). In their analysis, Tolkach and King (2015) and their participants identified the factors that will decide the fate of a community-based tourism project. They include: funding; information access and capacity building; involving actors and achieving consensus; leadership; mentality and attitudes of communities; government support; and power balance. Many of these factors highlight the importance of collaboration and the interconnectedness of all aspects of community life. But not all societies have strong formal institutions, not all members share the same interests and agendas and not all norms and values in society treat individual members fairly and equally. Also, not all local leaders are generous, professional and altruistic, not all governments are competent and not all of the powerful persons in a community have the interest of the wider community at heart. In this context, community-driven tourism has its merits, but the reality of its practice is more messy and complex than is frequently acknowledged. As mentioned already, local ways of doing tourism may actually aggravate the moral limits of the market. The response, however, is not to reject the local turn. Instead, it is important to be more sophisticated in going place based. This requires understanding local society holistically, which means knowing the implications of at least three intrinsic and interrelated universals of any community: each community is complex, is heterogeneous and changes.

First, all societies are complex (Ooi, 2019). A holistic understanding is necessary, as social, cultural, economic and political structures and activities are intertwined. In some societies, tourism activities are more intertwined with the community than others. A local turn strategy must account for the interlocking institutions and intertwined aspects of social life. An introduction of new tourism practices may not fit well in the complex web of interdependent and interlocking social forces. For example, achieving the fifth UN Sustainable Development Goal of gender equality and empowerment for all women and girls (UN World Tourism Organization, n.d.) is tough because of the entrenched complex of social relations in local society. The processes of allowing women to take better control, have a greater say in their community and to drive community-based tourism development entail addressing many sociocultural factors that constrain and enable both women and men in their roles in society. It may require significant changes to established relationships in society (Tucker & Boonabaana, 2012). Gender roles are manifested through norms and relations, even though women have agency in negotiating power relations. The negotiation process can be slow and may stutter. While a tourism development strategy will concentrate on women's role in that sphere, there are many other roles and functions that women play in society that need to be re-framed. Resistance to change and going back to old ways occur when men perceive threats from women and when women themselves are afraid of the uncertainty with the change (Tucker & Boonabaana, 2012).

Second, all societies are made up of a diverse myriad of individuals. Individuals may belong to different ethnicities, social statuses, occupations, genders and sexualities. They also have different income levels, hold different interests and carry different values. A society is inevitably heterogeneous and that entails at least three challenges in garnering views and support from local stakeholders (Ooi, 2013). One, identifying relevant stakeholders is a demanding exercise. For instance, who constitutes a stakeholder group? Or more specifically, which groups of locals should be consulted? Actively reaching out to stakeholders is important but with their varying levels of engagement and interests, members may not be responsive. Two, stakeholder groups may not want to or may not be able to cooperate. They may not agree because of their competing interests and desires. Certain cultural contexts may pose significant challenges to working and interacting with each other, and the local market system may actually discriminate against some locals. Three, consultation also necessitates coordination. Building up a broad consensus is time consuming and requires resources. In reality and practice, the diversity embedded in all societies means that any local turn strategy must also address the question of *whose* and *which* local perspectives, interests and agendas are embraced.

Third, no society is an island. Society evolves and outside influences are often internalised. Foreign influences have become part of local societies. English as the global *lingua franca* has become the second language in many societies. Technological advancements such as the internet, medical sciences and manufacturing have transformed almost all communities in the world. Tourism is a global phenomenon, and any society that engages with it must largely embrace internationally accepted institutions, structures, regulations, technologies, practices and norms, such as border control, health and safety standards, currency exchange facilities and means of communication (Ooi & Tarulevicz, 2019). The decentralisation of economic control and the celebration of the local often marginalises the importance of outside or global influences in communities. Going local is also about asking residents what they want and what influences they want to adopt and embrace. All societies change. When it comes to tourism development, it also involves changing local society in a more desired direction.

In acknowledging that society is complex, is heterogeneous and changes, what do these mean for taking the local turn in tourism development? In advocating a local turn strategy, there are some complicated realities to navigate. In accepting that any society is complex, heterogeneous and changes, locally repugnant transactions may still persist in the future, partly because those economic transactions are too lucrative to forgo for some (e.g. illicit drugs, poaching) and partly because there are disagreements on what transactions are acceptable or not (e.g. sex work, ecotourism development in a nature reserve). The distribution of benefits from the tourism market may be redirected by tweaking market mechanisms and through regulations but every market discriminates access and distributes benefits unevenly – the issue is whether those ways are considered fair and acceptable. That is always in contention. As mentioned earlier, the local approaches and principles may not bring about more social equity and equality. The moral limits of the market may be aggravated. Table 4.1 points to these challenges and the implications in practice.

The local turn, in principle, is essential in moving towards a more sustainable and regenerative tourism. This chapter supports this premise, but the discussions so far have highlighted the necessity to be realistic and not romantic about the local. The other chapters have rightfully criticised wanton and disrespectful tourism development. It is important to learn, listen and support the local community, while they/we develop a place-based tourism development strategy. Based on the complex, diverse and emerging reality of any local society, the listen, learn and support approach should be extended to include dialogue, proactive engagement and developing allies for desired changes. This may mean challenging some local players and also some of their practices and values.

Table 4.1 Society is complex, heterogeneous and evolves: Implications for a local-led change strategy

	Local-led tourism development strategy
Society is complex	Respecting established interlocking institutions, practices, norms and values. But some local social institutions may have to be dismantled and age-old practices, norms and values may have to change.
Society is heterogeneous	Respecting that society is diverse in terms of members having different interests, agendas and values. A feasible local-led strategy will not necessarily please everyone.
External influences in society	Local experiences, heritage and tradition define the place, but society also changes. These changes must be local-led or largely accepted. Discussion at the local level must be encouraged and future change in direction must be possible.

For instance, corruption is prevalent in many developed and developing countries. And corruption is bad for the community and the industry (Alola *et al.*, 2021; Das & Dirienzo, 2010). It makes an economy less competitive and less attractive for investments, including in tourism (Das & Dirienzo, 2010). A small group of persons will benefit from such practices, and the community-at-large suffers. Corruption is most prevalent when formal institutions are weak, as people find solutions to navigate and to benefit from the system (Jancsics, 2014). Non-governmental organisations in some societies are inadvertently caught in the web of unacceptable practices, as local elites have access to them and to public resources. As a result, a more sustainable and responsible way of supporting the local will be to dismantle unhelpful and anti-social behaviour and practices. Such an approach may mean being proactive in following the lead and supporting the changes that more righteous local members want.

We are cognisant of the dangers of continuing the colonisation of local communities through the introduction of 'modern' practices. Capitalism and the modern markets have shown how many local societies and social causes have been subjugated, marginalised and exploited (Chambers & Buzinde, 2015; Ooi, 2005, 2021; Said, 1979; Tucker & Zhang, 2016). And at the same time, we observe the corruption and abuses of various forms of capitalism, democracy and bureaucracy in developed western countries. The evils of tourism, including overtourism, climate change and the touristification of culture, heritage and society, are damning. Global domination of multinationals and the prevalence of economic and political ideas such as democracy and neoliberal governance have generated discontents (Chandler, 2019; Higgins & Larner, 2017). But still turning totally to the local ways of doing things is not necessarily the best way forward.

The discussion on the moral limits of the market highlights the reality that markets function locally, and they are shaped by locals too. There are local players – business people, policymakers, consumers – that embrace, perpetuate and possibly also lament about their own market systems. The market needs to be better regulated and possibly redesigned; this may mean that unhelpful and established rules would have to change. The rules should change to create a market that is fairer, more transparent and more accessible to all. Values of transparency, accountability and the rule of law are essential for democracy. These are also the same values for us to manage the market, and to address the moral limits of the market.

Democracy is not usually convenient and efficient. It is messy and mistakes are made. However, the principles of accountability, transparency and the rule of law ensure that leaders can be booted out when they fail, and the diverse interests of the population are factored into policymaking processes (Ooi, 2013; Weymans, 2016). This chapter is not the place to debate the imperfections of the variety of democracy. If we see tourism development as part of social change, there must be a mechanism and the room to correct its direction when undesirable consequences result. This implies that local residents must have access to information, know how tourism development projects are managed and have the chance to direct changes in the future. Community engagement in the projects is necessary and must be sustained.

Also, while we must respect all societies on their own terms, we must also be aware that many members of a local society may disagree with their own leaders and are unhappy with their own situation. While outsiders have their own values, many locals share the same values too. Should outsiders impose their values on locals? No, but there are sections of local society that want change, are challenging the status quo and share our 'universal' values. They do not want to continue to be marginalised and abused and tourism must develop in ways that will support these members of the local community.

Conclusions

The local turn refers to a form of tourism that is respectful and more inclusive of the local community, ecology and to its past and future generations. Many chapters here have convincingly made the case for tourism to be more place-based and local-focused. This chapter follows the same trajectory, but its focus is on the complicated reality of local communities.

Many of the challenges of tourism arise from the two moral limits of the tourism market. Moral limit one points to how economic exchanges transform many products, services and/or experiences in ways that denigrate and even destroy their intrinsic values. The market allows for repugnant transactions. Moral limit two refers to how the market distributes the benefits of market exchange in unequitable ways. Accessibility to goods and services is also based on people's ability to pay rather than their needs, and the benefits of the market do not always go to more deserving parties. Tourism issues of authenticity, worker exploitation, climate change and touristification are examples of repugnant transactions, discriminatory access to markets and the unfair distribution of benefits from the market. While these moral limits of the market are in-built and universal, each market is locally embedded. There is a variety of markets.

Respecting and listening to the local is a common and welcomed practice in tourism for a long time. The TBL, PPP, regulating and designing the market and community-focused tourism development strategies incorporate some forms of listening, learning and supporting the local. The challenge however is the actualisation of the local-turn: which and whose local turn to take?

There are some messy local realities to navigate. Local approaches and principles may not necessarily bring about more social equity and equality. The moral limits of the market may even be aggravated by local norms and values. But in accepting that society is complex, heterogenous and constantly evolving, then we might be able to use some common values we share with locals in supporting local tourism development in the society. We should be mindful not to impose our values on locals. There are many groups in local society who are struggling and wanting to change their society for the better – we should listen, learn and support local groups while remaining aware of local complexities. Additionally, we must stay attentive and be critical and reflexive on the local turn discourse to not fall into romanticism that hinders efforts.

References

Alola, U.V., Alola, A.A., Avci, T. and Ozturen, A. (2021) Impact of corruption and insurgency on tourism performance: A case of a developing country. *International Journal of Hospitality & Tourism Administration* 22 (4), 412–428. https://doi.org/10.1080/15256480.2019.1650686.

Andersson, T.D. and Getz, D. (2009) Tourism as a mixed industry: Differences between private, public and not-for-profit festivals. *Tourism Management* 30 (6), 847–856. https://doi.org/10.1016/j.tourman.2008.12.008.

Angelo, P. and Maria, C. (2010) Analysing tourism stakeholders networks. *Tourism Review* 65 (4), 17–30. http://dx.doi.org/10.1108/16605371011093845.

Arguea, N.M. and Hawkins, R.R. (2015) The rate elasticity of Florida tourist development (aka bed) taxes. *Applied Economics* 47 (18), 1823–1832. https://doi.org/10.1080/00036846.2014.1000519.

Asker, S., Boronyak, L., Carrard, N. and Paddon, M. (2010) *Effective Community Based Tourism: A Best Practice Manual*. Singapore: Asia Pacific Economic Council. https://www.apec.org/Publications/2010/06/Effective-Community-Based-Tourism-A-Best-Practice-Manual-June-2010

Ateljevic, I. (2020) Transforming the (tourism) world for good and (re)generating the potential 'new normal'. *Tourism Geographies* 22 (3), 467–475. https://doi.org/10.1080/14616688.2020.1759134.

Budeanu, A., Miller, G., Moscardo, G. and Ooi, C.-S. (2016) Sustainable tourism, progress, challenges and opportunities: an introduction. *Journal of Cleaner Production* 111, 285–294. https://doi.org/10.1016/j.jclepro.2015.10.027.

Burns, S. (2010) Local authorities, funding tourism services and tourist taxes. *Local Economy: The Journal of the Local Economy Policy Unit* 25 (1), 47–57. https://doi.org/10.1080/02690940903545398.

Castellani, V. and Sala, S. (2010) Sustainable performance index for tourism policy development. *Tourism Management* 31 (6), 871–880. https://doi.org/10.1016/j.tourman.2009.10.001.

Cave, J. and Dredge, D. (2020) Regenerative tourism needs diverse economic practices. *Tourism Geographies* 22 (3), 503–513. https://doi.org/10.1080/14616688.2020.1768434.

Chambers, D. and Buzinde, C. (2015) Tourism and decolonisation: Locating research and self. *Annals of Tourism Research* 51, 1–16. https://doi.org/10.1016/j.annals.2014.12.002.

Chandler, D. (2019) Resilience and the end(s) of the politics of adaptation. *Resilience* 7 (3), 304–313. https://doi.org/10.1080/21693293.2019.1605660.

Chen, Y.Y.B. and Flood, C.M. (2013) Medical tourism's impact on health care equity and access in low- and middle-income countries: Making the case for regulation. *Journal of Law, Medicine & Ethics* 41 (1), 286–300. https://doi.org/10.1111/jlme.12019.

Currie, R.R., Seaton, S. and Wesley, F. (2009) Determining stakeholders for feasibility analysis. *Annals of Tourism Research* 36 (1), 41–63. http://www.sciencedirect.com/science/article/pii/S0160738308001138.

Das, J. and Dirienzo, C. (2010) Tourism competitiveness and corruption: A cross-country analysis. *Tourism Economics* 16 (3), 477–492. https://doi.org/10.5367/000000010792278392.

Decrop, A., del Chiappa, G., Mallargé, J. and Zidda, P. (2018) 'Couchsurfing has made me a better person and the world a better place': The transformative power of collaborative tourism experiences. *Journal of Travel & Tourism Marketing* 35 (1), 57–72. https://doi.org/10.1080/10548408.2017.1307159.

Denny, L., Shelley, B. and Ooi, C.S. (2019) Education, jobs and the political economy of tourism: Expectations and realities in the case of Tasmania. *Australasian Journal of Regional Studies* 25 (2), 282–305. https://www.anzrsai.org/assets/Uploads/PublicationChapter/AJRS-25.2-pages-282-to-305.pdf.

Deville, A., Wearing, S. and McDonald, M. (2016) WWOOFing in Australia: Ideas and lessons for a de-commodified sustainability tourism. *Journal of Sustainable Tourism* 24 (1), 91–113. https://doi.org/10.1080/09669582.2015.1049607.

Fennell, D.A. (2018) *Tourism Ethics* (2nd edn). Bristol: Channel View Publications.

Fligstein, N. (2002) *The Architecture of Markets: An Economic Sociology of Twenty-First Century Capitalist Societies*. Princeton, NJ: Princeton University Press.

Freeman, R.E. (2010) *Strategic Management: A Stakeholder Approach*. Cambridge: Cambridge University Press.

Geoffrey Deladem, T., Xiao, Z., Siueia, T.T., Doku, S. and Tettey, I. (2021) Developing sustainable tourism through public-private partnership to alleviate poverty in Ghana. *Tourist Studies* 21 (2), 317–343. https://doi.org/10.1177/1468797620955250.

Germann Molz, J. (2013) Social networking technologies and the moral economy of alternative tourism: The case of couchsurfing.org. *Annals of Tourism Research* 43, 210–230. https://doi.org/10.1016/j.annals.2013.08.001.

Guttentag, D.A. (2009) The possible negative impacts of volunteer tourism. *International Journal of Tourism Research* 11 (6), 537–551. https://doi.org/10.1002/jtr.727.

Hales, R., Dredge, D., Higgins-Desbiolles, F. and Jamal, T. (2018) Academic activism in tourism studies: Critical narratives from four researchers. *Tourism Analysis* 23 (2), 189–199. https://doi.org/10.3727/108354218X15210313504544.

Higgins, V. and Larner, W. (eds) (2017) *Assembling Neoliberalism: Expertise, Practices, Subjects*. London: Palgrave Macmillan.

Higgins-Desbiolles, F. (2020) Socialising tourism for social and ecological justice after COVID-19. *Tourism Geographies* 22 (3), 610–623. https://doi.org/10.1080/14616688.2020.1757748.

Higgins-Desbiolles, F. and Bigby, B.C. (2022) A local turn in tourism studies. *Annals of Tourism Research* 92, 103291. https://doi.org/10.1016/j.annals.2021.103291.

Higgins-Desbiolles, F., Scheyvens, R.A. and Bhatia, B. (2022) Decolonising tourism and development: From orphanage tourism to community empowerment in Cambodia. *Journal of Sustainable Tourism* 1–21. https://doi.org/10.1080/09669582.2022.2039678.

Iossa, E. and Martimort, D. (2016) Corruption in PPPs, incentives and contract incompleteness. *International Journal of Industrial Organization* 44, 85–100. https://doi.org/10.1016/j.ijindorg.2015.10.007.

Jamal, T. (2019) *Justice and Ethics in Tourism*. London: Routledge.

Jancsics, D. (2014) Interdisciplinary perspectives on corruption. *Sociology Compass* 8 (4), 358–372. https://doi.org/10.1111/soc4.12146.

Joseph, J. (2013) Resilience as embedded neoliberalism: A governmentality approach. *Resilience* 1 (1), 38–52. https://doi.org/10.1080/21693293.2013.765741.

Khieng, S. and Dahles, H. (2015) Commercialization in the non-profit sector: The emergence of social enterprise in Cambodia. *Journal of Social Entrepreneurship* 6 (2), 218–243. https://doi.org/10.1080/19420676.2014.954261.

Lai, S. and Ooi, C.-S. (2015) Branded as a World Heritage city: The politics afterwards. *Place Branding and Public Diplomacy* 11 (4), 276–292. https://doi.org/10.1057/pb.2015.12.

Lasso, A.H. and Dahles, H. (2021) A community perspective on local ecotourism development: lessons from Komodo National Park. *Tourism Geographies* 1–21. https://doi.org/10.1080/14616688.2021.1953123.

Liu, Z. (2003) Sustainable tourism development: A critique. *Journal of Sustainable Tourism* 11 (6), 459–475. https://doi.org/10.1080/09669580308667216.

Mietzner, A. and Storch, A. (eds) (2019) *Language and Tourism in Postcolonial Settings*. Bristol: Channel View Publications. https://doi.org/10.21832/MIETZN6782.

Muganda, M., Sirima, A. and Ezra, P.M. (2013) The role of local communities in tourism development: Grassroots perspectives from Tanzania. *Journal of Human Ecology* 41 (1), 53–66. https://doi.org/10.1080/09709274.2013.11906553.

Nepal, R. and Nepal, S.K. (2019) Managing overtourism through economic taxation: Policy lessons from five countries. *Tourism Geographies* 1–22. https://doi.org/10.1080/14616688.2019.1669070.

Nguyen, D.T.N., D'Hauteserre, A.-M. and Serrao-Neumann, S. (2021) Intrinsic barriers to and opportunities for community empowerment in community-based tourism development in Thai Nguyen province, Vietnam. *Journal of Sustainable Tourism* 1–19. https://doi.org/10.1080/09669582.2021.1884689.

Nilsson, P.Å. (2007) Stakeholder theory: The need for a convenor. The case of Billund. *Scandinavian Journal of Hospitality and Tourism* 7 (2), 171–184. http://www.informaworld.com/10.1080/15022250701372099.

North, D.C. (1991) Institutions. *Journal of Economic Perspectives* 5 (1), 97–112. https://doi.org/10.1257/jep.5.1.97.

Nyaupane, G.P., Morais, D.B. and Dowler, L. (2006) The role of community involvement and number/type of visitors on tourism impacts: A controlled comparison of Annapurna, Nepal and Northwest Yunnan, China. *Tourism Management* 27 (6), 1373–1385. https://doi.org/10.1016/j.tourman.2005.12.013.

Okazaki, E. (2008) A community-based tourism model: Its conception and use. *Journal of Sustainable Tourism* 16 (5), 511–529. https://doi.org/10.1080/09669580802159594.

Ong, F., Pearlman, M., Lockstone-Binney, L. and King, B. (2013) Virtuous volunteer tourism: Towards a uniform code of conduct. *Annals of Leisure Research* 16 (1), 72–86. https://doi.org/10.1080/11745398.2013.769402.

Ooi, C.-S. (2021) Gay tourism: A celebration and appropriation of queer difference. In O. Vorobjovas-Pinta (ed.) *Gay Tourism: New Perspectives* (pp. 15–33). Bristol: Channel View Publications.

Ooi, C.-S. (2022) Sustainable tourism and the moral limits of the market: Can Asia offer better alternatives? In A. S. Balasingam and Y. May (eds) *Asian Tourism Sustainability* (pp. 177–197). Singapore: Springer Nature.

Ooi, C.-S. and Tarulevicz, N. (2019) From Third World to First World: Tourism, food safety and the making of modern Singapore. In E. Park, S. Kim and I. Yeoman (eds) *Food Tourism in Asia* (pp. 73–88). Singapore: Springer. https://doi.org/10.1007/978-981-13-3624-9_6.

Ooi, C.-S. (2005) The Orient responds: Tourism, Orientalism and the national museums of Singapore. *Tourism* 53 (4), 285–299.

Ooi, C.-S. (2013) Tourism policy challenges: Balancing acts, co-operative stakeholders and maintaining authenticity. In M. Smith and G. Richards (eds) *Routledge Handbook of Cultural Tourism* (pp. 67–74). London: Routledge.

Ooi, C.-S. (2019) Asian tourists and cultural complexity: Implications for practice and the Asianisation of tourism scholarship. *Tourism Management Perspectives* 31 (July), 14–23. https://doi.org/10.1016/j.tmp.2019.03.007.

Papathanassis, A., Katsios, S. and Dinu, R.N. (2017) 'Yellow Tourism' – Crime and corruption in tourism. *Journal of Tourism Futures* 3 (2), 200–202. https://doi.org/10.1108/JTF-09-2017-060.

Pollock, A. (2019) Regenerative tourism: The natural maturation of sustainability. Activate the Future. See https://medium.com/activate-the-future/regenerative-tourism-the-natural-maturation-of-sustainability-26e6507d0fcb (accessed 4 March 2022).

Ringham, K. and Miles, S. (2018) The boundary of corporate social responsibility reporting: The case of the airline industry. *Journal of Sustainable Tourism* 1–20. https://doi.org/10.1080/09669582.2017.1423317.

Roth, A.E. (2015) *Who Gets What and Why: The Hidden World of Matchmaking and Market Design*. London: William Collins.

Ruhanen, L. (2013) Local government: Facilitator or inhibitor of sustainable tourism development? *Journal of Sustainable Tourism* 21 (1), 80–98. https://doi.org/10.1080/09669582.2012.680463.

Said, E.W. (1979) *Orientalism*. New York: Vintage Books.

Sidelsky, E. and Skidelsky, R. (2015) The moral limits of markets. In E. Skidelsky and R. Skidelsky (eds) *Are Markets Moral?* (pp. 77–102). London: Palgrave Macmillan. https://doi.org/10.2491/jjsth1970.6.153.

Simmel, G. (1978) *The Philosophy of Money*. London: Routledge & Kegan Paul.

Sin, H.L. (2009) Volunteer tourism: 'Involve me and I will learn'? *Annals of Tourism Research* 36 (3), 480–501. https://doi.org/10.1016/j.annals.2009.03.001.

Sofield, T.H.B. (1993) Indigenous tourism development. *Annals of Tourism Research* 20 (4), 729–750. https://doi.org/10.1016/0160-7383(93)90094-J.

Tolkach, D. and King, B. (2015) Strengthening community-based tourism in a new resource-based island nation: Why and how? *Tourism Management* 48, 386–398. https://doi.org/10.1016/j.tourman.2014.12.013.

Tucker, H. and Boonabaana, B. (2012) A critical analysis of tourism, gender and poverty reduction. *Journal of Sustainable Tourism* 20 (3), 437–455. https://doi.org/10.1080/09669582.2011.622769.

Tucker, H. and Zhang, J. (2016) On Western-centrism and 'Chineseness' in tourism studies. *Annals of Tourism Research* 61, 250–252. https://doi.org/10.1016/j.annals.2016.09.007.

Tun, Z.T. (2020, March 5) Top cities where Airbnb is legal or illegal. Investopedia. See https://www.investopedia.com/articles/investing/083115/top-cities-where-airbnb-legal-or-illegal.asp (accessed 4 March 2022).

UN World Tourism Organization (n.d.) Women's empowerment and tourism. See. https://www.unwto.org/gender-and-tourism (accessed 3 March 2022).

Weymans, W. (2016) Radical democracy's past and future: Histories of the symbolic. *Modern Intellectual History* 13 (03), 841–851. https://doi.org/10.1017/S1479244315000141.

Williamson, O.E. (1998) Transaction cost economics: How it works; where it is headed. *De Economist* 146, 23–58. https://doi.org/10.1023/A:1003263908567.

Zapata, M.J. and Hall, C.M. (2012) Public–private collaboration in the tourism sector: Balancing legitimacy and effectiveness in local tourism partnerships. The Spanish case. *Journal of Policy Research in Tourism, Leisure and Events* 4 (1), 61–83. https://doi.org/10.1080/19407963.2011.634069.

Part 2

Case Studies of Local Community (Dis)/(Re)/Empowerment

5 Unheard Voices: Youth Activism for Social and Environmental Justice

Antonia Canosa

Introduction

This chapter builds on ongoing ethnographic research in the communities of Byron Shire, the location of Byron Bay – a popular tourism destination on the East Coast of Australia. The popularity of Byron Bay as a surf, wellness and alternative destination with national and international tourists has caused tensions and challenges for local residents, including impacting their quality of life and well-being (Wray *et al.*, 2010). The abundance of natural resources coupled with the alternative lifestyle and cultural diversity of the communities in the Byron Shire have contributed to the popularity of the area both as a domestic as well as an international tourist destination. The unprecedented growth of visitors to the area in the last decade has created problems like overcrowding, environmental pressures, loss of ambience and impacts on residential areas (Canosa *et al.*, 2019a).

Research shows that the social and emotional well-being which comes from having a special connection to favourite places in the community is important for young people (Thomson, 2007). Overcrowding due to tourism influx may jeopardise young people's connection to place and their sense of belonging, as such towns shift to accommodating the needs of tourists (Canosa *et al.*, 2021b). This chapter focuses on an important but often overlooked dimension of intersectional exclusion – that of being young (Josefsson & Wall, 2020). I draw on interdisciplinary approaches in childhood and tourism studies and the ontological lens of 'childism' (Wall, 2019) to advocate for child-centred and child-inclusive tourism research, policy and planning. The chapter responds to calls for more research on marginalised and unheard participants like children and young people in tourism (Canosa *et al.*, 2019b; Schänzel & Smith, 2011; Yang *et al.*, 2020), providing research evidence of local youth activism for social and environmental justice.

Three examples of children and young people's activism pre- and post-COVID-19 are analysed to provide a critical commentary about these issues and reflect on the possibilities for more socially and environmentally just forms of tourism. In the first example, a group of 14 young people aged 7–16 years living in the communities of the Byron Shire were involved in a participatory filmmaking project in 2015 to explore their views on tourism and their concerns about living in a tourist destination. In the second example, the chapter explores how young people between the ages of 18 and 24 years ($n = 9$) co-researched and co-designed an intervention strategy to educate tourists about marine litter and the impacts of tourism on the environment of Byron Bay. The third example presents findings from a participatory social theatre research project which engaged a group of young people between the ages of 15 and 21 years ($n = 12$) to research tourism-related issues in the communities of the Byron Shire during and after COVID-19 and resulted in the creation of a responsible tourism pledge.

Building on this research evidence and on the theoretical concepts of 'childism' and 'relational activism', I discuss how localised activism is embedded in children and young people's everyday lives to resist dominant forms of unsustainable tourism development and the hyper-consumption of people and places in the Byron Bay area. Ultimately, a case will be made to reorientate tourism based on the rights of residents of host communities (Higgins-Desbiolles, 2020; Higgins-Desbiolles *et al.*, 2019) including the rights of children and young people, as important but often overlooked stakeholders in the tourism industry (Canosa *et al.*, 2021b).

Background

Sustainable development has been defined by the UN World Commission on Environment and Development ('the Brundtland Commission') in 1987 as 'development that meets the needs of the present without compromising the ability of future generations to meet their own needs' (WCED, 1987, ch. 2, sec. 1, para. 1). Since this time, a growing body of scholarship has focused on sustainable tourism development exploring the complex relationship between tourism and the sociocultural, economic and environmental contexts of tourism destinations and host communities. Much has been written on how residents perceive tourism and the changes to their quality of life (Andereck *et al.*, 2005; Choi & Sirakaya, 2005; Moyle *et al.*, 2010; Sharpley, 2014), as well as the positive and negative impacts of visitation on the environment (Hall *et al.*, 2015; Weaver & Lawton, 2014). The concepts of sustainable, eco- and nature-based tourism have all stemmed from a cautionary platform in tourism studies (Jafari, 2005), where concerns emerged regarding the 'carrying capacity' of tourism destinations, their natural environments and the communities that live within them (Lawton & Weaver, 2015; McCool &

Lime, 2001; Saarinen, 2006). While the triple bottom line (TBL) three pillars of sustainability – economic viability, ecological preservation and societal well-being – are often referred to when discussing sustainable tourism development, the degree to which a tourism destination is pursuing sustainable growth can be difficult to ascertain (Slaper & Hall, 2011).

COVID-19 has prompted a time of reflection in which many critical tourism scholars have advocated for rethinking and redirecting tourism away from neoliberal growth models to a rights-based approach focused on the benefits, interests and rights of local populations (Canosa *et al.*, 2021b; Cheer, 2020; Everingham & Chassagne, 2020; Higgins-Desbiolles, 2020; Higgins-Desbiolles & Bigby, 2022; Lew *et al.*, 2020; Rastegar *et al.*, 2021). Higgins-Desbiolles (2020) advocates for 'socialising' tourism by focusing first and foremost on the rights of local communities. This can only be achieved by redefining tourism 'in order to place the rights of local communities above the rights of tourists for holidays and the rights of tourism corporates to make profits' (Higgins-Desbiolles *et al.*, 2019: 1936).

Discussions concerning alternative approaches to the linear progrowth approaches to tourism development are ongoing, with degrowth, circular economy, regenerative tourism and community-centred tourism frameworks put forward in the hope of promoting sustainable development and the responsible consumption and production of tourism products and services as advocated by Sustainable Development Goal 12 (Boluk *et al.*, 2019; Higgins-Desbiolles *et al.*, 2019; Higgins-Desbiolles *et al.*, 2022). Moving away from 'business-centric' tourism development, Weaver *et al.* (2021: 2) argue for a 'resident-centric' approach embodied in alternative and community-based forms of tourism which recognise residents of host communities as 'the stakeholders with the most to gain or lose from local tourism activity' and as such in need of 'special moral rights and privileged status'.

What is missing, however, are considerations of intergenerational equity and research on how children and young people understand concepts of sustainability in regards to tourism development (Canosa *et al.*, 2020; Seraphin *et al.*, 2020). In advocating for new approaches to understanding and reconceptualising tourism, we must consider the perspectives of our youngest citizens who often remain unheard stakeholders in tourism. If, as Higgins-Desbiolles (2020: 618) argues, tourism must be harnessed 'for the empowerment and well-being of local communities', children and young people need to be engaged in conversations surrounding tourism planning and development, particularly at the local level.

Children are defined by the United Nations Convention on the Rights of the Child (UNCRC) as 'every human being below the age of eighteen years unless under the law applicable to the child, majority is attained earlier' (United Nations, 1989: n.p.). This definition acknowledges the social construction and cultural relativity of the term 'childhood', which may differ according to particular contexts, cultures or environments (Morrow, 2011). The research evidence discussed in this chapter is based on ongoing ethnographic fieldwork in the communities of Byron Shire with children and young people aged 7–24 years of age. The broad spectrum of ages is important given the transition from childhood to adulthood is not clear-cut. As Sibley (1995: 34–35) argues 'the act of drawing the line' in the construction of discrete categories between childhood and adulthood is an arbitrary act which 'interrupts what is naturally continuous' and may be felt 'as unjust by those who suffer the consequences of the division'. Thinking of young adults coming of age at 18 and their ability to vote or drive as a clear threshold into adulthood limits our understanding of young people and our ability to conduct research on youth. The projects presented in the following sections provide evidence to suggest children and young people are social and cultural agents in their communities, working towards sustainable solutions to tourism development.

Youth Participatory Filmmaking

In the first example, a group of 14 young people aged 7–16 years, living in the communities of the Byron Shire, were involved in a participatory filmmaking project in 2015. The young people explored their concerns about living in a tourist destination in pre-COVID-19 times, when tourism influx at peak times was considerable (Canosa *et al.*, 2017). The young people participated in the film project as 'co-researchers', exploring their experiences of growing up in a tourism community and creatively representing these views through the medium of stop-motion animation. The group met once a week for a period of six weeks to discuss what they liked, what they disliked and what they would like to change in their communities. The young people employed a technique known as 'clay-motion' whereby they created plasticine figures and moved them in small increments. These were photographed to create the illusion of movement when played in a sequence. In collaboration with a local filmmaker and the Byron Youth Services (a not-for-profit youth organisation), the young people produced three animations which were showcased in the community.

Two of the three stop-motion animations created by the young people addressed environmental concerns and the problem of marine litter. 'Just One Piece' was created and produced by a group of six young girls who were particularly eager to feature in a news flash to warn visiting tourists that littering in marine environments is unacceptable and that they must respect nature (Canosa *et al.*, 2020). The following quote is evidence of their environmental stewardship and activism:

> We have to help now, not next year, not the year after but now…so please put your rubbish in the bins. No more walking along the beach and throwing your rubbish in the ocean anymore! (7-year-old girls, 'Just One Piece' animation)

Similarly, 'Rubbish Run' describes what the community of Byron Bay would look like 20 years into the future. With overtourism and increasing beach and street litter, the animation produced by this group of five young boys centred on the sanctions that could be put in place to curb environmental degradation:

> Rising numbers of tourists are throwing rubbish on the beautiful beaches of Byron Bay. New laws have been announced that if you get caught littering you get one chance and if you get caught again you will receive 2 weeks in jail. So stop littering and make Australia beautiful! (10-year-old boys, 'Rubbish Run' animation)

Young people's strong connection and care for the environment is evidence of their agentive role in the community where they grow up and of a keen civic engagement for nature conservation purposes and more sustainable forms of tourism. Due to its inherent pedagogic nature, filmmaking was particularly suited to initiating processes of change. Despite the young age of some of the participants, the children identified as 'activists' and took on the important role of educating other young people and adults during the screening event which took place in the community. In this case the stop-motion animations (or the final product of the filmmaking activity) were not as important as the process of filmmaking, which initiated children's critical reading of their worlds (Freire, 1970) and facilitated the emergence of their voices in the community (Blazek & Hraňová, 2012; Parr, 2007).

Youth4Sea

In this research, a group of nine young adults between the ages of 18 and 24 years living in Byron Bay co-designed an intervention strategy to curb marine littering during the peak tourist season of Schoolies in 2017. Schoolies is a popular milestone in the lives of young people in Australia as it marks the end of high school and the beginning of their independent adult life. As such, young school leavers flock to popular tourist attractions to celebrate the end of school (Canosa *et al.*, 2019a). The participatory project provided the young people with the necessary tools to explore understandings of marine pollution and the attitudes and behaviours of visiting tourists. The young people conducted video interviews with their peers, created visual diaries, took photographs and designed and implemented an intervention plan.

The young people took on an active role in advocating for change in their community and came up with the idea of creating 'butt-feeders' to reduce the issue of marine pollution and encourage visiting tourists to place their cigarette litter in the appropriate 'butt-feeder' containers (see Figure 5.1). The humorous pun proved to be popular with tourists and ignited casual but important discussions between the co-researchers and visitors. The young people took on the role of activists and educators creating and installing the 'butt-feeders' on the beach and subsequently talking to tourists about the issues of marine pollution. The youth-framed participatory research design (Barratt Hacking *et al.*, 2012; Cutter-Mackenzie *et al.*, 2015) channelled young people's activism for more environmentally just forms of tourism in their community.

There is now growing academic scholarship that explores the role of environmental education in increasing young people's knowledge and awareness of global environmental problems (Biswas & Mattheis, 2021; Cutter-Mackenzie & Rousell, 2019; Malone, 2016); however, this is rarely in relation to marine debris (Canosa *et al.*, 2021a) nor in relation to tourism activity at the local level (Seraphin *et al.*, 2020). A recent scoping review of academic literature shows that environmental education programmes developed *by* young people *for* young people are scarce but show promise in promoting pro-environmental behaviours among young people (Canosa *et al.*, 2021a). This is important in the context of tourism given environmental education is essential in fostering more responsible behaviours among tourists (Hehir *et al.*, 2021).

Our Home Holiday Town

The final example presented in this chapter, is a participatory social theatre research project which engaged a group of young people between

Figure 5.1 Cigarette 'butt-feeder' containers employed by young co-researchers in the Youth4Sea project. Credit: Author

the ages of 15 and 21 years ($n = 12$) to explore tourism-related issues in the communities of the Byron Shire during and after COVID-19 (Canosa *et al.*, 2021b). Specifically, young members of the Byron Youth Theatre (a non-for-profit youth theatre company) actively participated as co-researchers by interviewing their peers about their lived experiences of life in a tourist destination. The co-researchers participated in a series of workshops to develop their skills as researchers and engaged in discussions around the different phases of the project (data collection, analysis and development of the theatre production).

The data collection phase of this project coincided with the Australian COVID-19 lockdown in 2020 and the temporary end of both domestic and international travel. Hence the lived experiences of the young people living in the Byron Shire were significantly different from their 'normal' experiences in the tourist town. The co-researchers reflected on the heightened connection to place they experienced due to the lack of tourists as a positive outcome:

> During COVID there were a lot more people on the beach but locals, because everyone was coming out and it was a legal area to be in because of the need to exercise…and it was really nice not seeing the tourists. (Aiyana, age 17 years)

Not only did young people's ecological consciousness transpire from the interviews and group discussions, but also their sensitivity to social problems (e.g. housing affordability, homelessness and disrespect for Aboriginal sacred sites) was evident and connected to 'overtourism' in their community pre-pandemic.

> The homeless problem in the Byron Shire is definitely spiking more and more because of that [rising property prices]. People come into the Byron Shire and they are not able to afford anything and just live really low or live on the streets, that makes it hard. I'd love to see that being dealt with as well. (Grace, age 16 years)

Findings suggest young people wanted more socially and environmentally just forms of tourism. The halt to domestic and international tourism was a good point in time to reflect on what tourism could look like post-pandemic. In advocating for change, these young people wanted to be active citizens in making decisions to ensure a sustainable future. Mission Australia's (2019) recent report shows that young people's concerns for the environment have tripled in Australia since 2018. Recent research also shows that youth activism is on the rise and young people want to have a say in the governance of our communities, cities and nations (Canosa *et al.*, 2020; Jourdan & Wertin, 2020).

As a result, the young people in this research collaborated with several other key stakeholders to create the Byron Way Pledge (Byron Way Pledge, n.d.). This is a set of responsible tourism principles which aim to educate visitors about the need to respect nature, culture and people. Visitors can sign the online pledge and promise to respect the fragile natural environments, recognise the significance of the Arakwal – Bundjalung country on which they find themselves and behave in ways that do not infringe upon locals.

Responsible tourism pledges are emerging in other tourist destinations at national level to positively influence visitor behaviour (Albrecht & Raymond, 2022). Iceland, Hawai'i, New Zealand and Palau are all examples of countries that have implemented responsible tourism pledges with success. Albrecht and Raymond (2022) argue that this 'soft approach' to visitor management encourages tourists to connect at an emotional level with the destination they visit and build a sense of ethical responsibility. When visitors publicly sign the pledge and share this commitment via social media channels, they are making a promise to act responsibly and ethically. This is a relatively recent visitor management strategy in effect in other countries since 2017 but it has the potential to educate visitors and change behaviours. While most of these destination pledges are implemented at a national level and as top-down processes (Albrecht & Raymond, 2022), the Byron Way Pledge is significant because of the inherently grassroots way it was established.

The workshops and discussions with the young co-researchers during the 'Our Home Holiday Town' research project were the catalyst which sparked critical reflection on the impacts of tourism on locals in the Byron area. As in many instances, doing research with children and young people causes us, as researchers, to reflect on the outcomes of our projects which cannot be measured solely in publications and metrics, but which need to have positive impacts on the lives of our collaborators/participants in the research processes. Hence the idea of a responsible tourism pledge was put forward and wider consultation with other young people in the communities initiated by the young co-researchers at Byron Youth Theatre. These consultations involved asking young people how they would want visitors to behave when they visit their town; what they would want the government to do about the impacts of tourism in their town; and to propose one pledge item that tourists could make when they arrived in Byron Shire. In collaboration with Byron Youth Theatre, local Arakwal woman Delta Kay and local ecotourism operator Vision Walks, a website was developed by a local young person, student at the SAE Creative Media Institute and mentored by an experienced web designer at Hubway. All stakeholders involved donated their time freely to see this project launched, including the beautiful photos donated by local photographers. Although still in the early days, having only been launched in December 2021, this youth-driven intervention strategy has proven promising in the promotion of more socially and environmentally just forms of tourism at a local level.

Discussion and Conclusion

The research presented in this chapter provides evidence of children's and young people's agentive role in advocating for and promoting more sustainable forms of tourism. Being socialised in an increasingly interconnected world due primarily to the media, and in particular, social media, young people are sensitive to local and global social and environmental problems (Boulianne & Ohme, 2021). The global school strikes instigated by young climate activist Greta Thunberg, the political activism by working children in Latin America and the ongoing campaigns against refugee and migrant deportation laws are all examples of youth-led initiatives which start at the local level, but which have global impact because they appeal to the moral consciousness of youth worldwide (see also Josefsson & Wall, 2020). It is thus in the everyday moments that youth express their desire for sustainable futures and through their interdependent and interconnected actions they seek change. Josefsson and Wall (2020) argue that children's struggles for social and environmental justice lie in their 'deep interdependence' with each other and adults in their lives. They propose that the powerful youth activism initiatives, which are having a ripple effect worldwide, 'do not rise from independent agency, but rather from within interdependent networks of relationships' (Josefsson & Wall, 2020: 1049). The concept of 'relational activism' is thus useful in the context of childhood research because it posits that change generally commences and flourishes in the everyday lives of individuals as part of a collective community (Metler, 2021). O'Shaughnessy and Kennedy (2010: 566) argue that 'relational activism locates agency in the collective and uses relationships as the locus for change'. In discussing women's participation in environmental activism, the authors suggest the relationship-building work typical of 'relational activism' that happens in everyday life is often more important than the actual public-sphere of activism itself. O'Shaughnessy and Kennedy (2010: 566) argue that women, in particular, see themselves as acting as a collective and are 'potentially more accountable to, and have more influence upon, each other'. These principles are applicable to children and young people who rely on each other and significant adults to make their voices heard and take action for social and environmental change in small steps in their everyday lives.

Three distinct characteristics of relational activism are applicable to the research evidence presented in this chapter: (a) children and young people are members of a community; (b) actions are taken in daily practices to change norms of high consumption of people and places through tourism; (c) the private spheres of children's and young people's everyday lives are utilised for public purposes to enact change for sustainable tourism (see also O'Shaughnessy & Kennedy, 2010). By participating in regular beach clean ups, educating visiting tourists and making their voices heard in rallies and protests aimed at curbing tourism overdevelopment, children and young people are actively contributing to protecting their communities.

In previous work, I have discussed the importance of child-centred tourism research and the applicability of 'childism' as an ontological lens to explore the complexity of childhood experiences in the tourism industry in the global context (Canosa *et al.*, 2019b, 2020). 'Childism' has emerged as an analogy to other 'ism's such as feminism, postcolonialism and environmentalism in an effort to disrupt normative assumptions about childhood and children and promote more child-centred and child-inclusive perspectives (Biswas, 2021; Biswas & Mattheis, 2021; Wall, 2019). Childism is defined as a 'critical lens for deconstructing adultism across research and societies and reconstructing more age-inclusive scholarly and social imaginations' (Wall, 2019: 1). Given children have historically been marginalised from many spheres of life, they are generally precluded from participating in important decisions about the sustainability of their futures including tourism policy and planning. This chapter provides evidence to suggest that localised activism is embedded in children's and young people's everyday lives to resist dominant forms of unsustainable tourism development. While often these acts go unnoticed, I suggest there needs to be a concerted effort to reorient tourism based on the rights of residents of host communities (Higgins-Desbiolles, 2020; Higgins-Desbiolles *et al.*, 2019), including the rights of children and young people as important, but often overlooked, stakeholders in the tourism industry (Canosa *et al.*, 2021b).

Acknowledgements

I would like to express my gratitude to all the participants in these research projects; thank you for making fieldwork so much fun! I would also like to acknowledge the contribution of my colleagues and co-investigators on two of these projects, including Dr Catharine Simmons, Associate Professor Peter Cook and Lisa Apostolides (Our Home Holiday Town); and Professor Amy Cutter-Mackenzie-Knowles, Professor Alexandra Lasczik, Dr Marianne Logan and Marie-Laurence Paquette (Youth4Sea).

References

Albrecht, J.N. and Raymond, E.M. (2022) National destination pledges as innovative visitor management tools – social marketing for behaviour change in tourism. *Journal of Sustainable Tourism* 1–18. https://doi.org/10.1080/09669582.2022.2037620.
Andereck, K.L., Valentine, K.M., Knopf, R.C. and Vogt, C.A. (2005) Residents' perceptions of community tourism impacts. *Annals of Tourism Research* 32 (4), 1056–1076.
Barratt Hacking, E., Cutter-Mackenzie, A.N. and Barratt, R. (2012) Children as active researchers: The potential of environmental education research involving children.

In R.B. Stevenson, M. Brody, J. Dillon and A.E.J. Wals (eds) *International Handbook of Research on Environmental Education* (pp. 438–458). New York: Routledge.
Biswas, T. (2021) Letting teach: Gen Z as socio-political educators in an overheated world. *Frontiers in Political Science* 3 (28). https://doi.org/10.3389/fpos.2021.641609.
Biswas, T. and Mattheis, N. (2021) Strikingly educational: A childist perspective on children's civil disobedience for climate justice. *Educational Philosophy and Theory* 1–14. https://doi.org/10.1080/00131857.2021.1880390.
Blazek, M. and Hraňová, P. (2012) Emerging relationships and diverse motivations and benefits in participatory video with young people. *Children's Geographies* 10 (2), 151–168.
Boluk, K.A., Cavaliere, C.T. and Higgins-Desbiolles, F. (2019) A critical framework for interrogating the United Nations Sustainable Development Goals 2030 Agenda in tourism. *Journal of Sustainable Tourism* 27 (7), 847–864. https://doi.org/10.1080/09 669582.2019.1619748.
Boulianne, S. and Ohme, J. (2021) Pathways to environmental activism in four countries: Social media, environmental concern, and political efficacy. *Journal of Youth Studies* 1–22. https://doi.org/10.1080/13676261.2021.2011845.
Byron Way Pledge (n.d.) See https://www.byronpledge.com.au/about/ (accessed 28 February 2022).
Canosa, A., Graham, A. and Wilson, E. (2019a) My overloved town: The challenges of growing up in a small coastal tourist destination. In C. Milano, J. Cheer and M. Novelli (eds) *Overtourism: Excesses, Discontents and Measures in Travel and Tourism* (pp. 190–204). Wallingford: CABI.
Canosa, A., Graham, A. and Wilson, E. (2019b) Progressing a child-centred research agenda in tourism studies. *Tourism Analysis* 24 (1), 95–100.
Canosa, A., Graham, A. and Wilson, E. (2020) Growing up in a tourist destination: Developing an environmental sensitivity. *Environmental Education Research* 1–16. https://doi.org/10.1080/13504622.2020.1768224.
Canosa, A., Paquette, M.-L., Cutter-Mackenzie-Knowles, A., Lasczik, A. and Logan, M. (2021) Young people's understandings and attitudes towards marine debris: A systematic scoping review. *Children's Geographies* 19 (6), 659–676. https://doi.org/10.1080 /14733285.2020.1862759.
Canosa, A., Simmons, C., Cook, P.J., Apostolides, L., Wall, A. and Evington, R. (2021) 'Reclaiming place in a tourist town': Preliminary findings from a social theatre research project run by young people during COVID-19. *Annals of Tourism Research Empirical Insights* 2 (1), 100008. https://doi.org/10.1016/j.annale.2020.100008.
Canosa, A., Wilson, E. and Graham, A. (2017) Empowering young people through participatory film: A postmethodological approach. *Current Issues in Tourism* 20 (8), 894–907. https://doi.org/10.1080/13683500.2016.1179270.
Cheer, J.M. (2020) Human flourishing, tourism transformation and Covid-19: A conceptual touchstone. *Tourism Geographies* 1–11. https://doi.org/10.1080/14616688.2020. 1765016.
Choi, H. and Sirakaya, E. (2005) Measuring residents' attitude toward sustainable tourism: Development of sustainable tourism attitude scale. *Journal of Travel Research* 43 (4), 380–394.
Cutter-Mackenzie, A., Edwards, S. and Quinton, H.W. (2015) Child-framed video research methodologies: Issues, possibilities and challenges for researching with children. *Children's Geographies* 13 (3), 343–356. https://doi.org/10.1080/14733285.201 3.848598.
Cutter-Mackenzie, A. and Rousell, D. (2019) Education for what? Shaping the field of climate change education with children and young people as co-researchers. *Children's Geographies* 17 (1), 90–104. https://doi.org/10.1080/14733285.2018.1467556.
Everingham, P. and Chassagne, N. (2020) Post COVID-19 ecological and social reset: Moving away from capitalist growth models towards tourism as Buen Vivir. *Tourism Geographies* 22 (3), 555–566.

Freire, P. (1970) *Pedagogy of the Oppressed* (2000 edn). New York: Continuum International Publishing Group.

Hall, C.M., Amelung, B., Cohen, S., Eijgelaar, E., Gössling, S., Higham, J., Leemans, R., Peeters, P., Ram, Y. and Scott, D. (2015) On climate change skepticism and denial in tourism. *Journal of Sustainable Tourism* 23 (1), 4–25.

Hehir, C., Stewart, E.J., Maher, P.T. and Ribeiro, M.A. (2021) Evaluating the impact of a youth polar expedition alumni programme on post-trip pro-environmental behaviour: A community-engaged research approach. *Journal of Sustainable Tourism* 29 (10), 1635–1654. https://doi.org/10.1080/09669582.2020.1863973.

Higgins-Desbiolles, F. (2020) Socialising tourism for social and ecological justice after COVID-19. *Tourism Geographies* 22 (3), 610–623. https://doi.org/10.1080/14616688.2020.1757748.

Higgins-Desbiolles, F. and Bigby, B.C. (2022) A local turn in tourism studies. *Annals of Tourism Research* 92, 103291. https://doi.org/10.1016/j.annals.2021.103291.

Higgins-Desbiolles, F., Carnicelli, S., Krolikowski, C., Wijesinghe, G. and Boluk, K. (2019) Degrowing tourism: Rethinking tourism. *Journal of Sustainable Tourism* 27 (12), 1926–1944. https://doi.org/10.1080/09669582.2019.1601732.

Higgins-Desbiolles, F., Doering, A. and Bigby, B.C. (eds) (2022) *Socialising Tourism: Rethinking Tourism for Social and Ecological Justice*. Abingdon: Routledge.

Jafari, J. (2005) Bridging out, nesting afield: Powering a new platform. *Journal of Tourism Studies* 16 (2), 1–5.

Josefsson, J. and Wall, J. (2020) Empowered inclusion: Theorizing global justice for children and youth. *Globalizations* 17 (6), 1043–1060.

Jourdan, D. and Wertin, J. (2020) Intergenerational rights to a sustainable future: Insights for climate justice and tourism. *Journal of Sustainable Tourism* 28 (8), 1245–1254.

Lawton, L.J. and Weaver, D.B. (2015) Using residents' perceptions research to inform planning and management for sustainable tourism: A study of the Gold Coast schoolies week, a contentious tourism event. *Journal of Sustainable Tourism* 23 (5) 660–682.

Lew, A.A., Cheer, J.M., Haywood, M., Brouder, P. and Salazar, N.B. (2020) Visions of travel and tourism after the global COVID-19 transformation of 2020. *Tourism Geographies* 22 (3), 455–466. https://doi.org/10.1080/14616688.2020.1770326.

Malone, K. (2016) Reconsidering children's encounters with nature and place using posthumanism. *Australian Journal of Environmental Education* 32 (1), 1–15.

Metler, D. (2021) The promise of childism for imagining transformation: Reimagining the future of activism, blog. See https://www.kindredmedia.org/2021/05/the-promise-of-childism-for-imagining-transformation-reimagining-the-future-of-activism/ (accessed 14 November 2021).

McCool, S.F. and Lime, D.W. (2001) Tourism carrying capacity: Tempting fantasy or useful reality? *Journal of Sustainable Tourism* 9 (5), 372–388.

Mission Australia (2019) Youth survey report, online report. See https://www.missionaustralia.com.au/what-we-do/research-impact-policy-advocacy/youth-survey (accessed 3 May 2021).

Morrow, V. (2011) *Understanding Children and Childhood*. Centre for Children and Young People Background Briefing Series, no. 1. (2nd edn). Lismore: Centre for Children and Young People, Southern Cross University.

Moyle, B.D., Croy, G. and Weiler, B. (2010) Community perceptions of tourism: Bruny and Magnetic Islands, Australia. *Asia Pacific Journal of Tourism Research* 15 (3), 353–366.

O'Shaughnessy, S. and Kennedy, E.H. (2010) Relational activism: Re-imagining women's environmental work as cultural change. *Canadian Journal of Sociology* 35 (4), 551–572.

Parr, H. (2007) Collaborative film-making as process, method and text in mental health research. *Cultural Geographies* 14 (1), 114–138.

Rastegar, R., Higgins-Desbiolles, F. and Ruhanen, L. (2021) Covid-19 and a justice framework to guide tourism recovery. *Annals of Tourism Research* 103161. https://doi.org/10.1016/j.annals.2021.103161.

Saarinen, J. (2006) Traditions of sustainability in tourism studies. *Annals of Tourism Research* 33 (4), 1121–1140.

Schänzel, H.A. and Smith, K.A. (2011) Photography and children: Auto-driven photo-elicitation. *Tourism Recreation Research* 36 (1), 81–85.

Seraphin, H., Yallop, A.C., Seyfi, S. and Hall, C.M. (2020) Responsible tourism: The 'why' and 'how' of empowering children. *Tourism Recreation Research* 1–16. https://doi.org/10.1080/02508281.2020.1819109.

Sharpley, R. (2014) Host perceptions of tourism: A review of the research. *Tourism Management* 42, 37–49.

Sibley, D. (1995) *Geographies of Exclusion*. London: Routledge.

Slaper, T.F. and Hall, T.J. (2011) The triple bottom line: What is it and how does it work. *Indiana Business Review* 86 (1), 4–8.

Thomson, R. (2007) Belonging. In M.J. Kehily (ed.) *Understanding Youth: Perspectives, Identities and Practices* (pp. 147–179). London: Sage/The Open University.

United Nations (1989) *Convention on the Rights of the Child (UNCRC)*. New York: Office of the High Commissioner for Human Rights.

United Nations World Commission on Environment and Development (WCED) (1987) Report of the World Commission on Environment and Development: Our Common Futures. Oxford: Oxford University Press.

Wall, J. (2019) From childhood studies to childism: Reconstructing the scholarly and social imaginations. *Children's Geographies* 1–14. https://doi.org/10.1080/14733285.2019.1668912.

Weaver, D.B. and Lawton, L.J. (2014) *Tourism Management* (5th edn). Milton, QLD: John Wiley & Sons.

Weaver, D.B., Moyle, B. and McLennan, C.-l.J. (2021) The citizen within: Positioning local residents for sustainable tourism. *Journal of Sustainable Tourism* 1–18. https://doi.org/10.1080/09669582.2021.1903017.

Wray, M., Laing, J. and Voigt, C. (2010) Byron Bay: An alternate health and wellness destination. *Journal of Hospitality and Tourism Management* 17 (1), 158–166.

Yang, M.J.H., Yang, E. and Khoo-Lattimore, C. (2020) Host-children of tourism destinations: Systematic quantitative literature review. *Tourism Recreation Research* 45 (2), 231–246.

6 Enhanced Food Security Through Localised Community Cryptocurrency: Experiences of a Costa Rican Tourism Town

Mary Little

Introduction

The COVID-19 pandemic created an immediate and almost complete disruption in the tourism industry. The World Tourism Organisation (UNWTO) (2021) has estimated that international tourism dropped by 74%. This interruption has caused the termination of between 100 and 120 million direct tourism jobs. Job losses and reduced work hours have drastically reduced the flow of revenue to tourism sites throughout the world (UNWTO, 2021). An immediate impact of unemployment is food insecurity. The extensive interruptions in the tourism industry caused by COVID-19 are a catalyst for international evaluation of food security and sustainable agriculture (IPES-Food, 2020; Stephens et al., 2020).

Food insecurity is a by-product of commercial tourism, anchored in the inequalities of tourism as a product of global capitalism (Fletcher & Neves, 2012). As part of the neoliberal model, the tourism industry seeks low-wage labour and cheap natural resources to sustain profits (Patel & Moore, 2018). Galvez explored the connections between these neoliberal operations and local food systems. 'For locals, it does not matter if they work in or around ecotourism or mass-market commercial tourism; greater market dependency has meant a decline in dietary and nutritional status in low-income communities' (Galvez, 2018: 32). Tourism business closures during the COVID-19 lockdown resulted in the loss of salaries to buy food combined with the cumulative impacts of land sales and loss of farming knowledge. The connection between tourism and food insecurity is surprising for an industry that touts increased jobs and earning power for locals. Yet labour and land commonly move out of agricultural production and into tourism. The link between tourism and displacement from farmlands has been recorded for decades (Bélisle, 1983: 498) as people move away from agricultural jobs and lose farming skills. The imported products that fill the food void cost much more than local goods (Cantor, 2016: 3). When combined with the seasonality of most tourism, communities often face food security issues even in fertile areas.

Amid the upheaval caused by COVID-19, a space to reflect on the injustices in the current tourism system and seek a new model that promotes economic and ecological equality has emerged. Some scholars have identified this pause as a 'transformational moment opening up possibilities for resetting tourism' on a more resilient and regenerative path (Higgins-Desbiolles, 2020: 612; see also Jamal & Higham, 2021). In 'COVID-19 and a justice framework to guide tourism recovery' Rastegar *et al*. (2021: 2) proposed using the opportunity to redefine tourism through 'restorative action for a responsible and just recovery'. Food insecurity in rural tourism exemplifies how restorative action can create a new model that promotes economic and ecological equality.

The pandemic presents an opportunity to enhance communities' ability to reimagine tourism to support environmental and social interests. Creating local currencies is one way to reduce financial dependency on the growth-led, resource-intensive model that is prevalent in tourism (Bianchi, 2022). This moment of tourism disruption has created the space to redraw and implement new tourism economies. Local complementary currencies offer an option to decouple, fulfilling local needs from tourism revenue. Built on crypto-currency blockchain technology, community cryptocurrencies (CCC) offer a framework for creating resilient economies by implementing circular economies that encourage the use of local products that can be reused or recycled to reduce waste.

Monteverde is a mountainous cloud forest town in central Costa Rica and the site of this study. The pandemic raised issues of over-reliance on tourism and consideration of alternative models to create local livelihoods. Recognising tourism over-reliance as a real danger to well-being led to questions about how to share plenitude and reduce scarcity to ensure that all community members' basic needs are met, particularly during tourism downturns. Verdes, a CCC that runs on blockchain technology, is one response to reactivate and diversify the local economy in an environmentally sound way.

This chapter examines the question of whether the Verdes CCC functions as a resilience-building tool, particularly in response to food vulnerability. This question will be assessed by: (1) presenting the role of supplemental currencies through the lens of the doughnut economic model; (2) exploring the development and implementation process of community currency; and (3) examining the community currency's role in strengthening local food security in a tourism community. Finally, I discuss the benefits and challenges of the CCC as a tool for building resilience and enhancing food security during the pandemic as tourism returns. I conclude by presenting constraints and solutions that might enable Monteverde to reduce its reliance on tourism revenue and create local abundance.

Community Currencies as a Resilience-Building Strategy

The economic upheaval created by the pandemic has incentivised communities to envision and implement self-directed development to enhance just resource distribution tactics. They follow in the steps of Indigenous communities that have used community currency (CC) to break with government institutions that maintain colonial paradigms (Alcántara & Dick, 2017: 3). Tourism communities in the Global South share the experiences of reliance and exclusion from financial institutions and can use CC to achieve self-determination and decolonisation. After experiencing the impacts of the 2008 financial crisis and now COVID-19, a growing number of tourism communities are using CC to reduce dependence on international monetary systems and government development plans with CCC. CCC's circular flow of resources reduces leakage or extraction of wealth away from tourism areas that is common in conventional tourism models (Suryawardani *et al.*, 2016: 9402). Expanding on the concepts of the circular economy, some regions are using CCs to implement what has been called a 'doughnut economy' model. The doughnut economy incorporates circular principles and also emphasises ways to meet the basic needs of all community members while living within environmental boundaries, the centre ring that gives the theory its name (Raworth, 2017).

These currencies can additionally amplify financial security by creating monetary diversity by linking 'unused resources with unmet needs within a community, region, or country' (Lietaer *et al.*, 2009: 1). If standard, or *fiat*, currency is weakened or inaccessible, another form of currency can be used (Zeller, 2020). Just as with ecological systems, diversified monetary systems can enhance resilience (Seyfang & Longhurst, 2013a). They are known as 'complementary because they do not replace the conventional national money, but rather, operate in parallel with it' (Lietaer *et al.*, 2009: 1; see also Telalbasic, 2017). As of 2015, there were estimated to be over 4500 complementary currency systems (Place & Bindewald, 2015: 153). This chapter will examine a case of a CCC in Costa Rica in order to uncover ways CCs and CCCs help empower local communities through economic self-determination and reduction of tourism dependency.

CCs promote localisation that can potentially moderate the negative environmental effects of global transportation and introduction of products within disposable packaging (Douthwaite, 1996). Additionally, CCs represent tools for raising people's awareness about environmental issues and tools to promote eco-friendly behaviour (Seyfang & Longhurst, 2013b). For example, Verdes are earned by completing environmental actions. It is no coincidence that Monteverde, a rural community with strong nature conservation and a tradition of cooperatives, has embraced a model of financial diversification which replicates and enhances biodiversity.

Researchers have, however, documented significant negative environmental impacts of cryptocurrencies, with one study demonstrating that each \$1 of Bitcoin value created was responsible for \$0.49 in health and climate externalities from carbon emissions when mined in the US (Goodkind et al., 2020). The type of technology used to run a cryptocurrency determines the amount of energy, and therefore, CO_2 emissions released (Schinckus, 2020). The two largest cryptocurrencies, Bitcoin and Ethereum, use Proof of Work verification to validate blockchain exchanges which uses large amounts of energy to authenticate every transaction. Verdes CCC's operational framework is based on EOS, a third-generation technology developed in 2018. EOS runs on Delegated Proof of Stake verification that uses a fraction of the energy compared to Proof of Work (Cordoba Brenes, 2021). EOS technology is 66,455 times more energy-efficient than Bitcoin and 17,236 times more energy-efficient than Ethereum (EOS Authority, n.d.) thus aligning with Verdes environmental objectives.

Community cryptocurrencies

The financial crisis of 2008 resulted in popular awareness that the monetary system was not working for the interest of many. Blockchain technology was launched soon after. Barinanga described: 'a wave of monetary experimentation ensued that took a most concrete form in two entrepreneurial spaces: crypto-currencies with global ambitions and local currencies based on communal democracy' (2020: 1).

Digital or cryptocurrencies are virtual currencies that are secured by cryptography. Fiat currencies, such as dollars, are issued by governments, while cryptocurrencies rely on the computing power of a decentralised network of computer nodes that track and validate accounts, balances and transactions (Alcántara & Dick, 2017: 2). Cryptocurrency platforms have been adopted to host many community currencies. Diniz et al. point out that 'solidarity crypto-currencies evolved from the combination of two types of currencies that share similarities and differences' (2021: 5). Dapp explained that 'local communities can design and create currencies and incentives according to their local needs, tokenize what is agreed to be relevant, and collectively track the parameters in question' (2019: 167). For example, Cambiatus, the Costa Rican/Brazilian CCC platform that hosts Verdes, uses blockchain to help organisations create social currencies to achieve common social and environmental goals (Cambiatus, 2021). In practice, a community sets objectives that target the negative externalities they would like to control or positive externalities they would like to encourage (Cristine, 2021).

Conceptual Framework – The Doughnut Economy

Monteverde's circular economy committee has embraced the doughnut economic model to visualise and implement a just COVID-19 recovery. Verdes emphasises circular economic goals of meeting local economic needs of all community members in a regenerative and distributive manner while living within environmental limitations and planetary boundaries. Kate Raworth, architect of the doughnut model, proposed an embedded economic model, 'one that situates the economy within the greater sphere of society and earth's biosphere' (Monbiot, 2017). It is called a 'doughnut' because the inner edge of the ring represents all basic human needs being met while the outer edge of the ring represents the resource ceiling we cannot overshoot (see Figure 6.1):

Figure 6.1 The doughnut economy model has helped the Monteverde community envision and enact strategies to meet human needs without overshooting the ecological ceiling. Doughnut model imposed over coordinators planning a community garden site in a disused bullring. Photo source: Paula Vargas, used with permission

The hole in the middle is the space where people are falling short on life's essentials, be it food and water, health care, or political voice. But we mustn't go over the outer crust because there we are overshooting pressure on the planet, causing climate change, biodiversity loss, and acidifying the ocean. Our well-being lies in personal health, but also planetary health, the space within the doughnut itself. (Raworth, 2017: 34)

Monteverde organisations have identified the doughnut model as a useful model to visualise sustainable community efforts (Van Dusen, 2021). The model applies to the community's food security vision of meeting local nutritional needs at all times with culturally appropriate food while supporting a healthy planet. Verdes CCC align with Raworth's doughnut model movement away from extractive models of wealth creation to localised, needs-based production without overshooting local resources.

Research Site: Monteverde, Costa Rica
Development history

Ecological and economic factors have shifted Monteverde from a primarily agricultural-driven economy to one focused primarily on ecotourism. Some community members have identified tourism reliance as a threat to long-term prosperity after facing multiple crises that interrupted tourism, including the 2008 economic downturn, impacts of a devastating storm that closed off the community for weeks and now the COVID-19 pandemic. These experiences have sparked collective rethinking of traditional economic growth models and thoughts on ways to encourage local and ecologically responsible food production and consumption.

The area generally referred to as Monteverde is a seven square kilometre region that consists of small neighbourhoods including Santa Elena and Cerro Plano, with a population of approximately 6000 residents. The area also includes two private nature reserves, the Monteverde Rainforest Reserve and the Children's Eternal Rainforest that attract tourists and biological researchers. Traditionally, the area relied on subsistence farming and coffee production. In 1949, a group of American Quakers relocated to Monteverde when Costa Rica abolished its army which aligned with the Quakers' pacifist ethics. They opened the Monteverde Cheese Factory and dairy cooperative (Weinberg *et al.*, 2002: 373). The cooperative also provided set prices for milk and encouraged farms to hold shares in the company stock to create a collective decision-making model. Coffee, the main export crop, was also sold and processed by the local cooperatives.

The Quakers reforested a third of their ranch land as a watershed at a time when government policy encouraged farmers to clear forests for pasture. Realising the dangers that surrounding deforestation created, the Tropical Science Centre was established which later oversaw the Monteverde Cloud Forest Reserve. Biologists and naturalists began to visit the Reserve in the 1970s and in 1986, community members created the Monteverde Conservation League and the Monteverde Institute.

Several private hotels, restaurants, nature tours and adventure activities have accompanied the growth of ecotourism. Most tourism in the area is locally owned and developed to promote positive environmental and economic impacts; however, ecotourism has grown exponentially. In 1972, there were less than 100 visitors to the Monteverde Cloud Forest Reserve. In 1999, nearly 60,000 visitors entered the Reserve (Weinberg, 2002: 374) and there were estimated to be 250,000 annual visitors in 2019 (Guillermo Vargas, pers. comm., 18 November 2021).

This transformation has incentivised many farmers and their children to choose tourism as a more lucrative path. Weinberg *et al.* (2002: 374) asserted that increased tourism employment and business opportunities have been accompanied by challenges. First, tourism is seasonal. From November to May, unemployment is high, and many accumulate debt to survive the low season. The cyclical nature of tourism, even when it is strong, creates economic vulnerability. Second, tourism expansion has meant an increase in external investors acquiring land to build businesses and rental homes. Migrant workers also come to fill job vacancies. An inflow of investment and workers places price pressure on the local housing market. Some residents are no longer able to find or afford local housing and must move to neighbouring communities. This creates geographic separation between residents, disrupts communities and damages people's sense of belonging. Third, increased tourism, often accompanied with a greenwashed 'eco' representation, leaves communities to address the environmental impacts of tourism. Negative impacts have included an increase in solid waste, sewage run-off and traffic. One community organiser explained 'we struggle to correct all the problems caused by this never-ending tourism. Maybe it is time for us to direct tourism instead of playing catch up to all the problems it has caused' (Maria Vargas, pers. comm., 18 October 2021).

Food security in Monteverde

Food insecurity is defined as: 'limited or uncertain availability of nutritionally adequate and safe foods, or limited or uncertain ability to acquire acceptable foods in socially acceptable ways' (Matheson *et al.*, 2002: 210). Income instability created by dips in tourism leads to food insecurity in what can appear to be a thriving economy. One comprehensive study by the Monteverde Institute analysed how the shift from an agricultural economy to tourism impacted local food security for local women. Food insecurity was high in both communities, ranging from 67% to 73% (Himmelgreen *et al.*, 2006: 305). The survey found a connection between food insecurity and tourism with 60% of all the interviewees reporting that tourism and limited business competition were responsible for the high cost of food (Himmelgreen *et al.*, 2006: 307).

The viability of sustainable food production in Monteverde has received less attention. One study on sustainable food production identified challenges farmers encounter (Andia *et al.*, 2002). These included economic challenges such as the inability to support their family on farming income alone and the inability to compete with supermarket prices. Study participants also explained that most people prefer other opportunities than farming, especially preferring tourism (Andia *et al.*, 2002: 6).

COVID-19 and Monteverde's Food Security Responses

Due to COVID-19, all international flights to Costa Rica were suspended in mid-March 2020. The roads to Monteverde were closed shortly thereafter to create another level of protection from the virus. Monteverde's mix of collective Quakerism and environmental protection action, combined with a history of agricultural cooperatives, has created a community with relatively high ecological awareness and shared values of collective action. Before COVID-19, the area had created strong institutional resilience through mechanisms such as the Monteverde Institute, Monteverde Fund and Corclima. Five areas of need were identified during strategy meetings to create support during the COVID-19 crisis: (1) statistics, (2) health, education and recreation, (3) communications, (4) social assistance and (5) the circular economy. These five sub-committees formed the Enlace Commission (Comisión Enlace), with 'enlace' translating from Spanish as 'network' but also meaning to weave together and intertwine, hold and connect the community so that no one falls through (Wilkins, 2021).

The community collected data on the immediate needs of households, including employment and need for direct food assistance. The statistics committee collected data from 1050 people of the 6000 households. The initial survey conducted in April 2019 found that 65% of the population experienced a change in employment status and 30.6% were already unemployed within a month of the tourism restrictions. At that time, almost 13% of the survey participants said they were not sure if they would be able to purchase food in the next week. Others expressed concerns about longer-term food security (Enlace Statistics Committee, 2019). Data from this survey generated food security responses including production of food in household and community gardens, financial management workshops and a form of bartering or exchange for goods, products and services. To address immediate needs, a food bank was organised that distributed between 50 and 100 food baskets to families for the next year. These initial food programmes utilised locally produced food previously directed to tourism. A disused bullring was converted to a community garden and now provides fresh vegetables to the neighbourhood elementary school (see Figure 6.1). Experienced farmers and gardeners were recruited and paid to set up and teach people how to manage community gardens. Using this moment for knowledge transfer was especially important as many unemployed people had time available to grow their own food.

Verdes community cryptocurrency

The Enlace Circular Economy Committee determined that a digital community currency could promote the goal of local food production without reliance on tourism revenue. The committee collaborated with the Cambiatus blockchain platform to create the Verdes CCC. This initiative aimed to create abundance by harnessing local products and skills to meet needs. Cambiatus guided the articulation of the mission of the Verdes CCC in a series of workshops. As a result, Verdes' mission was determined to be the reactivation of the economy through activities that continue to promote environmental sustainability and conservation in line with the core tenets and values held by the people of the Monteverde region.

The Cambiatus CCC creation process began with community-articulated key objectives. These objectives underpin actions that community members can undertake to earn Verdes. While the objectives are permanent, actions can change over time, especially as goals are met or change. Based on this mission, the Verdes committee established the objectives and actions set out in Table 6.1. Once an action is completed, the action is verified by the Verdes platform administrator and Verdes are deposited into the participant's account once verified.

Verdes are valued at 1 Verde (VRD) to 1 colon, the national currency, to facilitate pricing and exchange. People can buy into the currency using dollars or colones to make it easier for tourists to participate, who would find it difficult to earn Verdes. Verdes cannot be 'cashed out' for colones in order

Table 6.1 The creation of community cryptocurrencies begins by determining objectives and developing specific actions to support those objectives. Community members complete actions to earn Verdes currency

Verdes objectives	Verdes actions
1. Reactivate the economy and facilitate local exchanges	Buy and sell using Verdes via: • Verdes digital sales platform • Verdes market (*feria*) • La Tilichera second-hand store
2. Encourage actions related to sustainability, responsible consumption and environmental conservation	• Riding bikes instead of driving • Walking to work or school instead of driving • Composting
3. Promote solidarity and inclusion	• Volunteer with local organisations • Share information about Verdes on social media

to encourage people to spend Verdes instead of hoarding them. To participate, users create an account on the digital platform that can be accessed via mobile phones and computers. Before launching via a digital platform, the Verdes committee conducted research and found that 88% of people surveyed had access to a smartphone, making web access a viable option.

Methodology

The insights of this chapter are based on case study research (Yin, 2014) gathering qualitative data in the community. An exploratory qualitative community perception and participation survey was conducted in November 2021 to better understand community awareness and uptake of Verdes CCC one year after the project was launched. The geographic parameters of the survey aligned with the coverage area of Verdes and included a sample of five neighbourhoods in the Monteverde region. Neighbourhoods were selected to represent a variety of economic realities and distance from the town centre. All homes in the survey areas were visited once and asked to participate. The survey tool was collaboratively designed with Verdes CCC coordinators. A central aim of the study was to collect data that Verdes CCC can use to inform future decision-making. The survey consisted of closed-ended questions on demographic information, basic needs, knowledge about and use of Verdes and interest in various information and sales platforms. The survey asked whether participants are active on that account, what they would like to buy and sell with Verdes and which environmental actions they would take to earn Verdes. The survey instrument also included open-ended questions on suggestions to facilitate Verdes uptake. Responses were collected from 121 respondents. Survey data were organised and analysed using Google Sheets software and open-ended responses were coded manually.

Additionally, semi-structured interviews were conducted with eight expert informants, including the co-founder of the Cambiatus platform, the current and two previous Verdes CCC coordinators and four additional community development coordinators. These interviews inform this presentation of Monteverde's COVID-19 response strategies, the creation of the Verdes platform and food security adaptation strategies. They also facilitated community input into the research design, including the objectives and application of the community survey.

Verdes uptake survey findings

The Verdes uptake survey reveals three main findings: (1) while there was strong initial creation of Verdes accounts, the uptake and use of Verdes has not been sufficient to reduce food insecurity; (2) despite rebounding tourism visitation in the Monteverde regions, almost 80% of participants reported unmet household needs; and (3) high levels of interest in Verdes demonstrates its continued potential as a tool to decouple food security from tourism revenue. The results indicated that the community currency has the potential to help residents meet their goals of living within the 'doughnut' depending on Verde CCC's ability to adapt and meet remaining needs as the pandemic subsides.

The 2021 uptake survey showed that 46% of survey participants are aware of Verdes CCC. Of those, 20% said they had an account and 11.5% had sold or purchased items on the platform. These findings indicate that people are interested in using Verdes; however, there are obstacles to joining or using the platform. Obstacles included lack of Verdes to buy products or lack of products they want to purchase. In either case, the Verdes electronic platform is not being used extensively enough to have sufficient impact on food security.

Although tourism was on the rebound at the time of the survey, 79% of respondents reported seeking more resources most or some of the time. The survey showed that the items respondents can offer for sale align with items people want to buy. Two open-ended survey questions asked participants if there was anything they would like to sell on the Verdes platform and also what they would like to buy. Responses to both questions were: food items, clothes and other products. This finding indicates alignment between locals needs and resources.

Residents can use Verdes to access local goods through the electronic platform, the *feria* market and La Tilicheria store, a second-hand clothing and household store that accepts Verdes. This analysis does not include La Tilicheria, as its operations are not directly related to food security. While participants report interest and some have used Verdes, making electronic and physical platforms more accessible to users is the main barrier to Verdes' uptake. To be effective, Verdes CCC needs to modify platforms to bring buyers and sellers together to meet basic needs.

Verdes CCC launched on a web-based platform which requires internet access to function. In addition to setting up a user account, people can learn about Verde's mission and how to earn Verdes on the platform. A central feature of the Verdes site is access to the sales platform, where people can offer goods and services for Verdes. The Verdes electronic platform shows that over 1000 accounts have been created since its launch in 2021 in a community with a population of 7000. Over 550 'actions' have been taken to earn Verdes. Active Verdes users have predominantly spent them on food and beverages (30% of transactions), followed by handmade items and clothing (14% of transactions). This indicates that resources and skills to make food items exist locally. The question remains whether people have earned Verdes to buy these products.

Initial uptake of the CCC was swift. However, some accounts are likely to be duplicates. To enhance security, the cryptocurrency platform assigns a 12-word set of passwords that cannot be changed. When people lose the passwords, it may be easier to create another account if they do not have many Verdes associated with the first account. Another issue surfaced when the platform was launched. People were creating additional accounts to collect a bonus for new users that they would then transfer to themselves. The bonus has been eliminated but this demonstrates the potential for system abuse. The Verdes initiative is preparing to make the platform easier to use. This includes alerts to sellers when buyers inquire about products, a forum where buyers could post which products they want to purchase and possibly converting the platform to a downloadable application.

In the first months after the launch of Verdes, the coordinators organised a *feria*, or market, for people to buy and sell items using only Verdes. The Verdes CCC market was held in a community space on Saturdays. The physical market was a successful way for people to use their Verdes and become familiar with the digital platform. Food and drinks were the main items sold at the market followed by an assortment of locally made crafts and household items.

Despite the market's success, it has faced challenges as the CCC has evolved. The purchase of items with fiat currency is not allowed in the community space. Transactions with Verdes avoided restrictions on sales in colones or dollars. However, since some ingredients cannot be sourced locally (for example, flour for baked goods), vendors have to buy these ingredients with colones or dollars. For this reason, vendors usually do not want to accept full payment in Verdes.

To address this issue, a hybrid model of payment partially in Verdes and partially in fiat currency is being implemented but cannot be carried out at the community space. The Verdes committee is creating a more suitable, permanent home for the Verdes market. Starting in 2022, Verdes will hold a physical market around the perimeter of a disused bullring that also holds a community garden started during the pandemic (see Figure 6.2).

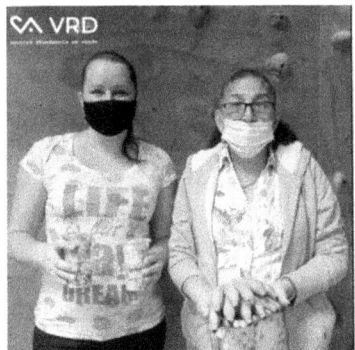 

Figure 6.2 Participants in the first Verdes market selling ceviche made of green plantains from a home garden (left, photo source: Irene G. Chen, used with permission) and the Verdes market's future home, a defunct bullring the market will share with a community garden (right, photo credit: Author)

The bullring is on municipal land and permits have been granted for this market. The bullring is located beside a public elementary school and vegetables from the community garden and food from the market will be donated for school meals. The market provides a new space that successfully adapts Verdes to a new physical platform to match food needs with local resources. The market further reduces emissions by supporting local production and reducing waste by requiring that all items are sold without packaging or alternatively, with reusable/compostable packaging. The market venue provides accessibility to a different segment of the population, which potentially promotes social capital building through knowledge and idea exchange.

Discussion

To be successful, a community currency must guarantee 'significance and benefits to their users by providing the commercial benefits of connecting underused assets with unmet needs' (Diniz *et al.*, 2021: 3). Harnessing underused assets offers a number of benefits including: local goods enter the food market more easily; excess perishable items are not wasted; and people exchange products they have on hand for other goods they need. During the economic crisis and food insecurity that COVID-19 sparked, Verdes met Diniz *et al.*'s requirement for success of 'connecting underused assets with unmet needs'. My findings show a continued role for Verdes to connect local resources to meet needs and interest in using Verdes. The first year of Verde exchange demonstrated how CCC can reduce reliance on tourism, moving the economy away from environmentally and culturally disruptive forms of tourism towards the regenerative sweet spot described in the doughnut economic model. The logistics surrounding the electronic platform, lack of knowledge of the how Verdes functions and how few businesses currently accept Verdes remain significant challenges, however.

The structural inequalities of tourism often leave basic food needs unmet, even as tourism returns. The Monteverde economic committee identified the local potential to grow and sell more local food to meet those needs. On the economic side, the complementary currency is a tool to unite local buyers and sellers, even when fiat money is scarce. Environmentally, consuming more local food reduces transportation emissions and revives small-scale agroecological production. Socially, community garden projects and volunteering opportunities to earn Verdes and Verdes skills training connect people to share knowledge and ideas about their community. These multiple features align with the goals of Raworth's doughnut economics model of creating ways to address everyone's basic needs without overshooting resource limits.

As tourism returns to this region, the CCC's biggest challenge is adapting to ensure user-friendly ways to encourage individuals and businesses to adopt this tool. While awareness of the Verdes initiative is relatively high, active use is low. This is attributed to difficulty or disinterest in using the digital platform, lack of awareness of the extent of products available on the Verdes platforms and how few businesses are accepting Verdes. The first solution derived from survey data and interviews is to make the digital and physical platforms more user-friendly. This can be addressed by making a downloadable Verdes application that can be used when not connected to the internet, reducing peer-to-peer Verdes payment friction and implementing physical tokens or a ledger at the Verdes market. The second issue of awareness about Verdes and its benefits can be addressed by using varied strategies to market Verdes to different groups. The current Verdes Committee Coordinator was selected partially based on her experience conducting social media outreach. The Monteverde Community Fund Communication Director is assisting with social media awareness and promoting Verdes information sessions. These sessions are essential to training people on how to create and use accounts. Sixty-eight percent of survey respondents reported an interest in neighbourhood information sessions. As the immediate dangers of COVID-19 recede, a point person could volunteer from each neighbourhood to conduct these sessions, answer questions and gauge interest in other activities such as small community gardens or neighbourhood mini-Verdes markets. Third, to make the circular economic model function, more businesses also need to accept Verdes as a form of payment. They must, however, have items they can spend their own Verdes on in order to accept them. The Verde coordinators could plan individualised solutions with local businesses to demonstrate what goods and services are available for purchase with Verdes. For example, the Life Monteverde Café accepts Verdes for coffee and lattes made from locally produced coffee and milk. The Cafe uses the Verdes they earn to pay for English lessons for employees, who need language skills to assist consumers. Once more businesses and producers accept Verdes, demand will drive the process.

Conclusion

Community currencies are an important means to address crises (Diniz *et al.*, 2021: 4). The COVID-19 crisis may have spurred the launch of this community currency but the underlying challenges of over-dependence on tourism and resulting food insecurity remain. A lesson of this crisis has been the unpredictability of economic incentives unmoored from ecological and social responsibilities. This study demonstrates that a decentralised, community currency that is based on collectively determined objectives and actions is an adaptive tool that can incentivise reliance on local resources in tourism communities. However, work is needed to better inform people on how to use Verdes, to make digital and electronic platforms more accessible and to incentivise more businesses to accept Verdes.

The COVID-19 pandemic has created opportunities for us all to rethink what we value and reimagine the future of tourism (Fountain, 2021: 1). Verdes CCC and other innovative strategies that place social and environmental interests before purely economic incentives are emerging from these reflections. The Monteverde community is applying new iterations of this regenerative economic model to fit changing needs as tourism returns. Through an adaptive process, Verdes serves as a tool for the local community to redefine their relationship with tourism. Verdes represent a way to localise food production, first to feed the community and then support farmers of all sizes to feed tourists.

Research interviews

Vargas, Maria. Development and Communications Specialist, Monteverde Foundation, 18 October 2021.
Vargas, Guillermo. President, Monteverde Chamber of Tourism, 18 November 2021.
Vargas, Paula. Community Garden Coordinator, Corclima, 25 November 2021.

References

Alcántara, C. and Dick, C. (2017) Decolonization in a digital age: Cryptocurrencies and Indigenous self-determination in Canada. *Canadian Journal of Law and Society, Revue Canadienne de Droit et Société* 32 (01), 1–17. https://doi.org/10.1017/cls.2017.1.
Andia, J., Davis, E., Klein, R. and Wirshing, E.A. (2002) Assessing food security in the Monteverde Zone: A multi-method approach. Unpublished study. See https://digital.lib.usf.edu/SFS0000206/00001?search=food+=security (accessed 11 November 2021).
Barinaga, E. (2020) A route to commons-based democratic monies? Embedding the governance of money in traditional communal institutions. *Front. Blockchain* 3. https://doi.org/10.3389/fbloc.2020.575851.
Bélisle, F.J. (1983) Tourism and food production in the Caribbean. *Annals of Tourism Research* 10 (4), 497–513. https://doi.org/10.1016/0160-7383(83)90005-1.
Bianchi, R. (2022) Tourism, COVID-19 and crisis: A case for a radical turn. In F. Higgins-Desbiolles, A. Doering and B.C. Bigby (eds) *Socialising Tourism: Rethinking Tourism for Social and Ecological Justice* (pp. 98–108). Abingdon: Routledge.
Carroll, J.M. and Bellotti, V. (2015) Creating value together. In *Proceedings of the 18th ACM Conference on Computer Supported Cooperative Work & Social Computing-CSCW'15*, New York, NY, March 14–18, pp. 1500–1510.
Cambiatus (2021) New organizations for planet regeneration. See https://www.cambiatus.com (accessed 18 October 2021).
Cantor, A. (2016) 'Costumbres, Creencias, y 'Lo normal': A biocultural study on changing prenatal dietary practices in a rural tourism community in Costa Rica. Graduate Theses and Dissertations. See http://scholarcommons.usf.edu/etd/619 (accessed 14 November 2021).
Cordoba Brenes, K. (2021) Cryptocurrency, mining and the environment, Crypto Week Honduras Conference. See https://www.cambiatus.com/videos-press (accessed 9 January 2022).
Cristine, J. (2021) Tutorial objectives and actions, Cambiatus. See https://medium.com/cambiatus-tutorials/tutorial-objectives-and-actions-cambiatus-5dfdc093f26d (accessed 18 October 2021).

Dapp, M.M. (2019) Toward a sustainable circular economy powered by community-based incentive systems. In H. Treiblmaier and R. Beck (eds) *Business Transformation through Blockchain* (pp. 153–181). Cham: Palgrave Macmillan. https://doi.org/10.1007/978-3-319-99058-3_6.

Diniz, E.H., Cernev A.K., Denis A.R. and Daneluzz, F. (2021) Solidarity cryptocurrencies as digital community platforms. *Information Technology for Development, Blockchain* 27 (3), 524–38. https://doi.org/10.1080/02681102.2020.1827365.

Douthwaite, R. (1996) *Short Circuit: Strengthening Local Economies for Security in an Unstable World*. London: Chelsea Green Pub Co.

Enlace Statistics Committee (2019) Unpublished report on Community need and COVID-19 in Monteverde, May.

EOS Authority (n.d.) The future of blockchain in sustainable. See https://eosauthority.com/green/ (accessed 9 January 2022).

Fletcher, R. and Neves, K. (2012) Contradictions in tourism: The promise and pitfalls of ecotourism as a manifold capitalist fix. *Environment and Society* 3 (1), 60–77. https://doi.org/10.3167/ares.2012.030105.

Fountain, J. (2021) The future of food tourism in a post-COVID-19 world: Insights from New Zealand. *Journal of Tourism Futures* No Vol. ahead-of-print, 1–14. https://doi.org/10.1108/JTF-04-2021-0100.

Galvez, A. (2018) *Eating NAFTA: Trade, Food Policies, and the Destruction of Mexico*. Oakland: University of California Press.

Goodkind, A.L., Jones, A. and Berrens R.P. (2020) Cryptodamages: Monetary value estimates of the air pollution and human health impacts of cryptocurrency mining. *Energy Research & Social Science* 59, 1–9. https://doi.org/10.1016/j.erss.2019.101281

Higgins-Desbiolles, F. (2020) Socialising tourism for social and ecological justice after COVID-19. *Tourism Geographies* 22 (3), 610–623. https://doi.org/10.1080/14616688.2020.1757748.

Himmelgreen, D., Romero-Daza, N., Vega, M. and Brenes Cambrenero, H. (2006) The tourist season goes down but not the prices: Tourism and food insecurity in rural Costa Rica. *Ecology of Food and Nutrition* 45 (4), 295–321.

IPES-Food (2020) Special Report: Covid-19 and the Crisis in Food Systems, April. See http://www.ipes-food.org/_img/upload/files/COVID-19_CommuniqueEN.pdf (accessed 1 October 2021).

Jamal, T. and Higham, J. (2021) Justice and ethics: Towards a new platform for tourism and sustainability. *Journal of Sustainable Tourism* 29 (2), 143–157. https://doi.org/10.1080/09669582.2020.1835933.

Lietaer, B., Ulanowicz R. and Goerner, S. (2009) Options for managing a systemic bank crisis, *S.A.P.I.E.N.S* [Online], 2.1. See http://journals.openedition.org/sapiens/747 (accessed 9 January 2022).

Matheson, D.M., Varady, J., Varady, A. and Killeen, J.D. (2002) Household food security and nutritional status of Hispanic children in the fifth grade. *American Journal of Clinical Nutrition* 76 (1), 210–217.

Monbiot, G. (2017, 12 April) Finally, a breakthrough alternative to growth economics – the doughnut *The Guardian* (Online). See https://www.theguardian.com/commentisfree/2017/apr/12/doughnut-growth-economics-book-economic-model (accessed 9 January 2022).

Patel, R. and Moore, J.A. (2018) *History of the World in Seven Cheap Things: A Guide to Capitalism, Nature, and the Future of the Planet*. Oakland: University of California Press.

Place, C. and Bindewald, L. (2015) Validating and improving the impact of complementary currency systems through impact assessment frameworks. *International Journal of Community Currency Research* 19 (Summer), 152–164.

Rastegar, R., Higgins-Desbiolles, F. and Ruhanen, L. (2021) COVID-19 and a justice framework to guide tourism recovery. *Annals of Tourism Research* 103161. Advance online publication. https://doi.org/10.1016/j.annals.2021.103161.

Raworth, K. (2017) *The Doughnut Economy: Seven Ways to Think Like a 21st-Century Economist*. London: Random House Business Books.

Seyfang, G. and Longhurst, N. (2013a) Desperately seeking niches: Grassroots innovations and niche development in the community currency field. *Global Environmental Change Part A: Human & Policy Dimensions* 23 (5), 881–91.

Seyfang, G. and Longhurst, N. (2013b) Growing green money? Mapping community currencies for sustainable development. *Ecological Economics* 86, 65–77. https://doi.org/10.1016/j.ecolecon.2012.11.003.

Schinckus, C. (2020) The good, the bad and the ugly: An overview of the sustainability of blockchain technology. *Energy Research Social Sciences* 69, 101614. https://doi.org/10.1016/j.erss.2020.101614.

Stephens, E.C., Martin, G., van Wijk, M., Timsina, J. and Snow, V. (2020) Editorial: Impacts of COVID-19 on agricultural and food systems worldwide and on progress to the sustainable development goals. *Agricultural Systems* 183, 102873. https://doi.org/10.1016/j.agsy.2020.102873.

Suryawardani, I.G.A.O., Bendesa, I.K.G., Antara, M., Nursetyohadi, D. and Wiranatha, A.S. (2016) Implementation of social accounting matrix in calculating tourism leakage of accommodation in Bali. *International Journal of Applied Business and Economic Research* (IJABER) 14 (13), 9377–9405.

Telalbasic, I. (2017) Redesigning the concept of money: A service design perspective on complementary currency systems. *Journal of Design, Business & Society* 3 (1), 21–44.

UNCTAD (2020) Covid-19 and tourism – assessing the economic consequences. See https://unctad.org/webflyer/covid-19-and-tourism-assessing-economic-consequences (accessed 1 August 2021).

UNWTO (2021) 2020: Worst year in tourism history with 1 billion fewer international arrivals. See https://www.unwto.org/news/2020-worst-year-in-tourism-history-with-1-billion-fewer-international-arrivals (accessed 1 August 2021).

Van Dusen, K. (2021) Let's create a local, circular, doughnut economy in Monteverde, *Seeds*. See https://monteverdequakers.org/economics/ (accessed 13 November 2021).

Weinberg A., Bellows, S. and Ekster, D. (2002) Sustaining ecotourism: Insights and implications from two successful case studies. *Society & Natural Resources: An International Journal* 15 (4), 371–380. https://doi.org/10.1080/089419202753570846.

Wilkins, J. (2021) Community weaving and the circular economy: How COVID-19 has prompted a new vision for community building in Costa Rica. Global Community Fund blog. See https://globalfundcommunityfoundations.org/blog/strengthening-the-fabric-of-civil-society-in-monteverde/ (accessed 9 January 2022).

Yin, R.K. (2014) *Case Study Research: Design and Methods*. Los Angeles: Sage.

Zeller, S. (2020) Economic advantages of community currencies. *Journal of Risk and Financial Management* 13 (11), 1–11.

7 An Ethnographical Study of Community Tourism: Seeking Alternative Tourism Options for Malta Through 'Meet the Locals'

Andrew Jones and Julian Zarb

Introduction

The importance of tourism as a tool for social, cultural and economic development has been widely emphasised in the past (Murphy, 1985; Pearce *et al.*, 1997). This chapter is focused on the evaluation of alternative approaches to tourism with a focus on the role that local communities have in their contribution to the visitor learning experience. It aims to demonstrate how such forms of tourism can help move away from the over reliance on more traditional forms of Mediterranean tourism based upon 'beach sun, sea and sand' destinations.

The research is based upon a current, ongoing project titled: 'Discovering Malta and Gozo through its People and Culture' (see Figure 7.1). The project was originally initiated by the Ministry for Tourism, the Environment and Culture of Malta in 2011, and has led to the development of a number of locally derived visitor itineraries for villages and towns which are considered to be on the 'island periphery' with regard to mainstream tourism, but which nonetheless can often 'showcase' the real and authentic experience of the Maltese Islands. The key aims and objectives of the project thus aim to produce a number of specialist tourist experiences, such as community heritage and cultural resource audits, cultural trails, pilot tours and, in turn, encourage strategies that develop local tourist craft enterprises and 'cottage' industries.

The study is based on an ethnographical study and action research utilising a qualitative research approach focusing on six island

Figure 7.1 Discovering Malta and Gozo through its people and culture. Source: Malta Tourism Society 2018 (used with permission)

communities based in six local villages on the Maltese islands of Mqabba, Safi, Lija, Attard, Balzan and Kirkop. The aim is to disseminate or spread the benefits of tourism, which in Malta are currently very much based upon major coastal tourism resorts, to smaller more outlying village communities and to often disparate local stakeholders who might otherwise be excluded from the current wealth and benefits that tourism brings to the Maltese Islands (Visit Malta, 2022).

The development of community-based tourism and associated tours of village communities aims to be very different to the mainstream 'holiday experience' that is presently offered across Malta and Gozo as part of mainstream tourism activities. This alternative approach is very much based on the development of local communities, where the emphasis and focus are on meeting and interacting with the host community and learning about traditions, history, legends and folklore developed by the locals themselves.

In this context, the research updates the research work already undertaken in 2015–16 and the assessments of Jones and Zarb at that time (2017, 2018). This current assessment provides a more contemporary review of the issues and challenges for 2022 and beyond. As such, this research evaluates the issues and challenges of the project thus far. It evaluates how effective current community-based tourism has been, particularly the successes and failures, in attempting to rekindle host–visitor interaction and in the development of local cultural tourism resources and enterprises since the initial findings reported from 2015–16. In broader terms it assesses the effectiveness of this more sustainable, interactive and responsible approach to encourage local tourism and continues to evaluate some of the key lessons that can still be learnt. In summary, the key objectives of the study were to:

(1) introduce the concept of community-based tourism to the islands of Malta and Gozo;
(2) work with local stakeholders to develop and promote a tourism experience that is culturally unique, culturally authentic and benefits all local stakeholders;
(3) manage and monitor the development of community-based tourism and evaluate the issues and challenges for local implementation and stakeholder engagement.

Community Tourism: The Research Context

The literature provides a discussion of critical issues for community-based tourism which tends to focus on issues and challenges associated with consultation, interactivity, communication, stakeholder theory and ultimately community ownership and capacity building. Such issues are, for example, discussed by authors such as Murphy (1985), Pearce *et al.* (1997), Richards and Hall (2000) and more recently, by Moscardo (2008), Messer (2010), Wiltshier and Clarke (2019) and Walia (2020). These studies have provided the main theoretical framework and backdrop for the study. The concept is a simple one; it embraces the principles of sustainable and responsible tourism and provides a method for implementing those principles effectively through involving the key local stakeholder groups who work together towards an integrated approach, such as explained by Zarb (2020) in the 'tourism planning triangle'. In this respect the literature suggests that tourism needs to be considered as a sociocultural activity rather than just an economic industry (Dodds & Butler, 2010). It is not just about quantifying the industry in terms of tourist arrivals, bed-nights and revenue, but rather it is about enhancing the host–visitor interaction through an inter-cultural dialogue, communication and sharing experiences. The prime concept suggests that community-based tourism is an alternative development approach to a destination, region or locality. As such, it is not a process which works from the top-down but from the bottom-up. Again, early approaches and assessments of such notions have also been identified by Veal (2010) and Murphy (1985) who referred to Arnstein's (1969) work regarding the 'the Ladder of Citizen Participation' which consists of 13 levels of stakeholder participation ultimately culminating in 'citizen power'; notions and concepts that were further explored by, for example, Zarb (2020).

Hall and McArthur (1998) described this as a shift from the 'expert' view and policies to a more integrated and holistic involvement of the local community who should be the real owners of local tourism, heritage and culture. Moscardo (2008) referred to this as 'building community capacity' and demonstrated the importance of linking community-based tourism to local stakeholder involvement. Moscardo also defined this community capacity as building 'the community's awareness of, and education in, tourism development' (Moscardo, 2008: 10). By understanding the community with its needs, culture, traditions and characteristics, Moscardo asserted that this will, in turn, assist in promoting a sense of pride and stimulate local tourism, innovation and enterprise within local communities. More recent work by, for example, Freeman *et al.* (2010), has referred to the concept of 'stakeholder management theory' and stated that tourism planning concerns stakeholder relationships together with the dynamic relationship that exists between stakeholder groups.

Local community tourism has achieved positive results in a number of localities over the last decade or so. Calvia, Mallorca (Spain) is a case in point. In Calvia, the Local Agenda 21 project has been based on four key principles: the principle of environmental sustainability, the principle of local economic development, the principle of quality tourism and the principle of citizen participation. The outcomes of this have been assessed by, for example, Williams and Lawson (2001) and Dodds (2007). Pedersen (2002) also supported such principles and suggested that community tourism needs to be seen as a process of working from within the community and together evolving a complete strategy that offers the visitor a holistic experience of living history, cultural integration and social interchange.

Hall and McArthur (1998), however, cited a number of issues that prevent full stakeholder involvement due primarily to institutional malaise and poor stewardship. Krutwayso and Bramwell (2010) also cited a number of implications for community tourism implementation and stakeholder engagement. Their research revealed various dialectical relations between policy implementation and the socioeconomic, political, governance and cultural contexts. Dodds and Butler (2010) also considered a number of critical themes that can be identified in the literature, ranging from power clashes between political parties at a national level to lack of stakeholder involvement and accountability at the local level (Dodds & Butler, 2010). The idea of 'citizen power', as Arnstein (1969) referred to in his last stage of the ladder of citizen participation, is pertinent in this context. The need to understand and avoid the repercussions that are so common today in top-down tourism planning or stalled efforts in community participation are key notions that have further scope for exploration and focus (Walia, 2020; Wiltshier & Clark, 2019). These are certainly pertinent points, especially if community-based tourism approaches are to provide pragmatic alternatives to mainstream tourism.

In summary, the literature indicates that the contexts and positives for community-based tourism are increasingly being understood and recognised by tourism practitioners and policymakers alike. In this respect, there are now many good examples of how community-based tourism can aid tourism diversification, help innovate local community capacity and enterprise and help contribute to alternative forms of sustainable tourism based upon local culture and heritage. There are nevertheless challenges to such approaches which have been clearly highlighted by several authors. In this respect understanding the challenge of sustaining community engagement in tourism activities is one that remains critical for the success of community-based approaches.

Tourism in Malta: Promoting Alternative Approaches to Traditional Tourism

Malta as a traditional Mediterranean package tour destination (sun sand and sea) has increasingly experienced pressure on its infrastructure (Jones, 2016; Lockhart, 1997). Over the last decade, tourism has attained year-on-year growth. Arrivals increased from 1.4 million visitors in 2012 to over 2.7 million in 2019, with an expenditure estimated to be above € 2.2 billion (Malta Tourism Authority (MTA), 2019). The imbalance between the increase in tourist arrivals (pre-COVID-19) and the rise in earnings off-set against increasing resource pressures (particularly societal and environmental) and the consequences of overtourism have increased debate on the future direction for tourism in Malta (Briguglio & Avellino, 2021; Dodds & Butler, 2010; Goodwin, 2017).

Pre-COVID-19 statistics illustrating continued growth in tourism numbers (5% average growth per annum), particularly in the peak summer period, were already creating environmental strains which were increasingly leading to carrying capacity issues, resource pressures, waste and pollution impacts, congestion, environmental degradation and local community unease. Indeed, up until late 2019 news headlines from across Europe including Barcelona, Dubrovnik, Rome, Prague, Mallorca and Venice highlighted growing disquiet, anger and often conflict between host communities and visitors due to overtourism. Primarily, these concerns have been attributed to mass tourism and the perceived lack of local benefits from tourism for such host communities (Kettle, 2017). These, unfortunately, are notions and beliefs now growing in Malta (Debono, 2019).

Tourism in Malta and Gozo has traditionally focused on traditional coastal resorts of Sliema, St Julians, Buggiba, Qwara and Mellieha and key cultural icons which include Mdina, Valletta, the Grand Harbour, the prehistoric temples and Gozo (MTA, 2019). This has been the mainstay of the tourist 'offer' over the past 50 years or more and has created a somewhat narrow tourism image of the Maltese islands. This, in turn, has been used by the various main tourism stakeholders and authorities as key symbols and the backbone of the 'traditional tourist package' for tourism on the Islands.

The Ministry for Tourism's *National Tourism Strategy 2015–2020* (Ministry for Tourism, 2015) focused on ensuring sustainable growth. It emphasised the development of tourism based upon cultural, community and resource assets placing the growth of community-based tourism and the development of associated local tourism industries at the forefront for new alternative strategies for tourism. This has now been recently enhanced and supplemented by a new strategy *Recover, Rethink, Revitalise: Malta's Tourism Strategy 2021–2030* (MTA, 2021). This is a Tourism Strategy in which 'recovery' is conditioned by the dual principles of 'rethinking' and 'revitalising' the tourism economy for the islands. Its focus is not merely a plan to return to the tourism activity prevailing pre-COVID-19 but one in which policy is designed to enhance tourism growth, which is stronger, more competitive and better equipped to handle the challenges of the next decade. Sustainability, niche tourism and inclusive stakeholder strategies significantly form the main focus in this respect (MTA, 2021).

The 'Discover Malta and Gozo through its People and Culture' project complements these new strategic approaches. Since its inception in 2011, the rationale of the project has been to provide visitors with tourism alternatives based upon tailor-made, locally sourced community and cultural tourism experiences. Despite much effort in recent support and, laudable as the new strategic tourism approach is for community-based initiatives, the project continues to demonstrate challenges and mixed results. These have already been discussed by, for example, Jones and Zarb (2018, 2019) in earlier assessments. Nevertheless, the refocusing of tourism strategies to include new niche developments in outlying villages and non-traditional tourism areas remains a worthy policy re-focus and, in this respect, one which continues to present real opportunities, as well as challenges for continued and future local community engagement.

The Research Approach and Project Delivery

As Freire stated, research can be attributed to: 'Learning to do it by doing it' (1982). In this context the methodology adopted for this project has and primarily continues to be based on a critical, participative action research approach supported by ethnographical qualitative techniques used for data collection. This has involved local stakeholder focus group participation, personal interviews with both locals and visitors, scoping meetings and local village-based seminars.

Kemmis and McTaggart (1988) described action research as a six-step process: plan, reflect, replace, act, observe and reflect. These descriptions express the benefit of applying a participatory action research process to this study, since this can take a continuous and cyclical format. Action research models, such as the work of Kumar (2012) as well as other authors such as Kemmis *et al.* (2014) and McIntyre (2008), indicate that participatory action research in practice can provide insightful data and can act as a key motivator and incentive to stakeholders engaged in research. This research approach works to develop the process and implementation of tourism community development at a local or 'grassroots' ethnographical level by using such models, while also focusing on the principles for sustainable development.

Critical participatory action research has two basic objectives according to Kemmis *et al.* (2014). These primarily focus upon opening up space for dialogue and bringing about change. It is a process that helps all the participants and stakeholders work together to 'meet the criteria of rationality, sustainability and justice' (Kemmis *et al.*, 2014: 22). Phillmore and Goodson (2004: 23) considered the advantages of action research from the point of view of a more reliable and representative study since it involves the actual participants in a direct and sincere discussion rather than the more rigid method of interviews. The framework for this research followed a participatory, four-staged approach for developing community tourism which has been, to date, primarily programmed between 2013–2021.

Between 2010 and 2019, a number of meetings were held with stakeholders. These meetings served two main purposes: to keep the stakeholders informed of the progress with the project and to listen to their suggestions and proposals for enhancing the value added by the project. The stakeholders included the local council mayors (or their designated representative, sometimes the Executive Secretary or the Councillor responsible for tourism in the locality); tour operators, destination management companies (DMCs) and non-governmental organisations

(NGOs) such as the local band clubs and businesses in the locality. These meetings were chaired by the project coordinator together with the research assistants and notes were taken to be included in minutes that served more as references to the project's progress and issues that can be considered as learning points for any future or similar projects. The quotations provided below are derived from these notes and protect the sources through anonymity. Outlined below is the research process described in four stages.

Stage 1: Scoping meetings: 2013–2015: This stage involved working with local councils to identify the strengths and weaknesses of tourism activity in each of the village localities. Here the project started by targeting local councils in Malta and Gozo and developing itineraries and tourist maps for their respective localities. The itineraries did not follow the traditional 'programme-based' set lists of sites and places of interest but were instead a compilation based upon local cultural ethnography, i.e. local knowledge of those historical, cultural, religious and social sites that each community felt important to promote. These were, in turn, developed and offered to visitors in the form of local visitor itineraries (Visit Malta, 2022). The intention was to provide a set of alternative and unique local experiences tailored by in-depth local community ethnographical knowledge at a given locality.

Stage 2: Pilot projects: 2015–2016: This stage aimed to identify those localities that would be prepared to develop the concept of a community-based experience for visitors through the development of local guides and village tours. The key objective was to develop local, authentic cultural experiences where the visitor and the host communities were given the chance of interacting, learning from each other and where opportunities availed, secure economic benefit from the experience. As a change from the original proposals in 2018, six localities (Mqabba, Safi, Kirkop, Lija, Attard and Balzan) were included in the final participant list for the pilot project stage.

Stage 3: Project implementation, promotion and initial feedback: 2016–2018: In collaboration with local destination management organisations (DMOs), this stage considered two objectives: (i) to provide the logistical framework to establish and kick-start community tourist guides/tours which aimed to provide visitors with authentic, local tourism experiences based upon local culture and heritage and opportunities to meet 'the locals'; and (ii) to analyse initial feedback from the pilot projects in terms of understanding key source markets (including demographics and qualitative data concerning the experiences and perceptions from visitors). In addition, analysis also focused on an assessment of initial project implementation issues and challenges.

Stage 4: Project reflection – Monitoring: 2017–2021: This aimed to analyse the promotion and implementation process for the community-based operations using both quantitative and qualitative analysis methods. This sought key data, to evaluate the current and future needs of the project. In this context, the monitoring and reflection process has focused particularly on current issues and challenges, for example, project implementation, project management and project sustainability. This stage remains ongoing for the foreseeable future.

Current Research Outcomes

The implementation of the community-based operations within the six pilot villages continues to present a number of rich research findings. The findings build on earlier findings from the initial research in 2018 (Jones & Zarb, 2018). The results are now clearly identified and can be primarily categorised in terms of both the positives and negatives of: (i) physical–environmental issues, (ii) ethnographical–sociocultural issues and (iii) economic issues.

Outcomes from focus group meetings, municipal community events and social-media responses have coalesced much current and topical sentiment in this respect. For instance, statements included:

'Tourists are changing in synchronization with the tourist sector and the community which is a good thing'.

'Luckily tourism has stopped to be seen as a stop gap, job opportunity during the summer. Now people recognise its importance and power in the development of a country and our communities'.

'Small groups of tourists and the local participation are essential to an enhanced tourist experience for the locality'.

Key opportunities: By and large the reaction by the key local stakeholders, as evidenced from some of the comments above, continues to be very positive and supportive with a number of positive synergies relating to coordinated approaches to planning, community actions and heritage conservation.

'Great idea! It should be started in Hamrun. The town retains a vintage feel about it and has lots of urban local to offer, especially since the urban landscape is unspoilt with a unique streetscape of entire rows of townhouses without blocks of flats in between'.

'The emphasis here should be on retaining the uniqueness of the locality in terms of character and culture'.

'The need for upmarket-quality tourism has been recognised for many years but now some good people are now making it happen. Bravo'.

In this respect, Freeman *et al.*'s work (2010) on stakeholder management theory and Zarb's (2020) integrated planning research highlighted the synergies that can occur when local community projects are executed successfully. Such an approach would still seem very pertinent for this project today. It is one that still encourages a more integrated, inclusive approach to support alternative forms of tourism development based upon ethnography, community, heritage and culture. In turn, one which also offers clear community benefits in terms of community inclusion, innovation, reinvention, revitalisation and enterprise.

'It is important to see the project preserve the culture, character and heritage instead of turning the destination into one concrete jungle!'.

These are now notions clearly expressed and fostered in the new tourism strategy for Malta (MTA, 2021). As such, evidence from the study, thus far, can also demonstrate that involving local stakeholders can stimulate an enhanced affinity between all interested parties. This can, in turn, produce consistent and unique local ethnographical, community and cultural experiences for the visitor and measurable tangible cultural and financial benefits for a local host community.

As an example, during recent meetings with the local councils and some industry stakeholders there was a fair amount of enthusiasm shown, particularly when these included travel agents and tour operators looking for new experiences to offer their clients. During the launch of the tour in Lija, there was a strong response from the press. This included the presence of two television stations and a primary national newspaper reporter. Following the launch project, organisers were contacted by media associates who were interested in covering the Lija tour for easyJet's inflight magazine; the article was included in the April 2020 edition.

Such examples have required robust and committed participation by a multitude of diverse local stakeholders that make up the local community. This, as a consequence, has strengthened local engagement, engendered community pride, widened local participation and built upon community inclusivity. In this respect, there continues to be clear evidence from the project that local communities can enhance the host–visitor interaction through the development and dissemination of knowledge for local or more intangible, ethnographical-based cultural heritage.

There have been several examples of synergy between stakeholders. The most prominent example of this was a meeting that was held between five local councils (a no mean feat) where strong support and enthusiasm was expressed for such community initiatives as a way of promoting village localities, local cultural products and places that are normally perceived as being 'off the beaten track'. This aspect has also been referred to by authors such as Norkunas (1993: 218) as preservation of local culture by promoting: 'a sense of themselves through orally transmitting family stories and through celebrations and rituals performed inside the group'. Encouragingly, the richness of such an approach has and still continues to contribute to a wealth of local innovation which provides for both a valued visitor cultural experience and significant community gain.

As well as these key opportunities, the study has also shown that benefits for the local community and visitors can accrue positive outcomes in a number of other ways. For example, the six-village project has provided opportunity to revive local heritage and culture and establish a unique brand potential for non-traditional tourism areas across Malta that are not currently perceived or recognised as tourism destinations This has included, for example wine and culinary tastings and tours, cookery courses, craft stone masonry classes, bee-keeping and honey-making demonstrations. The project has also had potential to create new tourist demand and encourage the development of new tourism economies and off-season opportunities. In this respect the project has stimulated growth for new community tourism businesses based on crafts, cultural trails, cultural events, accommodation and associated hospitality enterprises (see Figure 7.2).

There has also been some scope to encourage environmental enhancement, townscape improvements, better coordinated public transport access and small-scale infrastructure investment.

As often happens, the concept of promoting a locality for tourism brings with it an enthusiasm to improve the environment and infrastructure of that locality. In Safi, for example, the local council allocated funds for the improvement of an old garden (known as the Governor's garden – Il-Gnien tal-Kmand) so that the local DMC operator could feature this as one of the attractions and also host some of the craft demonstrations there, such as the production of honey. In the same village of Safi, one retail outlet started to sell souvenirs, some made locally, others made in Malta. This has been one encouraging outcome given that Safi has never been a locality on the tourist route. These benefits are very much in line

Figure 7.2 Community Tourism Malta: 'Meet the Locals'. Source: Malta Tourism Society 2018 (used with permission)

with findings and conceptualisations in Messer (2010), Moscardo (2008), Wiltshier and Clarke (2019) and Walia (2020) in their works focused on the benefits of community tourism.

Key challenges: At a less positive and more challenging level there have been some outcomes that need continued consideration. Especially when related, again, to the cultural, physical and economic concerns or challenges of the project. The concept of stakeholder scepticism, malaise and fatigue while maintaining consistent involvement and consultation throughout each stage of the project continues to be a challenge in this respect.

> 'First they ruined the tourist hotspots like Paceville, St. Julians ,Qawra and Bugibba with over building. Now they are Turning their beady eyes to what is left of the true Maltese village. Beware, you are next!!!'.

Stakeholder scepticism and fatigue did appear in this project as the previous quote illustrates. Initially, the southern villages of Safi, Kirkop and Mqabba included two other localities – Qrendi and Zurrieq – the former well known as the locality with the twin temples of Hagar Qim and Mnajdra while the latter was much larger and was renowned for the cave known as the Blue Grotto in the valley of Zurrieq (Wied iz-Zurrieq, an area that fell under the jurisdiction of Qrendi). It was this issue that led to the withdrawal of Zurrieq and Qrendi from the project because they felt 'politically' that they could continue the project independently without working as a group.

The perceptions of fatigue and scepticism have been described by Dodds and Butler (2010) as one of the key challenges of maintaining momentum and energy in local projects. This has and continues to affect project outcomes particularly by causing delays and interruption in achieving the project's goals. This, for example, continues to be the case for the roll out and development of the local village guides and itineraries and for the development of local cultural tours.

> 'We don't have the time or expertise to devote energy here we have to earn a living'.

> 'We need help money and support for this to be successful'.

Three key factors would seem apparent here: (i) accountability and stewardship; (ii) title and tenure of cultural assets; and (iii) consistency and reliability in participation. The political aspects that tend to govern the local council administration, not particularly unique to Malta, has led to a situation that 'stewardship' and 'ownership' have not been holistic or universal, which to some degree has failed to generate equality and parity of commitment across councils or municipal boundaries. In fact, intermunicipal rivalries have become more acute in this context.

As mentioned earlier, the political rivalry between Zurrieq and Qrendi led to stakeholder disputes which resulted in both these localities leaving the southern villages group. On the other hand, the initial other localities, Tarxien, Santa Venera, Zebbug, Gozo and Bormla, worked together in presenting a diverse example of experiences that were unique for each locality. This primarily occurred due to strong interventions from local drivers/facilitators. In each case, project actions and initiatives were carried out with all representatives from each of the localities being present for meetings and other activities.

To this end, collaboration still tends to be very dependent on individual drivers and individual's passion for the project. In this respect participation, ownership and stakeholder fatigue have been very much determined by changing local council priorities, local politics and functions and inter-municipality rivalries where gaining competitive advantage has, at times, negated collaborative effort. Sometimes, as suggested, this was political, while at other times it was a strong sense of parochialism. These were the main reasons why in the initial part of the project, at the creation of village itineraries and online guide maps for localities, there was a failure to engage with a wider community base. Only 16 of the 68 local councils presented their itineraries in four years.

Pedersen (2002) highlighted such problems for the continuity of similar projects. He stated that there is need to work inclusively from within and across communities in order to ensure that there is a complete inclusive strategy. Such an approach, he suggested, offers the visitor and local community a holistic experience of living history, cultural integration, social interchange and economic benefit. Zarb's (2020) work on integrated inclusive planning approaches and lessons that can be gained from such an approach are pertinent considerations as well. Outcomes to date show that the general patchwork of participation has sometimes challenged established local stakeholder relationships, alienated some individuals or groups and in some instances caused friction, anger and disenfranchisement.

In this respect, the Zurrieq and Qrendi issue is the clearest example where there were angry sentiments that sometimes manifested themselves at the regular meetings where collaborative agreement was often lacking. The lessons learnt from these instances have taught the project that it is important to ensure that all stakeholders get a chance to participate where there are signs of disagreement or intransigence, otherwise the project ultimately has to move on with other partners. Outcomes here suggest that it is not advisable to continue a project without all three key stakeholder representatives present – that is the local authority, the local businesses and the local community – as this defeats the idea of community-based project that is integrated and authentic. Otherwise, there is danger that initiatives just become another commercial sightseeing tour rather than a community-based tour.

This has certainly pointed to a need to maintain and encourage a stronger synergy between stakeholder partners in order to avoid delay or discord as Pederson (2002) suggested. Dodds and Butler (2010) also suggested that challenges to the implementation of community strategies include tensions that occur between how stakeholders perceive tourism either as a sociocultural activity or tourism as a socioeconomic industry. In this respect, most still perceive tourism growth usually in quantitative, commodity terms and within short-term financial return timeframes. In this respect, the project thus far has to some extent suffered from the 'short-termism' of the local political representatives, whose period of office is normally between three to four years.

> '*We need to demonstrate returns on our investments and hard work and show local people what they can earn and how they can benefit – otherwise it's a wasted effort*'.

This is also exacerbated by the fact that local councils and the local business community still tend to expect an immediate or quick economic return from time, effort and investment spent on the project. Inter-municipal rivalry to effect competitive advantage, as already stated, has also compounded such challenges as the project has matured. Hall and McArthur (1998) addressed these dynamics by suggesting that stakeholder involvement strategies cannot ignore the fact that prime motivations for involvement are driven by politics and popularity rather than a genuine desire to involve and achieve stakeholder participation and the community sense of ownership. Unfortunately for the project, this still remains a pertinent point.

Lai *et al.* (2006) discussed barriers that concern the issue of management and planning that are important for a balanced and successful stakeholder approach to project implementation. The study certainly has had its challenges in this respect. Action research focusing on three pilot areas experienced plenty of situations which involved 'trial and error' and presented a real learning curve for balancing the often diverging rather than converging interests between the multitude of diverse stakeholder groups. Mention has already been made to the Zurrieq–Qrendi disagreement, which encouragingly is probably the only really 'polarised' stakeholder disagreement so far in this project. The need to further promote and manage stakeholder participation which ensures collaborative goals being met remains a challenge in this respect too.

At a more operational level stakeholders have identified some unease and growing concerns regarding the development of the broader project objectives over time. These now increasingly relate to perceived future viability and sustainability of end goals in terms of physical, cultural and economic attributes. Concerns regarding village gentrification and the potential removal of traditional working and living practices through commercial exploitation and commodification of some facilities and operations have been reported.

> '*Concrete is already on the way. Boutique hotels, shops, bars, restaurants, and to hell with those old buildings.*'

Tourism is nothing but venom which destroys the fabric of all localities it touches. We will just never learn'.

'The big tour bus operators have taken over things and will take over everything! Just like all the other tourism'.

During 2018 and 2019 the local enterprises entrusted with promoting the community-based tour in Safi hosted some 3000 visitors to the small village in just six months. The concern was not about the extraordinary interest in a village that was previously off the tourist map of the islands, but whether the number of visitors was actually defeating the idea of a community-based tour where the emphasis should always be on the host–visitor interaction. This issue was taken into consideration in the production of the Lija itinerary in December 2021 where the intention is to host fewer visitors and offer better quality experiences.

The potential exploitation in promoting 'alternative' cultural tours through the continuing and all-pervasive institutional dominance of mass tourism has been another concern. The 'hijacking' of local tours by established mass tourism operators offer a case in point. In this context 'cultural greenwashing' has been levelled at some of the cultural enterprises that have emerged as a consequence from the project. This has also led to more divergence or fragmentation of stakeholder interests within each village community; leading in turn to more perceived levels of strategic dis-coordination rather than collaboration. Project outcomes suggest that the issues of 'exploitation' and 'commercialisation' mean that the project needs to ensure that there is a careful selection and focus on the tour operators and DMCs selected for such tours; the focus being on quality not quantity and the emphasis on the visitor profile not on visitor numbers.

Hospitality services and service delivery have also caused some concern from both visitors and hosts alike. This has been exacerbated by some stakeholders often criticising the poor quality of local infrastructure and, in part, often poor quality of the townscape. Comments included:

'There is no place anywhere for decent coffee'.

'The whole place is clogged with cars and lorries. We need a nice village plaza'.

'There is still an awful lot of litter and mess which needs to be cleared'.

'The building and digging noise from new construction is terrible and will affect tourists'.

'There are too many regulations to deal with'.

During the project there were certain examples that had been noted including environmental health hazards at a farm in one of the three southern villages, as well as at a stone crafting workshop near a quarry where there was no border wall or safety bars in place. Another health hazard has involved food hygiene. Challenges of allowing visitors into domestic and commercial kitchens (which happened in Kirkop and Lija) and ensuring that visitors have proper food handling attire are also a case in point.

These issues can potentially present key barriers for the future of the project and its continued sustainability. The need for continuing education, community capacity building and strategic collaborative delivery, in this context, continues to remain a real challenge and one that is still clearly relevant today. Insights provided by authors such as Moscardo (2008) and more recently by Wiltshier and Clarke (2019) also would support such views.

Conclusions

The research associated with this project remains ongoing and will continue through a monitoring process to understand and evaluate on-going stakeholder contributions, commitment and needs for community-based tourism in Malta. The research, thus far, contributes to developing a broader understanding of development and policy approaches for community tourism – its opportunities and its challenges at a local level (see Figures 7.3 and 7.4). The results from the project today, very much consolidate the discussions and outcomes initially advocated in the findings from the earlier study of Jones and Zarb (2018). In summary, the findings

Figure 7.3 Community Tourism Malta: Opportunities

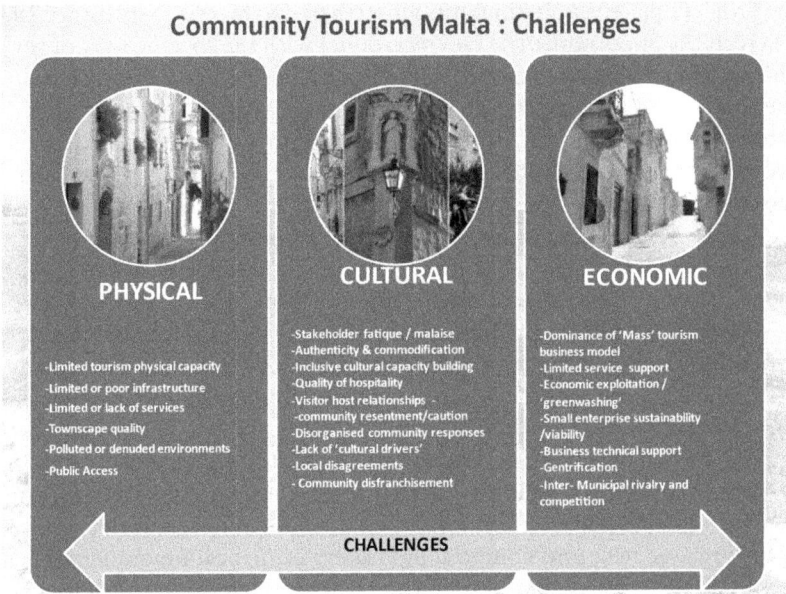

Figure 7.4 Community Tourism Malta: Challenges

support notions that community tourism can, indeed, be a useful tool in supporting a more sustainable, responsible and alternative form of tourism. It is one approach which primarily accrues benefits for local communities and in turn assists with diversifying and stimulating non-traditional tourism destination economies using cultural assets and local community enterprise.

Thus, key lessons can be learnt from the project to date. Lessons illustrate that community-based tourism can provide a strong basis for community integration and alternative forms of tourism development which are primarily focused on local and unique community and visitor cultural experiences. Indeed, the name given to this project – 'Meet the locals' – was actually chosen as an alternative to the idea of simply developing more mainstream sightseeing tours, by walking around and talking to people in the streets, providing opportunities to meet locals in their houses and local clubs and engage with local crafts.

That said, lessons have shown that the project has experienced, in terms of sustainability outcomes, both opportunistic and challenging results. In this respect the project, to date, illustrates that it has been largely dependent on good political will and key champions, drivers or individuals and that there is a risk that these can be short-lived, thus threatening project continuity, viability and ultimately sustainability.

These are pertinent points. COVID-19 has put the project on hold for almost two years and with increasing concerns that this could jeopardise the project objectives and impetus altogether. To this end, September 2021 has seen the appointment of new personnel with the specific task of revitalising this project once again. Already, this has succeeded in rekindling the interest of the key stakeholders in many places, especially in Lija. Interestingly, this has already included a tour operator interested in developing these experiences for the quality visitor from new markets such as the United States.

Overriding principles for success also suggest that building community tourism operations that meet with all stakeholders' needs is paramount to ensure both local ownership and stewardship. In this context lessons suggest that the promotion of community-based tourism activities generally needs a different approach to the more 'commercial' or traditional mass tourist market and one which needs to be much more sensitive to local needs such as local carrying capacity limits and sensitive approaches to support and encourage community engagement. In this respect, lessons also demonstrate that local municipality rivalries, commercial exploitation or commodification, stakeholder inclusivity, townscape management and potential damaging gentrification outcomes need to be further addressed and managed more strategically. In essence this will be essential for the well-being and long-term sustainability of the project and for similar initiatives that come thereafter. As one informant cautioned:

> 'This sounds like a good initiative. But only until the first application is made to turn one or two of Lija old villas into boutique hotels, adding a couple of floors and converting existing orchards into swimming pools with decks. Just look at what is happening at Saqqajja and Vjal Santu Wistin in Rabat. Beware'.

In summary, indictors from the project emphasise the need for involving all stakeholders in building community capacity for local tourism projects. This approach is identified as important in sharing knowledge and best practice that will enhance both the visitor experience and benefit local communities in physical, cultural and economic outcomes. The research also points to a need for a continuous and consistent process of consultation, monitoring and engagement with the key stakeholders in order to ensure that projects remain on a viable and sustainably sound footing.

The results indicate that future directions for community tourism should recognise opportunities for community enterprise and cultural development and how these can be maximised or capitalised upon. In turn, where challenges remain, strategies need to focus on how these can be minimised, managed and addressed. The project 'Discovering Malta and Gozo thorough its People and Culture' demonstrates that such an approach can form a platform or model that can be used as an effective technique to encourage medium to long term sustainable community development. This is one which delivers a real gain for local communities within the Maltese Islands and, in turn one which provides meaningful experiential cultural exchange for visitors. It is a project that can also provide broader lessons for community-based tourism initiatives at other locations and for other communities.

References

Arnstein, S.R. (1969) A ladder of citizen participation. *JAIP* 35 (4), 216–224.
Briguglio L. and Avellino, M. (2021) Overtourism, environmental degradation and governance in small islands with special reference to Malta. In J.L. Roberts, S. Nath, S. Paul and Y.N. Madhoo (eds) *Shaping the Future of Small Islands* (pp. 301–322). Singapore: Palgrave Macmillan.
Debono, J. (2019) Maltese showing signs of unease at overtourism, *Malta Today*. See https://www.maltatoday.com.mt/news/national/96828/Maltese_showing_signs_of_unease_at_overtourism#.YdQdf1nTXIU (accessed 12 August 2021).
Dodds, R. (2007) Sustainable tourism and policy implementation: Lessons from the case of Calvia, Spain. *Current Issues in Tourism* 10 (4), 296–322.
Dodds, R. and Butler, R. (2010) Barriers to implementing sustainable tourism policy in mass tourism destinations. *Tourismos, International Multidisciplinary Journal of Tourism* 5 (1), 35–53.
Freeman, R.E., Harrison, J.S., Wicks, A.C., Parmar, B.L. and De Colle, S. (2010) *Stakeholder Theory: The State of the Art*. Cambridge: Cambridge University Press.
Freire, P. (1982) Creating alternative research methods: Learning to do it by doing it. In B. Hall, A. Gillette and R. Tandon (eds) *Creating Knowledge: A Monopoly?* (pp. 29–37). New Delhi: Society for Participatory Research in Asia.
Goodwin, H. (2017) The challenge of overtourism. Responsible tourism partnership. Working Paper No. 4, pp.1–19. See https://www.millennium-destinations.com/uploads/4/1/9/7/41979675/rtpwp4overtourism012017.pdf (accessed 30 January 2022).
Hall, C.M. and McArthur, S. (1998) *Integrated Heritage Management: Principles and Practice*. Chichester: John Wiley & Sons.
Jones, A. (2016) Island tourism. In J. Jafari and H. Xiao (eds) *Encyclopedia of Tourism* (p. 526). Cham: Springer.
Jones, A. and Zarb, J. (2017) Developing community-based tours for greater stakeholder benefit and commitment. *International Journal of Tourism Policy* 7 (3), 250–267.
Jones, A. and Zarb, J. (2018) Paradoxically speaking: Community tourism: Discovering Malta and Gozo through its people and culture – diverging or converging sustainability issues? [online]. In T. Young, P. Stolk and G. McGinnis (eds) *CAUTHE 2018 Conference: Get Smart: Paradoxes and Possibilities in Tourism, Hospitality and Events Education and Research*. Newcastle, NSW: The University of Newcastle, pp. 46–50.
Jones, A. and Zarb, J. (2019) 'Meet the Locals': Community tourism- an approach to combat over-tourism in Malta and Gozo. In P. Wiltshier and A. Clark (eds) *Community-Based Tourism in the Developing World: Community Learning, Development and Enterprise* (pp. 112–124). Abingdon: Routledge.
Kemmis, S. and McTaggart, R. (eds) (1988) *The Action Research Planner*. Melbourne: Deakin University Press.
Kemmis, S., McTaggart, R. and Nixon, R. (2014) *The Action Research Planner. Doing Critical Participatory Action Research*. Singapore: Springer.
Kettle, M. (2017) Mass tourism is at tipping point but we are all part of the problem, *The Guardian*. 11 August. See https://www.theguardian.com/commentisfree/2017/aug/11/tourism-tipping-point-travel-less-damage-destruction (accessed 25 October 2017).

Krutwayso, O. and Bramwell, B. (2010) Tourism policy implementation and society. *Annals of Tourism Research* 37 (3), 670–691.

Kumar, R. (2012) *Research Methodology. A Step-by-Step Guide for Beginners*. London: Sage Publications.

Lai, K., Li, Y. and Feng, X. (2006) Gap between tourism planning and implementation: A case of China. *Tourism Management* 27, 1171–1180.

Lockhart, D. (1997) Islands and tourism and overview. In D.G. Lockhart and D. Drakakis-Smith (eds) *Island Tourism Trends* (pp. 228–246). London: Pinter.

Malta Tourism Authority (MTA) (2019) *National Tourism Statistics 2019*. Valletta. See 2022. https://www.mta.com.mt/en/file.aspx (accessed 3 January 2022).

Malta Tourism Authority (MTA) (2021) *Recover Rethink Revitalise-Malta Tourism Strategy 2021–2030*, Ministry for Tourism, Valletta. See file:///C:/Users/User/AppData/Local/Temp/maltatourismstrategy_2030_v7.pdf (accessed 3 January 2022).

McIntyre, A. (2008) *Participatory Action Research – Qualitative Research. Methods Series 52*. London: Sage.

Messer, C. (2010) *Community Tourism Development* (3rd edn). Minneapolis: University of Minnesota.

Ministry for Tourism and Culture (2015) *National Tourism Policy 2015–2020*. Ministry for Tourism and Culture, Valletta. See https://tourism.gov.mt/en/Documents/FINALBOOKLETexport9.pdf (accessed 3 January 2022).

Moscardo, G. (2008) *Building Community Capacity for Tourism Development*. Oxon: CABI.

Murphy, P.E. (1985) *Tourism: A Community Approach*. Oxon: Routledge.

Norkunas, M.K. (1993) *The Politics of Public Memory: Tourism, History and Ethnicity in Monterey, California*. Albany: State University of New York Press.

Pearce, P.L., Moscardo, G. and Ross, G.F. (1997) *Tourism Community Relationships*. Bingley: Emerald.

Pedersen, A. (2002) *Managing Tourism at World Heritage Sites: A Practical Manual for World Heritage Site Managers*. Paris: UNESCO.

Phillmore, J. and Goodson, L. (2004) *Qualitative Research in Tourism: Ontologies, Epistemologies and Methodologies*. Oxon: Routledge.

Richards, G. and Hall, D. (eds) (2000) *Tourism and Sustainable Community Development* (1st edn). London: Routledge.

Veal, A.J. (2010) *Leisure, Sport and Tourism, Politics, Policy and Planning* (3rd edn). Oxon: CABI.

Visit Malta (2022) *Discovering Malta & Gozo Through its People & Culture*, Visit Malta. See https://www.visitmalta.com/en/a/itineraries/ (accessed 3 January 2022).

Walia, S.K. (eds) (2020) *The Routledge Handbook of Community-based Tourism Management: Concepts, Issues and Implications*. London: Routledge.

Williams, J. and Lawson, R. (2001) Community issues and resident opinions of tourism. *Annals of Tourism Research* 28 (2), 269–290.

Wiltshier, P. and Clarke, A. (eds) (2019) *Community-Based Tourism in the Developing World: Community Learning, Development and Enterprise*. Abingdon: Routledge.

Zarb, J. (2020) The process of research in developing an effective local tourism planning strategy and policy: A community based tourism Model from Malta. PhD thesis, Pontypridd: University of South Wales. See https://ethos.bl.uk/OrderDetails.do (accessed 3 January 2022).

Part 3

Practitioners' Views and Insights

8 Localhood

Signe Jungersted

The Epilogue

In late 2016, I was writing emails to close partners and not-so-close network relations that all had one thing in common: English was their native language. I was asking them what kind of connotations the word '*localhood*' held for them? At that point, we had already formulated 'Localhood for Everyone' as the new vision of Copenhagen's 2020 Tourism Strategy, but we wanted to be absolutely certain that the word didn't just resonate with us, but also intuitively resonated with non-invested outsiders with a much stronger sense of the English language. This was even more important as we had decided not to include a clear definition of the invented word in the strategy document. Localhood was for everyone – to live, experience and define. We made up the word in an endeavour to catch the spirit of something new, something intangible, perhaps vaguely defined, but definitely something era-making and tourism-changing. Funnily enough and completely unrelated to our international email sounding board, around the exact same time, a start-up in Hong Kong launched a social network to connect neighbours and boost community spirit, with the inspired name of … localhood. In many ways, it all starts much like in those movies where people, who are continents apart, somehow have thoughts and destinies intertwined, and in the end, the big plot reveals itself to be an interconnected and meaningful story. What started as a reflection on the increased demand for more local experiences and a very early vision of a more balanced tourism development in Copenhagen, shifted over the years after the launch to a much stronger focus on and discussion of community involvement, social well-being and quality of life.

The power of words was not lost upon us when we first launched the strategy document in early 2017 announcing the 'End of Tourism – as we know it' and welcoming a new era of 'Localhood for Everyone'. But we had not anticipated the extent to which Localhood as a word and idea embodied and connected the thoughts and aspirations of the tourism community at the time, and how it would even be reinforced and given new meaning with the developments and events that followed. With this epilogue on Localhood, I want to share with you how the idea and strategy came to be, what were the strategic determinants that led us to declaring the end of tourism and welcoming localhood in early 2017, and what has been both the aftermath and afterlife of tourism and localhood.

Local Huh?

If you have never heard of Localhood or The End of Tourism, the above introduction must have been confusing. In that case, here is the introduction that would probably have been more helpful: In early 2017, Wonderful Copenhagen – the official tourism organisation of the Danish capital region – published its new 2020 Tourism Strategy (Wonderful Copenhagen, 2017). The strategy, which was developed throughout 2016, loudly declared 'The End of Tourism – as we know it' and welcomed a new era of Localhood and people-based growth. In many ways, the strategy invented its own language on tourism to Copenhagen, describing tourists or visitors as temporary locals, closely linked to the Copenhageners – or better yet, to the permanent locals of the city (as described in the introduction of the strategy, Wonderful Copenhagen, 2017). Published in English (and not Danish – which in itself was a controversial and conscious decision), the strategy became part of a global conversation on the future of tourism and destination management.

In this chapter, I will share the case of the Copenhagen strategy, more as a commentary and post-reflection than a detailed and chronological case study of what happened, when and how. My objective is to highlight and differentiate between the original intention and the context surrounding the idea of localhood from what quickly became the meaning given to localhood (by ourselves and our surroundings), as well as the criticism it met along the way.

Who am I to share this? At the time of developing and publishing the strategy in 2016 and 2017, I was the Lead Strategist and Development Director of Wonderful Copenhagen, the Destination Marketing/Management Organisation (DMO) of Copenhagen capital region. As of early 2019, I no longer represent Wonderful Copenhagen, and this chapter should therefore be read as my own epilogue on the strategic work and the afterlife (and in some instances, new life) of localhood.

Strategic Determinants of Localhood

In developing and defining the vision of 'Localhood for Everyone' and the 2020 Strategy for Copenhagen, a number of key determinants framed the context and outlook of tourism to Copenhagen at the time. In the strategy, we included eight of these, entitled 'Road signs to a new beginning'. Here, I am adding a few perspectives to these road signs that perhaps were not explicitly formulated in the strategy, but still strongly influenced the direction and vision.

The rise of the digital traveller. These are travellers who are always on, connected with their social circles and are able to create and access all kinds of information at any time and at any point in their travel. The digital travellers were now largely independent of official recommendations and looking to connect beyond the picture-perfect destination with an authentic and more personal experience of place and people. The rise of the digital traveller in many ways enabled a much broader, mass demand for authenticity, which was reflected in the increased popularity of market offerings to 'live like a local' with Airbnb, or similar services such as, 'eat like a local' or 'do as a local', essentially making accessible an experience of place that was if not entirely, then nearly-authentic and non-touristic. At the same time, this new level of access and demand for authenticity negatively impacted the experiences considered to be 'too touristic'. In short, in the identity play that is tourism, being a **tourist was becoming an undesired identity**, as digital and social media enabled travellers to blend in, to seek out the same bars and hang-outs as locals and even be mistaken for **(temporary) locals** themselves.

The other side of the 'temporary locals' coin was that of **permanent locals**, who were not only becoming a crucial part of sharing the 'destination experience', but also increasingly becoming the answer to a bigger question of: 'why tourism in the first place?' Why attract and develop tourism to the city, if not for the benefit and better quality of life for the permanent locals of the city – the Copenhageners? At the time of developing the strategy, we had not yet entirely grasped the full meaning and extent of this question. But we did foresee the importance of engaging and addressing locals as a core focal point in both process and strategy, as well as to start to talk about the value of tourism in terms different from the past – beyond bed nights and economic growth.

I am not hesitant to claim that today this would have been more extensive (I will get back to this point later), but nonetheless 'localhood for everyone' specifically and especially also meant 'for local residents' and refocused on the value and contribution of tourism to **local livability and well-being**. In the 2020 strategy, this was still primarily contextualised as the support of locals for tourism to avoid negative sentiment towards tourism, but also to ensure their continuous contribution to the destination experience in demand. As described in the strategy:

> Locals are not a nice little sideshow, but, rather, one of the major attractions of a destination. The Little Mermaid offers no emotional or personal connection to the destination, the locals do. The delivery of an authentic destination experience depends upon the support of locals, whereas the liveability and appeal of our destination – and thereby the advocacy of locals – depends on our ability to ensure a harmonious interaction between visitors and locals. (Wonderful Copenhagen, 2017: 5)

While describing locals as attractions, the core value message is actually one of human relations – across both the vision of localhood and the strategic framework; a framework where visitor growth is not the key goal in itself. The new goals are the somewhat vaguely defined around 'people-based growth' and the objective of increasing the value of visitors for everyone. As described in the strategy, we were envisioning: 'A future, where tourism growth is co-created responsibly across industries with the destination's sustainable development and the locals' well-being at heart' (Wonderful Copenhagen, 2017: 10).

From a DMO organisational perspective, we were addressing a shift in role and mission in the strategy as well. This was a shift away from the organisational identity and mission as the destination's promotional superstar, the prime broadcaster of destination branding as **the official destination marketing organisation**. Instead, we were looking to shift towards a stronger focus on enabling human connection, telling the stories through and with others, but also a shift beyond destination promotion towards that of destination management, co-creating and advocating sustainable and long-term value beyond bed nights in a much broader context of societal impact and holistic value creation.

The final, more inwardly strategic determinant was the urgent need for Wonderful Copenhagen as an organisation to rise like a Phoenix in the aftermath of crisis. Following the Eurovision Song Contest in 2014 and a massive budget deficit, Wonderful Copenhagen received a lot of negative media attention and suffered severe reputational damage in the public eye. With an entirely new management team in place, the organisation had been laying low and rebuilding from the inside. At the time of declaring the End of Tourism and welcoming the era of Localhood, this was really a first expression of the new voice of both the organisation and tourism to Copenhagen. In short, the time was ripe for change and Localhood was put forward as a strategy to guide that change.

The End Was Just the Beginning...

Once launched in early 2017, the vision of Localhood for Everyone became part of a much bigger discussion. In the following, I will briefly outline how the vision and strategy were received and how the idea of Localhood developed an unexpected afterlife, reaching far beyond Copenhagen.

- **First globalhood, then localhood.** When first launched, the strategy got quite a bit of international attention with its loud declaration of 'The End of Tourism…'. While shared and presented extensively in several different contexts across Copenhagen and Denmark, the vision was first primarily adopted among peers (other DMOs) internationally. For example, Sedona, Arizona in the USA implemented an 'End of tourism as we know it' strategy (Sedona, 2022). My assumption is that such DMOs were sensing the same shift, which the strategy and vision of Localhood put into words. Local adoption was slightly slower within Denmark, but happened over the coming years, where Localhood became a reference across the tourism stakeholders of Copenhagen and in several instances, it was also incorporated into their own strategies. This included hotels that integrated initiatives to also become places for locals to hang out, with neighbourhood clubs or running-with-locals initiatives. It was also widely adopted by the municipality of Copenhagen, who run the visitor service – and even outside Copenhagen in other parts of Denmark, where Localhood has now become an integral part of the tourism vocabulary.

A simple reason why the conversation around localhood went beyond Copenhagen and became international was the language of publication. We made a conscious choice to publish only in English, with no parallel Danish version. We wanted it to become part of a global conversation on the future of tourism, not just a Danish one.

- **Part of a new vocabulary for tourism.** In the beginning, we jokingly set an objective that Localhood should become part of the Oxford Dictionary's 'word of the year'. While that didn't happen (or at least, hasn't happened yet), the word was widely adopted across destinations globally. I have seen it applied in multiple other DMO strategy documents, content strategies and also used by private companies, most ambitiously with the launch of Localhood.com (by Crowdriff) as a community platform to engage local experts and ambassadors to share their favorite places and experiences (see Localhood, n.d.).

  But, even more importantly, localhood became part of a new vocabulary for discussing and describing tourism throughout 2017 and beyond. In the summer of 2016, US think tank Skift first coined the term 'overtourism'. In the summers that followed, this was among, if not *the* dominant summer discussion about tourism in Danish media. Quickly, localhood was seen as Wonderful Copenhagen's proactive response to this issue – as a way to ensure dispersal, inspiring visitors to go beyond the beaten path, but also to engage and ensure support for tourism from local residents. And while this was definitely part of the rationale behind the vision and strategy at the time of writing it, we probably had not anticipated that this would become more or less the sole framework of discussion in which to unfold the vision of localhood. This also gave Localhood as an idea and concept an afterlife that related much more to the one strategic coordinate of people-based and community-based tourism development, rather than any other of the strategy's strategic coordinates (the four other coordinates were: '*Once attracted, twice valued*', '*Shareability is King*', '*Tomorrow's Business Today*' and '*Co-innovation at heart*' (Wonderful Copenhagen, 2017).

- **Localhood as the solution and a problem.** Upon re-reading the strategy document again, I was pleased to be reminded of the insistence that the strategy itself would not be a definitive solution, but rather setting a course towards the future we envisioned. Localhood was not the cure. Localhood was a desirable future state, balancing and mutually reinforcing the livability and visitability for which we were striving. Nonetheless, one of the key criticisms of localhood was that rather than a solution, it was part of the problem – or at the very least, accentuating the problem. Often simply referred to as localhood – and not the full vision of 'Localhood *for Everyone*' – the vision was criticised: firstly for encouraging tourism to local neighbourhoods and communities not accustomed to, not prepared for, nor interested in welcoming (more) visitors; and secondly for commodifying what is local, while not fully or extensively providing answers to the ways in which the permanent locals, the residents of Copenhagen, were given voice or priority.

It is not my intent to defend the strategy nor the vision of localhood against this criticism. I think both add an important point and perspective to the implementation and afterlife of localhood. I will reflect on that as part of the implementation section below. But the criticism did reflect the more popular interpretations of the strategy and simple definitions of localhood: namely localhood as 'going local' – a display of authenticity made accessible and experiential to those otherwise not included in the shared narrative. These interpretations often overlook the fact that the vision also included 'for everyone' and that the balance was an important part of the desirable future envisioned. Meanwhile, the strategy document did not address the extent to which permanent locals would be given voice, would be involved in co-creating a shared experience of localhood and have influence on the livability of place. I think this analysis is right, and I believe the afterlife of localhood (Localhood 3.0) would need to be much more ambitious and clearly inclusive to a broader extent than reflected in the 2017 document we produced.

Easier Said than Done: Implementing Localhood

From the initial launch and receipt of Localhood as a new strategic vision, we went to work in implementing this across the DMO as an organisation, but also in our approach to working with stakeholders, both existing and new. Looking back, I will say that implementing a strategy like this and the vision of localhood was not without challenges, and in some respects (though certainly not in all) it was a little ahead of its time, mindset and skillset – both within and outside the DMO.

Among the approaches of implementation, and the challenges faced in doing so, there were three key learnings.

- **Implementing by projects and programmes.** About one and a half years into the new strategic period, a colleague approached me to ask: What are we actually doing to implement this new vision? I was taken aback, because I felt we had been doing little else other than implementing the new vision and strategy. True, we did not implement a large organisational re-structuring; departments were still called the same and as such, with or without the End of Tourism, Wonderful Copenhagen as an organisation still felt much the same. Instead, we took an approach of implementing by projects and programmes. We were testing and experimenting with new approaches, new partners and new skills in targeted projects that were intended to bring us forward along the course of localhood. There were a variety of projects. Firstly, was the **TourismX project**, a big collaborative project to make tourism and the visitor economy an appealing playground for innovative, forward-pushing start-ups. Out of the initial phases of these efforts came startups that adopted the localhood idea, among others in enabling visitors to venture 'off-the-beaten-track' and seek out the less-visited neighbourhoods of Copenhagen. Secondly, the **Tourism + Culture Lab** was an experimental collaboration to make tourism more valuable and instrumentally beneficial to the cultural institutions of Greater Copenhagen. Prior to this project, tourism had been branded almost as an invader that came with the risk of carnivalising the cultural institutions to fit into the experiential ambitions of tourism proponents. But with this project, the approach was to put culture first and experiment with the best approach with which tourism could support and empower the cultural institutions' existing priorities and goals. In addition, following the first years of the strategy, Wonderful Copenhagen launched a sub-strategy for sustainable tourism – **Tourism for Good,** to expand on the idea introduced in the Localhood vision around sustainable and people-based growth. Tourism for Good added four concrete focus areas and related these to the specific sustainable development goals, which Wonderful Copenhagen had committed to supporting. Finally, as one of the major follow-up projects on the Localhood vision, Wonderful Copenhagen developed and implemented the **10XCopenhagen project** (10xCopenhagen, n.d.), an extensive research project which sought to address much more broadly the local voice on tourism to Copenhagen and local resident perspectives on tourism. The purpose was not to identify how we could grow tourism to Copenhagen by tenfold, but rather how we could completely rethink tourism and the visitor economy of Copenhagen towards 2030.

- **Missing mandate to manage.** It became clear in the research of 10xCopenhagen that the balance envisioned with 'Localhood for Everyone' towards 2020 was challenged by the continued growth and ensuing increase in hotel capacity planned for Copenhagen. At the time of publishing the 10XCopenhagen research in early 2019, the increase in room capacity was expected to be a massive 49% by 2021 compared with 2018; almost half of which would be located within the inner-city areas. This made clear the challenge of achieving the 2020 Strategy's ambitions: shifting from a *Destination Marketing* to a *Destination Management Organisation.*

This is not a challenge exclusive to Copenhagen, but rather one experienced by DMOs across most of the world; because a critical question to consider is: what does the DMO actually have a mandate to manage? Decisions to approve new hotel investments and capacity expansion usually do not involve the DMO. DMOs are also typically not involved in decisions concerning the broader strategic planning of a place, whether rural or urban, regional or municipal. While the DMO often stands first in line to take the blame with regards to issues of overtourism, the organisation is very often limited both in mandate and capability to make the changes required. The DMO response therefore often limits itself to that of marketing over management: marketing off-the-beaten-track places, off-season visitation or simply 'turning off' marketing altogether. As such, taking on the management role as a DMO requires a much bigger change in capability within the DMO (i.e. in terms of data management, usage and understanding), a mindset change among important stakeholders within the visitor economy in terms of the key measures of success (from visitor growth to community and ecosystem well-being), as well as a different degree of involvement of the DMO in strategic decisions, which again begs the important question of the DMO's mandate to take part in such strategic discussions. We need to consider whose interests does the DMO represent: those of the tourism industry, the visitor economy or those of local residents? And if the latter, how does the DMO ensure a mandate to represent the public?

- **Localwashing vs Local Watchman.** There were and are critical tensions in the Localhood strategy. The Localhood vision was criticised, particularly in academic circles, for its uncritical invitation of visitors into the homes of locals and for commodifying the idea of local. The strategy was aiming to bring visitors to more local experiences – with the risk of being at the expense of local quality of life and well-being. While Localhood was envisioned to be for everyone, the strategy itself weighed the visitor experience if not higher, then equally high with that of the residents' own experiences of Localhood. The ambition was to keep residents happy with tourism and the objective set in the strategy was to ensure resident support for continued visitor growth. Today, based on the years that followed, I personally believe this to be insufficient. It is not about what residents can do in support of tourism; but instead, it is about what tourism can do in support of residents, local lives, community priorities and ecosystem well-being. The strategy does not clearly reflect the consequences of this balance, nor does it describe the concrete approach to involve residents in ensuring the future balance of people-based growth or 'Localhood for

Everyone'. We still referred to 'our shared destination', while in essence DMOs were promoting places where people live. With that follows an important responsibility of representation – in terms of *defining localhood* and in terms of *representing that localhood*.

(1) **Defining localhood**: In offering the 'local version' of a place, the DMO must be aware of the danger of 'localwashing' and commodification of authenticity. By localwashing, I mean the temptation to use the attractiveness of the local for marketing value without the real commitment to a change in values (as 'greenwashing' has occurred in ecotourism). In simple terms, destination marketers are not just responsible for attracting more people to a place with whatever stories work the best, but for adding value to a place and to the lives of people already there. Representing the 'authentic version of somewhere' involves representation of the values of place and the culture and diversity of its people. The narratives offered, sharing the values and localhood of place, cannot be defined by the DMO alone, but rather by giving voice to the people who live and have lived there. This goes beyond using a few local Instagram influencers, who often (and no doubt wisely) are used by DMOs in their neighbourhood or 'go local' content strategies. The Freedom Campaign of Helsinki offers an example of such, where firstly the core value of freedom was identified through research and surveys among residents, and secondly defined in nature and value narratives through resident involvement and representation (see Helsinki Freedom, n.d.). Coming out of the pandemic crisis, we see more places and destinations reclaiming their own narratives, taking them back from visitors and destination fantasies. This is occurring not only in narratives, but also in broader promotional activities, enabling and empowering the inclusivity of tourism-related activity within the destination. And so, it is now time to talk about the democratic imperative of authentically representing identity and values of people and place, and thereby adding value to place and community.

(2) **Representing localhood:** The DMO promotes places where people live; thus, the responsibility of representation is beyond that of promotional inclusivity and value definition. It is also about broader involvement and empowerment in defining the priorities of place and the contribution of tourism to those priorities. Tourism is a phenomenon in the public domain and shared space, and the whole idea of a destination is founded on the identity of a place, carried by the people who call it home. This means that a destination can welcome tourism, but tourism cannot claim the destination, its resources, culture, people or space. In many ways, the business of tourism requires a license to operate – a destination contract, if you will, or a mandate – from the people who live there. This is what I see as the core of Localhood 3.0 (as described further below).

From Localhood to Local Good

Even in hindsight, I think 'The End of Tourism' strategy and the vision of 'Localhood for Everyone' pushed forward an important discussion at the time of their launch and in the years that followed, even if that discussion turned out to be more about overtourism and sustainable tourism development than we had the foresight to expect. The vision and framework of the strategy probably even saw a surge in relevance as the pandemic hit and the absence of tourism made local connection, local relevance and community contribution of the DMOs an even higher priority. Localhood was given new meaning in this context and was adopted again beyond Copenhagen. Nonetheless, as an epilogue written almost exactly five years after the final comma was added to the strategy document, I am envisioning a new Localhood 3.0 version to push the discussion even further – a discussion that should ambitiously and extensively involve and empower the people of place (whether Copenhagen or beyond) in defining the vision and priorities of tourism development and contribution.

No longer a representative of Wonderful Copenhagen, I define Localhood 3.0 as 'DMOcracy' (Group NAO, 2021). This describes an era in which we shift from involvement as primarily centred on residents' acceptance of tourism and motivated by the negative ambition to avoid tourism discontent, to instead involvement as the actual empowerment of the local residents. Empowerment here refers to empowering local residents to actively shape the future of tourism to the place which they call home. Through this empowerment, we reinvent: the social contract and tourism businesses' licenses to operate; the value creation and contribution of tourism; and essentially the mandate of the DMO to invite and develop tourism to the destination. It is less about marketing versus management and more about building an entirely new governance model and structure and it is about the legitimacy and credibility of the DMO as part of this structure.

In conclusion, it is time to start a new discussion on Localhood as the empowerment of a more inclusive and democratic destination governance, developing the accountability and legitimacy of the DMO. In the end, it is not the customer who is king. Nor is the visitor economy a fantasy kingdom that serves at the temporary pleasure of the visitor. Tourism and tourists' access to the places we have called 'destinations' is not a right but a privilege. That privilege is best earned by involving the local community and ensuring that tourism supports the well-being of both people and place. The Localhood strategy discussed here has started important work on this pathway and DMOcracy is set to take it even further (see case study that follows). With these stronger foundations, we can build tourism futures that are more democratic, sustainable, widely beneficial and thriving.

References

Group NAO (2021) Time for DMOcracy. See https://groupnao.com/time-for-dmocracy/ (accessed 4 January 2022).
Helsinki Freedom (n.d.) See https://www.myhelsinki.fi/en/work-and-study/helsinki-freedom (accessed 4 January 2022).
Localhood (n.d.) Empowering local storytellers & passionate creators. See https://localhood.com/ (accessed 4 January 2022).
Sedona (2022) The End of Tourism as We Know It. See https://visitsedona.com/sustainable-tourism-plan/the-end-of-tourism-as-we-know-it/ (accessed 30 January 2022).
10xCopenhagen (n.d.) Rethinking Tourism in Copenhagen Towards 2030. See https://10xcopenhagen.com/ (accessed 4 January 2022).
Wonderful Copenhagen (2017) The End of Tourism as We Know It. See http://localhood.wonderfulcopenhagen.dk/wonderful-copenhagen-strategy-2020.pdf (accessed 4 January 2022).

Case Study: Transforming Relations Between DMOs and Communities

NAO Launching: Time for *DMOcracy*

22 Destinations in new joint initiative: Time for DMOcracy – a new collaborative project and a curious journey into citizen activation and empowerment, the challenges, and imperatives of dialogue, power-sharing and new modes of governance in tourism development. The project is relevant to any destination – urban or rural, national or regional – that wants to empower their local communities in the future of tourism and who puts value to the shift from tourism as a goal in itself to tourism as a means to build better societies and communities and increasing the liveability of the people, who live there.

The initiative is developed by Group NAO and launched in association with Global Destinations Sustainability Movement, The Travel Foundation, TCI Research and European Cities Marketing. Over the next 12 months, the partners and the so far 17 participating destinations will deploy a wide range of research, case studies, master classes, learnings labs, boot camping and conferencing before delivering a white paper on public engagement in tourism in the early fall of 2022.

What We Want To Achieve

The destination and project partners have signed up with four shared objectives:

- **To unfold the meaning and practice of people-based tourism in the city:** Many city agencies and DMOs talk about community involvement – this project will explore best practice and methodologies to put action behind words and engaging citizens for not just better tourism, but better local quality of life.
- **To map citizen participation and involvement in tourism:** The project will map existing citizen-involvement and participatory models in relation to tourism, develop a typology of current approaches, and discuss the role of the DMO and the challenges involved.

- **To understand democratic mandate and participation:** The project will identify new ways of empowering people-based, democratic tourism development and destination governance, including methods that diversify participation and empower actual influence and decision-making.
- **To prepare and upskill the DMO for new modes of governance**, and what this means in terms of functions, skills, and accountability.

Why DMOcracy? Why Now?

There are **four reasons** why now is the time to explore DMOcracy:

First, *the need for a destination contract*: Unless you are Disneyland, destination marketing is about selling experiences in places where people live. Neither the tourism industry nor the DMO can claim ownership of the destination.

Tourism is a phenomenon in public domain and shared space, the destination is founded on the identity of a place – carried by the people, who call it home. This means that a destination can welcome tourism, but tourism can't claim the destination, its resources, culture, people or space. In many ways, the business of tourism requires a license to operate – a destination contract – from the people, who live there. With *Time for DMOcracy*, we want to identify governance models that build trust and accountability – models that reflect and respect the real ownership of the destination.

Second, *it's who we are*: In promoting a place where people live, DMOs hold a special responsibility to that place. Destination marketers are not just responsible for attracting more people to the place, but for adding value to it. Coming out of the pandemic crisis, many places and destinations are reclaiming their narratives – taking them back from visitors and destination fantasies. And so, it is now the time to talk about the democratic imperative of representing identity and values of people and place, adding value to local community.

Third, *hospitality vs hostility*: Destinations are getting ready to welcome back visitors, but the return of tourism is not necessarily met with enthusiasm from locals. From an era, where tourism was increasingly perceived as invasive by residents, to a period of no tourism, we are most likely now looking ahead to a time of increased local sensitivity to the return of tourism. There are indicators already that resident sentiment will not exclusively pivot to wide-open and hospitable cities in the hope of fast recovery. Instead, it is likely that we will see elements of visitor-phobia with demands for restrictions and regulation.

This raises the urgency of a continuous involvement in shaping the long-term accountability and sustainability of tourism development over the coming years. Well-designed resident sentiment surveys are a good place to start, but there is a need to activate the data with open conversation and real involvement and actual influence on the issues raised.

Fourth, *M for mandate to bridge the great disconnect*: For years, DMOs have discussed what the M stands for – balancing between the role of marketing and management. In finding this new balance, new disconnections sometimes emerge between the main stakeholders of the visitor economy:

(1) political decision-makers (municipal, regional, national);
(2) the commercial tourism industry stakeholders;
(3) the local community and citizens;
(4) the DMO – often in the no-man's-land in-between.

Liveability Over Visitability

The disconnects are potentially worsened by the pandemic crisis. Pressure from industry associations to rapidly regain tourism growth might face opposition from hesitant local populations that seek liveability over visitability. The success of the DMO is thereby measured by its ability to ensure continued growth, while simultaneously balancing this with its ability to mediate the gap between the different sides, without the actual mandate to do so. Hence, the DMO risks losing support and license to operate from all sides. That's why now is the time for DMOcracy – to build trust, accountability, and legitimacy through popular mandate.

In conclusion, it is time to start **a new discussion on how we can empower more inclusive and democratic destination governance**, developing the accountability and legitimacy of the DMO in reflection of the many complex interests at play.[1]

From: https://groupnao.com/time-for-dmocracy/

Note

(1) There is also now a North American parallel project conducted with Miles Partnership (Signe Jungersted, pers. comm., 30 January 2022).

9 The Story of Cambodian Children's Trust: Evolving Development Practice From 'Doing For' Communities To 'Doing With' Communities

Tara Winkler

Introduction

The Global North's International Development Sector is a trillion-dollar industry (UNCTAD, 2014), yet it has yielded limited results in alleviating poverty in the Global South. More than 20 years have passed since world leaders established the Millennium Development Goals, focused in part on eradicating extreme poverty. The United Nations (UN) claims that we have made great progress towards the newest iteration, the sustainable development goals (SDGs), with the proportion of people living in extreme poverty halved at the global level (UN, n.d.). However, recent research shows that the international poverty line, which is set by the World Bank at $1.90 per day, at 2011 Purchasing Power Parity, is not sufficient for basic human health, or even survival (Hickel, 2019). This is a very low bar when it comes to aspirational aims for a more just and equal world. In order to achieve a normal human life expectancy of just over 70 years, people need about $7.40 per day (Edward, 2006). But even this is likely to be inadequate for people to achieve the full spectrum of basic human rights to food, water, shelter, health, education and child survival (Hickel, 2019). And yet, even at a meagre $1.90 per day, we are set to fall short of the SDG's first target (UN, 2021). Furthermore, the number of people living in extreme poverty today is about 1 billion, which is exactly the same as it was when measurements began in 1981. There has been no improvement in over 35 years (Hickel, 2017). This is consistent with the realities I've seen on the ground in Cambodia over the last 15 years, where families struggle to care for their children and communities remain dependent on external aid.

My work as co-founder of a grassroots child protection organisation called Cambodian Children's Trust (CCT) has granted me front-row seats to the confronting realities of the international development sector. CCT has been the recent focus of an academic study (Higgins-Desbiolles *et al.*, 2022); however, feedback on this work indicated the need for a more in-depth, personal narrative of CCT's evolution, which I offer here. By narrating my experiences working through CCT, I intend to show how development agencies perpetuate a system of dependency rather than achieve development goals. This occurs due to preferencing a 'downstream' approach that responds to symptoms of poverty rather than enabling an 'upstream' approach that addresses the structural root causes of poverty and places power in the hands of communities.

This chapter will narrate how I went from being a volunteer tourist in Cambodia to a co-founder of an orphanage that eventually transitioned into the CCT of today – one which models how organisations can work in solidarity with local communities and support them towards self-determined development. CCT evolved its practice from 'doing to families' to 'doing for families' and finally 'doing with families and communities'. I will contextualise this work in terms of a critique of downstream development practices and ideologies and turn to Paulo Freire's *Pedagogy of the Oppressed* (1970) for a theoretical basis for thinking through contrasting practices of working together in solidarity. This analysis shows that maintaining dependency on development agencies is not accidental and that fundamental shifts in power are needed to place communities in charge of their own development pathways. In the final sections of this chapter, I will explain how CCT has accomplished the transition to its upstream Village Hive approach, demonstrating an effective model for the localisation and decolonisation of international development practice.

My Start in Development: 'Doing to' Families

My journey towards establishing CCT began in 2005, when I was 19 years old, with a holiday to Southeast Asia. The first part of my trip started with a guided tour through the well-trodden loop of Thailand, Laos, Vietnam and Cambodia. In Cambodia, we went to see the Killing Fields, the Angkor Wat temples and a large orphanage in Siem Reap. We watched the children perform a traditional Apsara dance and made a donation, naively believing that we were being ethical tourists and doing a good thing to help disadvantaged children.

After the tour ended, I extended my holiday in Cambodia for another month and set off to explore northwest Cambodia. Battambang was a sleepy little riverside town with remnants of colonialism in the beautifully dilapidated French architecture. Away from Siem Reap's tourist bubble, I was less insulated from Cambodia's endemic poverty, which left me feeling uncomfortable about my overt privilege as I sipped cocktails by the hotel pool. I was so taken with my first visit to the orphanage in Siem Reap that I wanted to recreate that experience. I bought some clothes, books and toys from the local market to donate to several orphanages in Battambang.

One of the orphanages I visited, named Sprouting Knowledge Orphans (SKO), was desperately poor. They didn't have funds for enough food, clean water or healthcare. Some of the children were HIV positive and didn't have access to treatment. As a privileged white girl from the eastern suburbs of Sydney, this was very confronting. The director asked me to stay and volunteer, but I felt overwhelmed and promised to help raise funds for the orphanage when I returned to Australia. Back in Sydney, I ran a fundraising event and raised AUD20,000. Like many donors in the Global North, I didn't trust that the money would reach the children if I sent it from Australia, so I planned a trip back to Battambang. I spent three months volunteering at SKO orphanage, overseeing how my donation was spent. I set up water filters, hired a nurse, got the kids vaccinated, and took them to the dentist for the first time in their lives. The children were so hungry for connection and attention that I spent a lot of time showering them with love and playing the role of substitute parent. I can now see how problematic my mindset was back then, but I was young and naive, and all the feedback I was getting from friends and family reaffirmed that I had achieved something profound and important.

I continued supporting the orphanage when I returned to Australia with the help of a local man, Pon Jedtha. Jedtha had been a monk for 17 years and had disrobed to take a job at SKO orphanage so he could do more to help children in his community. I set up a reporting system with Jedtha and continued raising funds to support the orphanage from Australia. Over the six months that followed, I received some concerning messages from Jedtha and the kids informing me that there was trouble brewing in the orphanage. When I returned to SKO orphanage, Jedtha told me that the director of SKO was corrupt. He had been embezzling every cent donated to the orphanage from foreign donors and, as a result, the children were suffering such gross neglect that they were forced to catch mice, bullfrogs and pest fish to feed themselves. I also found out later that the director had been physically and sexually abusing the children. I couldn't bring myself to turn my back on children whom I had come to know and care about, so I worked with Jedtha and the local authorities to set up a new orphanage in Battambang. In August 2007, we rescued the children from SKO orphanage and gave them a safe, new home in CCT's orphanage.

Between 2007 and 2011, CCT operated a well-resourced orphanage with therapeutic programmes and a very high standard of care. However, despite our best efforts, the care plans we made for the children had limited benefit. Attachment disorders had developed in the children due to the traumatic separation from their parents. These disorders had been exacerbated by a constant rotation of staff and volunteers coming in and out of their lives. Worse still, living in an orphanage robbed the children of the most important resource they needed to grow into healthy and well-adjusted adults – a responsive parent who is devoted to their well-being. In almost every case, we found that young adults who had grown up in the orphanage could not transition into independence and remained reliant on welfare from CCT into adulthood.

Later, when I had learned to speak Khmer fluently and could communicate properly with the kids, I came to the rather startling discovery that the children we had rescued from SKO orphanage weren't orphans at all. All of them had parents and relatives still alive. My initial reaction to this discovery was that perhaps these children were better off without their parents. I couldn't imagine how a parent could abandon their child to an orphanage. However, once I met the families of the children, I realised how wrong and prejudiced my judgements were. I now understand that this thinking stemmed from a place of privilege and unconscious bias. I was raised to be appalled by racism. My grandma was a Holocaust survivor, and so the scars of racial prejudice are imprinted in my DNA, but the problem with privilege is that it's blinding. We are socialised not to see the ways in which unjust systems benefit some people and harm others, placing them in impossible situations where they are forced to make impossible choices. It was clear the parents loved their children, but they had entrusted them into the care of SKO orphanage in the hope that it would lead to a path out of poverty to a better life.

During the years we operated the CCT orphanage, our approach to helping vulnerable children overlooked and excluded the adults in their lives. We did it that way because the international development sector perpetuates downstream systems designed to 'save' children who are being harmed by investing the lion's share of funds and resources into end-stage crisis interventions, like orphanages.

Understanding Development: Investing in Downstream Systems

At first glance, a downstream system appears logical. If children are being harmed, it seems rational to focus all efforts on getting them out of harm's way. If the number of children in harm's way grows, it makes sense to double down and pour even more funds into crisis interventions to save them and help them heal from the trauma they've sustained. There is, however, a grave flaw in this design. If families cannot access universal prevention services, like health, education, housing and sanitation, they are plunged into a cycle of multidimensional poverty. Living in multidimensional poverty creates an environment of chronic stress that makes the job of parenting extremely difficult. Targeted early intervention services, such as case management, counselling, home nursing, parenting skills and financial literacy, help ameliorate chronic stress and equip parents to meet life's challenges and care for their children. Therefore, if communities lack early intervention services, problems grow into intractable crises such as domestic violence, abuse and neglect. It's at this point that children fall into 'the system' and require crisis interventions.

If NGOs and development agencies neglect to fund prevention and early intervention initiatives and instead pour all their funds and resources into downstream interventions, a self-reinforcing dynamic is created that causes increasing numbers of children to require crisis services and alternative care. When children are separated from their parents and placed in alternative care, it triggers a biological stress response that remains activated until the child is reunited with their parent. High and persistent levels of stress disrupt the architecture of the developing brain and other biological systems, with life-long negative impacts on learning and behaviour, as well as physical and mental health (Shonkoff, 2018). Stated simply, the trauma of family separation stays with children for life. But the worst and most damaging aspect of a downstream approach in child protection is that support is available *after* children are already harmed and traumatised. How can we call that child protection?

To illustrate the difference between a downstream and upstream system of child protection (see Figure 9.1), I use this analogy:

Every day, families navigate a fast-flowing river as part of their daily lives. Some families have old boats. When their boats break, their children fall into the river. Some people see the children being washed down the river and over a waterfall. They call an ambulance to provide emergency first aid to the children.

Meanwhile, further upstream, more families' boats are breaking, and more children are falling into the river and being washed downstream. The rescuers call more ambulances to join the escalating rescue effort at the bottom of the waterfall. In this scenario, more and more children are harmed, and more and more resources are poured into end-stage crisis interventions.

Instead, some people walked upstream to find out why so many children were falling in. They saw that families were having problems with their boats and couldn't afford to repair them. They worked with the families to repair their boats to make them strong and resilient. They also built a safety net in the river to catch any children who accidentally fell in. In this scenario, the children were getting across the river safely with their families, and much of the costly rescue effort at the bottom of the waterfall is no longer required.

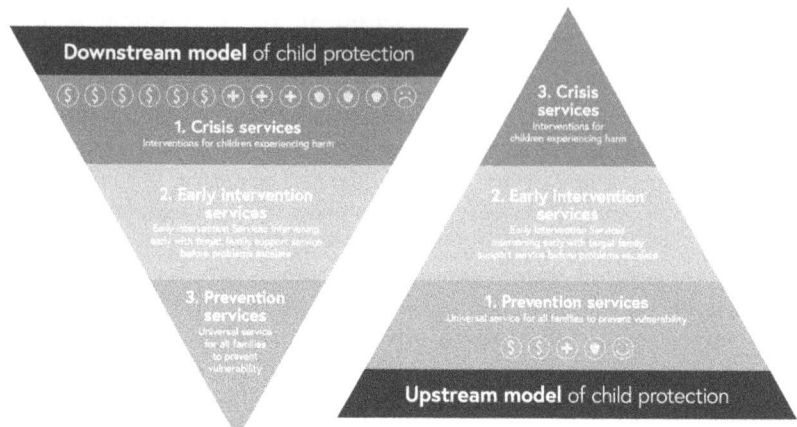

Figure 9.1 A downstream approach versus an upstream approach to child protection in development

A real-world example of the Global North's over-investment in downstream development is the orphanage boom seen in Cambodia between 2005–2010. During these five years, the number of orphanages in the country increased by 75% and the number of institutionalised children doubled (Government of Cambodia & UNICEF, 2011). It is no coincidence that the proliferation of orphanages matches the explosion of tourism in Cambodia; foreign arrivals increased 250% in the same period (Hartley & Walker, 2013: n.p.), and 100% of orphanages in Cambodia are funded by foreign donations (Government of Cambodia & UNICEF, 2011).

The children in these orphanages were being labelled as orphans, even though 80% had at least one living parent and the remainder mostly had other relatives still alive. It may appear as though families make a conscious choice to place their child in an orphanage, but in the absence of prevention and early intervention services and faced with multidimensional poverty, entrusting their children into the care of an orphanage becomes their only choice. Removing children from their family and community of origin so they can access essential services in orphanages may solve one problem, but it creates a bigger one. No matter how high the standard of care, institutional environments are harmful to the development of children. One study has shown that young adults raised in institutions are ten times more likely to fall into sex work than their peers, 40 times more likely to have a criminal record and 500 times more likely to commit suicide (Lumos, 2017b).

The orphanage boom is a phenomenon caused by the Global North that impacted many countries throughout the Global South (Lumos, 2017a). It could have been avoided altogether if the Global North had given sufficient funding to upstream development that addressed the root causes of family separation. Instead, we are left picking up the pieces, trying to reunite traumatised children with their families. In recent years, networks of NGOs, including CCT, have been working together to solve the orphanage crisis and transform the dominant model of care in Cambodia (FCF-REACT, 2019). However, these networks allocate all funds and resources to other downstream interventions like reintegration and foster care. Attempting to transform the dominant care system in Cambodia by shifting from one downstream intervention to another does nothing to address the root cause of child–family separation. It may reduce the number of children living in orphanages, but it increases the number of children in foster care. It can also create other unforeseen problems like incentivising families to place their children in orphanages to access the essential support services through reintegration programmes. If child protection systems fail to address the root causes of child–family separation and merely redirect efforts from one end-stage intervention to another, the only possible outcome is the ongoing perpetuation of harm.

Another example in Cambodia is the controversy concerning the Somaly Mam Foundation. Somaly Mam is the founder of a Cambodian anti-trafficking organisation. From 1996 to 2014, she and young girls in her care appeared on global media platforms including Oprah, PBS and CNN, sharing horrifying stories of trafficking and sex slavery. Millions of dollars were donated to the Foundation to assist with brothel raids and fund an institutional crisis shelter, which became heavily touristed. In 2012, Somaly Mam was exposed in a *Newsweek* article for instructing the girls, who were not in fact victims of sex slavery, to fabricate horrifying stories in order to raise funds (see Marks, 2014). What the *Newsweek* article failed to recognise is that Somaly did what she did because donors in the Global North are obsessed with donating to sensationalist downstream charities, and sex trafficking is one of the most popular and lucrative of all downstream causes (Haynes, 2014; Heynen & van der Meulen, 2021; O'Brien, 2018). The girls in Somaly's residential shelters may not all have been survivors of sex trafficking, but they were undoubtedly poor and in real need of support services that weren't available. It is important to recognise that these aid and development disasters are not the failings of a few immoral individuals; they are the inevitable result of downstream systems.

CCT First Evolution: 'Doing for' Families

I learned about Cambodia's orphanage boom in a report released by the Cambodian Government and UNICEF in 2011. By this stage, I understood that even high-quality institutional care was harmful to children and realised CCT's orphanage was a part of a wider problem. Knowing better comes with a moral imperative to do better, so Jedtha and I began the process of transforming the CCT orphanage and reuniting the children with their families. By 2012, five years after we started CCT, we finally closed the orphanage and shifted to a family-based care model that supported the families of the children we'd reunited. This was a step in the right direction, but we still didn't have it quite right.

Around this time, there was a significant increase in tourists passing through Battambang. With this uptick in tourism, we saw a dramatic increase in the number of children on the streets begging from tourists. Children of all ages were out late at night, hanging around the riverfront where tourists were dining. A concerning number of these kids were also sniffing glue to evoke sympathy from tourists who would then give more money. When we spoke to their families, it was clear they were struggling with complex, multidimensional poverty. Many parents said they were working several jobs to keep food on the table, which often meant their children were left unsupervised. As a result, these children followed their friends to the streets. Many parents said they were worried about their children, but they were having trouble managing their behaviour (see CCT, 2019). The kids weren't going to school and some were sniffing so much glue that they had become seriously ill and malnourished.

We set up a drop-in centre in town in an attempt to help these kids. We had a 'no glue' policy at the drop-in centre so the kids could only come when they were sober. The drop-in centre offered a nutritious breakfast and lunch, toilets and showers to wash and brush their teeth, a clean set of clothes and washing facilities. It also provided remedial tutoring and fun activities like music and art. The drop-in centre opened with just two boys attending, but their friends soon followed. Before long, we had the entire population of 'street kids' regularly dropping into the centre. We'd usually only have about 10 to 20 kids turn up on any given day, but we were building important relationships and gaining a deeper understanding of their world. We were also providing material support to their families, including rental assistance, essential food supplies, healthcare and, for families living in abject poverty, we were providing monthly support payments.

We wanted to get the children off the streets and back with their families, but we quickly realised that running a drop-in centre inadvertently enabled them to stay on the streets and engage in high-risk activities. They would turn up when they were hungry and disappear again when they wanted to sniff glue, so we changed the approach. The drop-in centre became a youth centre with compulsory six-day per week attendance. If the kids didn't show up every day, they were no longer allowed to participate. This was a game-changer. The solid relationships we had developed meant they wanted to continue coming, so they agreed they would quit sniffing glue and attend the youth centre daily. A cohort of children living on the streets told us they couldn't come every day because sometimes they were responsible for looking after their younger siblings while their parents were working. To address this, we set up a preschool and creche within the youth centre, which had the added benefit of setting these infants and young children on healthy trajectories. We also introduced a pick-up and drop-off service to encourage them to go to school and return home at the end of the day. Suddenly we saw dramatic behaviour change. Their health drastically improved, they started putting on weight and they were much more engaged and attentive in classes.

With 80 children attending every day, we outgrew the little four-by-four metre space in town, so we moved to a new location with several classrooms and an onsite kitchen much closer to the slum community where most of the children came from. Within six months, all the children attended school every day and slept at home every night. The wider Battambang community was amazed that the 'street kid problem' had disappeared. The owners of the restaurants were delighted. They assumed we had rounded them all up and put them in an orphanage, as commonly happened to Cambodian children living on the streets. They were surprised to learn that they were no longer roaming around the streets and begging from tourists because they were at home and busy with school.

We ended up opening several more youth centres to support hundreds of families living in slum communities in Battambang. Over 95% of these children remained enrolled in the youth centre for the next decade, with most continuing to high school, vocational training and higher learning. The youth centres proved to be a powerful intervention in these children's lives, and they still play a vital role in preventing child–family separation.

The one problem we had was that the families remained dependent on the youth centre service and the other financial aid from CCT. We had stopped excluding the adults from CCT's interventions, but we were still operating as if we knew what was best for them and their children. We were approaching families as if they were helpless victims who needed 'fixing' and we were 'the experts' who knew how to fix them. We would step in and over them, designing interventions to solve all the problems in their lives. These interventions seemed effective initially, but families remained dependent on CCT. Months turned into years, and the same families were still reliant upon monthly support. When we tried to push some of these families towards independence, the plans we made for them would inevitably unravel. The small businesses we had set up would fail; they would return to live on the streets, and, in some cases, the children would be swooped up by orphanages in need of vulnerable children.

The interventions were never sustainable because they were *our* plans. Families didn't have ownership or agency over the decisions we were making on their behalf, so it was no surprise that these plans didn't make sense to them. We weren't doing things 'to' families anymore, but we were doing things 'for' families, undermining their abilities and cultivating a culture of dependency.

CCT Second Evolution: 'Doing with' Families and Communities

The best way to ensure children grow into healthy, well-adjusted adults is to prevent them from experiencing trauma in early childhood by utilising an upstream development approach. The CCT team and I never set out to create an upstream model of child protection. It grew organically out of a lot of trial and error. Building an upstream model involved redirecting our focus, funds and resources towards prevention and early intervention. However, the real magic began when we realised that the best outcomes were achieved when we partnered 'with' families, working alongside them as allies and advocates towards a common goal. When families were leading the work, co-designing solutions that made sense to them, the transformations were profound and long-lasting.

Once we put families in the driving seat, recognising them as the experts in their own lives, we began seeing families and youth voluntarily exit from CCT's support. This was a profound shift from the hand-to-mouth dependency we had seen previously, but some of these families still struggled when they encountered setbacks down the track and would return to CCT for further assistance. We also realised that some of the most vulnerable children and families in the community were experiencing such significant social isolation that they were being overlooked and never referred to CCT for support.

These findings made us realise that we had major gaps in our model. We needed to find a way to build healthy ecosystems in which no vulnerable child could slip through the cracks and families could become resilient enough to overcome any future setbacks they encountered. In 2016, with the assistance of families, youth, social workers, nurses, teachers, local authorities and the wider community, we co-designed an upstream model of child protection called the Village Hive that successfully mobilises entire communities to protect children and strengthen families.

The Village Hive has two primary activities. It brings prevention and early intervention services into communities and cultivates community networks. Multi-sector services – such as education, healthcare, safe housing, sanitation, collaborative case management, counselling, youth centres, family finance and livelihood coaching – play an essential role in untangling the complex web of challenges that keep families trapped in multidimensional poverty. Village Hive social workers, recruited from within the community, partner with families to co-create case plans. They do this by helping families, and their extended network of kin, identify their existing strengths, think through their problems and arrive at their own solutions. They use a strengths-based, three-column framework called the Signs of Safety (n.d.) and a technique called appreciative inquiry to engage families in self-determined change.

To foster long-term resilience, we knew that it was essential to build healthy ecosystems around vulnerable families through a process of co-creation with the wider community. In Cambodia, the department responsible for delivering social protection services at a commune level is the Commune Council for Women and Children (CCWC), making them a key community stakeholder. We invited all the members of Battambang's CCWC to join a co-creation session with the CCT Village Hive team. We used the same three-column framework to facilitate these co-creation sessions. Together, CCT and CCWC thought through all the strengths and protective factors that kept children safe in Battambang, all the gaps and risks that made children unsafe, and then we workshopped how we could build on the existing strengths to devise an action plan. One of the primary action items that came out of this initial co-creation session was the need to establish community-run Child Protection Networks (CPNs). The primary aim of the CPNs was to give the community a voice in designing and evolving the Village Hive. CPNs would also help map risks and services gaps in their community and ensure that vulnerable children and families had a tribe of voluntary kin around them to provide support and, importantly, make sure that no vulnerable child was overlooked.

Next, we conducted a co-creation workshop with the Commune Chiefs and Village Leaders. We used the three-column framework to think through how we would establish the CPNs to ensure that the whole community was represented and had a voice in the network. Some villages decided that a self-nomination process would work best for their community, holding a quota for people from marginalised groups such as disability, ethnic or religious minorities and the LGBT+ community. Other villages decided to conduct a village election to ensure the most trusted members of the community were voted in. All villages conducted background checks on all new CPN members.

Once the CPNs were established, we then conducted a co-creation workshop with the CPN members to develop the terms of reference for their network, including who would lead, how often they'd meet and the process of referral to Village Hive social workers. The CPN members tailored a risk-assessment chart to the individual circumstances in their community and participated in workshops on child rights, child protection risks and awareness-raising.

The mobilisation of the CPNs resulted in the communities coming together to support one another to protect their most vulnerable and ultimately rely less on institutions and support services. These relationships of neighbours, extended family and voluntary kin have been central to creating a robust system of safety for children. The practical support wrapped around vulnerable families serves essential life functions, such as providing financial and emotional relief and, importantly, a sense of connection and belonging.

CCT's Development Insight: Empowerment is a Shift in Power

The lessons we had learnt from 'doing with families and communities' made us rethink what it means to empower. We realised that empowerment is not a matter of building capacity, as it is commonly understood. Capacity building may, in some cases, play a role, but it's not the endgame. To empower is to *give* power. If the Global North is to assist with the empowerment of the Global South, a radical rebalancing of roles is required. The Global North's role must shift from one of leadership to allyship. These lessons about power dynamics in international development also echoed through other projects we were operating around that time.

In 2013, we had established a social enterprise restaurant called Jaan Bai to provide training and employment for marginalised Cambodian youth. The restaurant was a roaring success, taking the number one spot on Tripadvisor for many years. However, while CCT managed the restaurant, it never turned a profit. Jaan Bai was not established for profit-making, but it was less than ideal that the training the youth received was not equivalent to the best practices of a profitable enterprise. In 2016, we approached the youth working in Jaan Bai with a proposal to transform the business model of Jaan Bai into a profit-share arrangement, with the employed youth all becoming shareholders of the business. They agreed unanimously and were eager to step up to the challenge. The agreement involved taking over day-to-day operations and management. The restaurant began turning a profit in the first year following the handover. The youth were becoming business savvy because they had ownership over the project, were invested in its success and had the opportunity to develop strong leadership skills. A percentage of proceeds was also coming back to support CCT's Village Hive work for the first time.

The other project that reinforced the importance of shifting power began in 2015 after noticing that children in Battambang were graduating high school without any computing skills. We won a grant from the Atlassian Foundation to establish a computer school. However, instead of setting up our own privately-run school, we partnered with a large public high school called Monivong with over 6000 students.

Computing programs delivered by NGOs are notoriously unsustainable. When we first partnered with Monivong High School, we found a graveyard of broken-down computer labs left there by another NGO. To avoid our project suffering the same fate, we co-designed the project *with* the school's faculty and teachers. We made sure it was integrated *within* the school's existing systems and was aligned with the national Master Plan for ICT education. We set up energy-efficient computer labs that were cost-effective to run and supported the school's administration to upgrade their administrative systems, moving them away from analogue record-keeping so they could oversee the programme efficiently. We co-designed all teaching resources, which were delivered from a cloud-based platform. The auto-marking system on the platform, developed by Prof Curran's Grok Learning, meant that teachers didn't need expertise in computing to teach the course. We also trained the students and school administration to manage the ongoing maintenance of the labs.

By 2016, we had co-designed additional courses that gave students an introduction to advanced computing. We then partnered with all four public schools in Battambang and the city's teacher training college to scale the project, so every high school student had the opportunity to gain essential computing skills. By 2017, all four schools and the teacher training college had the knowledge required to deliver the course. Sustainability was achieved when the schools and the teacher training college incorporated the computing course into their budgets, ending any dependency on CCT and Atlassian Foundation funding.

Seven years after funding from CCT ceased, the programme is still fully operational, funded independently by the schools and the training college. Over 25,000 students and over 500 teachers have completed the course, and thousands of students continue to enrol each year. Students from these schools have been the national STEM champions each year since the course began. Some students have even gone on to study computer science and engineering university degrees in Cambodia and abroad.

These projects achieved something rare in international development – true sustainability. The international development sector promotes the idea that solving poverty is a herculean challenge that is all but impossible to surmount. But in my experience, sustainable development isn't as difficult as it's made out to be. All that's required is an upstream approach with a shift from 'power-over' to 'power-with' (Lediard, 2016).

Why Fund Downstream?

It is not a mistake that the Global North is obsessed with funding downstream projects. The system has been designed that way with neo-colonial aims. The 'doing to' approach stems from a subconscious belief that adults in the Global South are inherently immoral and corrupt (King, 1968), and by removing their children, they can be 'saved' from those corrupting influences. This vilification of black and brown adults has led to atrocities throughout history, such as the stolen generations in settler-colonial Australia, Canada and the USA, and is still being perpetrated against those populations today.

The 'doing for' approach stems from a subconscious belief in the incompetence of people in the Global South. This belief helps to justify the Global North stepping in to do it for them – building their houses, digging their wells and saving their children. Treating people as if they are incapable of solving their own problems instils an internalised inferiority complex, which cultivates a culture of dependency. This culture of dependency reinforces the white saviour narrative that 'rescuers' are needed and necessary to save black and brown people from their circumstances.

The Global North encourages white saviourism by offering valuable reputational rewards to people who try to save people from the Global South. White saviours are viewed as morally superior to other white people and are hero-worshipped as living saints. Gaining white saviour status does not require any specific knowledge, expertise or even any success in helping people of colour. All that matters is the superficial intention to help.

There are countless examples of this throughout history. Mother Teresa is the quintessential white saviour. Everyone knows who she is, while few know anything about the work she did. Many are surprised to learn that Mother Teresa's main crusade was to abolish abortion and contraception and convert people to Catholicism (Hitchens, 2012). To save the souls of unborn children, she founded several dozen orphanages. If Mother Teresa's reputation was judged by her actions, I don't think she would be so widely lauded as a hero and humanitarian.

I have also received reputational rewards delved out from the white saviour hero-worshipping machine. When I was still running the orphanage, a documentary (ABC, 2010) was made about me – a young, unqualified white girl – for setting up an orphanage in Cambodia, something I would never have been allowed to do back at home in Australia. From that documentary, I was very quickly elevated to the status of a national hero, winning the New South Wales Young Australian of the Year Award. I was applauded in countless media stories that paid little attention to the work I was actually doing. The Global North has known that orphanages are a bad model for raising children for over 100 years. It was 1909 when Theodore Roosevelt held the first conference on child welfare and heralded the deinstitutionalisation of children in America (Michael & Goldstein, 2010). Today, there are no longer any orphanages in the Global North because we know that they are detrimental to childhood development. Yet, I was being celebrated for starting one in Cambodia.

If empowerment of the Global South is what the international development sector aims to achieve, then we need to unpack the concept. The term 'empowerment' is so overused by NGOs that it has lost its meaning. It is often used to describe activities that 'build capacity,' i.e. the adage of 'teaching a man to fish'. The problem with capacity building is that it infers an inherent *lack* of capacity and is thereby used as an instrument to instil an inferiority complex in the Global South.

One consequence of this focus on capacity building is that it leads to local NGO staff being caught up in incessant cycles of training. Paradoxically, all this training and capacity building rarely leads to any career progression for local staff. This is because an endless cycle of expatriate white saviours continues to assume managerial and supervisory roles. This feeds into a self-fulfilling system of dependency where local staff come to believe that they are never sufficiently skilled or trained for their jobs and thus always dependent on foreign leadership. Whenever I've asked CCT's local staff what additional resourcing they might require for their roles, they predictably answer 'more training'. For better or worse, people tend to live up to expectations. If local staff are never given the opportunity to step into senior roles and cultivate leadership skills, they are left dependent on a never-ending cycle of expatriates.

However, it is an illusion that expatriates from the Global North are more skilled and therefore more suited to leadership roles in the Global South. Foreigners lack an adequate grasp of local culture and the unique strengths that exist within local communities. They often cannot speak the language fluently, so English is frequently adopted as the dominant language, placing local staff at a disadvantage in their own country. There is simply no accounting for how much time, funds, meaning and progress is lost through the process of translation. Despite these shortcomings, expatriates are paid anywhere between 2–20 times as much as their local counterparts. The international development sector justifies this overtly racist double standard by invoking the ideology that people in the Global South are inherently less capable and therefore less valuable.

In the Global North, white saviours are viewed as altruistic, self-sacrificing people who act with the best of intentions. However, the motivations of expatriate staff in NGOs are rarely purely altruistic. Expatriates may be on a lower salary than they were in their home country, which gives the illusion that a major sacrifice has been made. The living allowance for volunteers in the Australian Volunteers International (AVI) programme, which is supposed to emulate a 'modest local lifestyle' (AVI, 2019), is more than eight times the salary of an average factory worker in Cambodia (Sovuthy, 2021). This means that expatriate salaries provide a standard of living that far exceeds what most could afford in their country of origin. Renting a luxury villa, enjoying frivolous consumption and hiring personal staff, including drivers, maids, cooks and nannies, is part and parcel of the expatriate lifestyle. Such a blatant demonstration of white privilege while working to alleviate poverty suggests expatriates benefit from the oppression they are supposed to be combating.

CCT's Shift to an All-Khmer Leadership Team In-Country

Following the closure of the CCT orphanage in 2012, CCT went through a period of growth, emulating the most established and prominent NGOs in Cambodia. This meant that we went from a small Khmer team to an organisation with expatriates in all management roles and technical advisors to assist with the capacity building of the rest of the local team. For several years we had highly skilled expatriate staff in every department of the organisation, including social work, medical, youth centres, communications, human resources and finance. Most of our expatriate managers were on one-year contracts, so we spent a lot of time recruiting. It would take several months for new expatriate managers to settle in, by which time we would be recruiting again for their replacement. It wasn't just time we were wasting. The expatriate staff were also paid twice as much as their local counterparts. I am embarrassed that it took me so long to question this overt racism.

In 2019, after I had gone through an intense period of self-examination and reflection regarding the ways in which I was unconsciously complicit in propping up oppressive neo-colonial systems in my personal and professional life, CCT developed a strategic plan to transition to Cambodian leadership by 2022. We drafted an affirmative action policy mandating only one salary scale for the organisation and rules for recruitment to ensure that local staff were given first preference for all roles and all expatriate staff required sufficient justification, with supporting evidence demonstrating that the specific skill set is unavailable in Cambodia. This was a big deal for me personally. I had spent my entire adult life in Cambodia. I spoke Khmer fluently and, in many ways, felt more at home in Cambodia than I did in Australia. Stepping aside in 2022 to allow CCT to be wholly led by Cambodians meant giving up the part of the job I loved the most. But CCT and its mission are bigger than me. The evidence I had seen was clear; shifting power to local leadership resulted in more efficient and effective outcomes. The plan was that I would focus on building a more sustainable fundraising body in Australia to support the Khmer team to scale up local grassroots innovation. This strategic plan gave us a three-year window to prepare Jedtha and our Khmer staff to take the reins.

At the time, I thought a three-year window was an ambitious goal, and I had some reservations. I knew Jedtha deeply understood the work. Like me, he had lived and breathed CCT since its inception. However, by nature, Jedtha is one of the most gentle, relaxed and easy-going people I've ever met. These are wonderful qualities that I admire greatly, in part because I am the exact opposite with a quintessentially Type-A personality. These differences in temperament, coupled with colonial racial dynamics, meant that Jedtha would never have stepped over me to assume the role of executive director. In all honestly, I wasn't sure whether he was up for the task of leading on his own.

Under the affirmative action policy, CCT's local team thrived. They jumped at the opportunity to step into leadership positions. In the departments led by Cambodians, operations began working much more efficiently and effectively. By the beginning of 2020, we only had four expatriates on the team – Keir Drinnan and myself working alongside Jedtha as directors, Samantha Besgrove in the communication team and Erin Kirby working as a technical advisor to the social work team. We kept one technical advisor to oversee the social work because we had struggled to complete an overhaul of our case management process. We had moved away from the paternalistic approach of 'doing for' families but still had hundreds of old 'set and forget' cases on our books. These were families who had been accessing Village Hive services and receiving support payments for years with no plan to transition to self-reliance. We had hired more than ten highly qualified expatriates over the years to transform our case management system and support families in setting their own goals and working towards independence. None of the expatriate experts from Australia, the UK and America achieved this goal.

Then, in 2020, the pandemic hit. I happened to be in Australia renewing my passport when the Australian borders closed. I decided to bring Keir, Sam and Erin back to Australia as border closures would have left them stuck in Cambodia away from their families. So, just like that, our 2022 plan was brought forward to 2020.

The ability to effectively lead, motivate and direct a group of people requires complex skills, mainly acquired through hands-on experience. Over the years that Jedtha had been leading CCT with me, he had quietly cultivated these skills, and when he stepped into the role of sole director of CCT, an extraordinary leader emerged.

Within a year of being the sole director of CCT, Jedtha had achieved some astonishing feats. First, he navigated CCT through the operational pivots required to respond to the community's changing needs through the COVID-19 pandemic. With schools closed for a total of 14 months, underprivileged children were unable to access home learning and were at risk of falling so far behind that they would have been forced to drop out of school. Under Jedtha's guidance, the CCT team acted as a conduit between students and their schools by hand-delivering all home-learning materials, assisting with home tutoring and ensuring year 12 students had access to computers to sit the exams. Throughout the pandemic, CCT also coordinated the delivery of life-saving equipment to the Battambang public hospital and vital personal protective equipment materials, emergency health care and vaccinations for thousands of families.

Second, in addition to navigating through a pandemic, Jedtha successfully overhauled our case management system. He did it by emulating the process himself. He sat with Village Hive social workers and families and patiently guided them through the practice of goal setting, utilising appreciative inquiry with the Signs of Safety framework. By supporting families to set achievable goals, Village Hive social workers now help families achieve robust and long-lasting resiliency.

Seeing Jedtha flourish in the Executive Director role revealed another blind spot. I had underestimated his ability and the ability of the entire Khmer team. I assumed that paternalistic hand-holding was required to achieve a successful and smooth transition to local leadership. The Khmer team have proven that not only have they got what it takes to lead CCT, but they are also the best people for the job, and Jedtha is the best person to lead them.

CCT's Next Evolution: Shifting the Power by Working Within Local Systems

The lessons we'd learnt about doing 'with' and not 'for', in a spirit of allyship and not leadership, set the foundation for the next phase of CCT's evolution. It is here that Paulo Freire's *Pedagogy of the Oppressed* (1970) and his articulation of mutuality presents a path forward for working collaboratively for development. His theorisation of conscientisation (conscientização) is key. Conscientisation refers to the mutual, critical awareness work that the oppressed do in understanding their conditions. It leads to a determination that 'the concrete situation that begets oppression must be transformed' (Freire, 1970: 50). The fundamental basis of this work is the appreciation of the oppressed's own knowing and abilities to transform their situation (Freire, 1970: 68). Freire explained that this work is achieved through 'praxis: reflection and action upon the world in order to transform it' (1970: 51). Higgins-Desbiolles *et al.* explained:

> Allies must work in solidarity with the oppressed, but they are cautioned to not fall into the trap of trying to lead and thereby return to oppressing. The method for this work is dialogical action where allies (revolutionary leadership in Freire's terms) support the oppressed in conscientisation and their praxis towards liberation (p. 67). This transformative work is not something the oppressed can do alone nor something others can do for them; it is derived from working with and not for… . (2022: 6)

The CCT team was effectively working 'with' families and communities, ensuring local stakeholders were leading the design of every project and every case plan. However, the Village Hive services were still being delivered by CCT staff out of private facilities. This is typically how most NGOs operate – delivering programmes in parallel to local systems. For CCT to be fully congruent with Freirian values, we needed to find a way to hand over the control and ownership of Village Hive services to local stakeholders. The only mechanism by which this shift in power could occur would be to integrate the Village Hive services into existing public systems. Once the services were operating out of community spaces, communities could finally take the lead.

We approached the District of Battambang with a proposal. We invited all ten communes and 62 villages in the district to partner with CCT and conduct an ambitious pilot to integrate the Village Hive services into public facilities. The communes in Battambang District had already played an active role in the design and evolution of the Village Hive, so we knew they were already familiar with the services and invested in the project. However, handing over the services meant they would be taking on a significant administrative burden, and the viability of the pilot rested on the willingness of all community stakeholders in the communes – the schools, the health clinics, the CCWCs, the commune chiefs and village leaders – to step up to the task. Despite our past successes in shifting power to local stakeholders, I imagined it would be a tough sell. Once again, I sold them short. Every stakeholder responded with resounding and overwhelming enthusiasm.

The biggest challenge we encountered was that international donors are reluctant to fund upstream work due to concerns over the lack of control when local stakeholders are leading development. This lack of trust is corrosive to sustainable development. It traps the Global South in binds of dependency by preventing development initiatives from being fully integrated into public systems and facilities where communities can assume control and ownership. In response to the lack of confidence from institutional donors, the local stakeholders decided to develop a zero-tolerance corruption policy and conduct anti-corruption training for all managers and frontline workers. Unfortunately, this has so far failed to allay the concerns of most institutional donors who can't see past the stereotype of 'corrupt Cambodians'. Fortunately, CCT's unrestricted donors and one institutional donor, World Childhood Foundation, have supported the vision and agreed to fund a six-year rollout partially. The pilot will only cost USD10 per person, per annum, and Battambang District has a population of 120,000. This cost will initially be covered by CCT donors but can eventually be integrated into the commune investment plan budgets and funded by the Cambodian Government. The cost per person is expected to decrease as CCT's involvement reduces and the resilience of the community increases, resulting in less need for support services.

Following the anti-corruption training, CCT and the local stakeholders co-developed the terms of reference and a child protection policy, established fiscal management protocols and signed a Memorandum of Understanding (MoU) with the relevant ministries, clearly outlining roles and responsibilities. Each co-creation meeting has offered opportunities to practice conscientisation, where informal round table discussions take place that critically unpack issues relating to the broader context of the Village Hive, including structural injustices and power dynamics between NGOs and communities.

The next stage of the pilot rollout involved integrating the Village Hive youth centre service into the communes' public schools. In 2020, the first Village Hive youth centre was embedded into a public school in Ou Char Commune, where it is being co-run and managed by the public school and the commune council, in partnership with CCT. After one year of operation, we conducted a focus group evaluation that revealed some unexpected findings. Every stakeholder felt confident that the Village Hive youth centre operating out of the public school was delivering value to the community above and beyond the expected outcomes. They identified improvements in school enrolments and a significant decrease in domestic violence within the wider community. They put this down to the indirect effect of greater community cohesion and stress reduction in families. In 2021, operations commenced integrating another two youth centres into public schools in Battambang District. As I write this, CCT and local stakeholders are co-creating the action plan to integrate the remainder of the services into all 10 communes and strengthen the existing universal services in the community.

Once the services are integrated into all 10 communes, ongoing oversight and governance will be managed at the district level, aligned with the Cambodian Government's National Child Protection Plan. This will create a clear exit pathway for CCT. Exit strategies should, after all, be the logical end goal of every NGO. This is why the process of *how* we are designing and implementing the Village Hive is so important. The activities that comprise the Village Hive will vary from commune to commune due to each community's unique assets, needs, ideas and resources. The upstream process, however, is something that can be scaled and replicated.

The upstream process must work within these five fundamental parameters:

(1) *Trust*: An essential prerequisite to commencing work as allies and equal partners is mutual and radical trust in the capacity and intentions of all stakeholders.
(2) *Conscientisation*: The practice of conscientisation is vital to project design and rollout. Ongoing dialogue transforms both coloniser and colonised mindsets and supports local stakeholders to step into their role as leaders of the work.
(3) *Co-creation*: All stages of the work, from ideation and planning to design and execution, are to be carried out as partners with collaborative and co-created engagements led by local stakeholders.
(4) *Prevention*: The work must be focused on tangible upstream actions that address the root causes of poverty, oppression and inequality.
(5) *Local integration*: Projects must work within existing public systems and operate out of community spaces with a clearly defined exit strategy for the partnering NGO.

Once the pilot is complete, 120,000 people across Battambang District will have access to a community-run, cost-effective, upstream child protection and social protection system. They will also have the power to adapt and evolve the system into the future. We are venturing into uncharted territory with this pilot, blazing a trail with the aim of building an evidence base to prove it is possible to dismantle neo-colonial systems of development, so the inherent wisdom that lives within local communities can be ignited and local stakeholders can rise to the challenge of transforming their world.

I sincerely hope that this work's success will invite an openness of heart and mind necessary to fuel a revolution that will decolonise development. Sometimes, it feels like an impossible dream. Then I remember that every major social justice milestone – the end of slavery, democracy, gender equality, same-sex marriage – would have once seemed like an impossible dream. Humanity is capable of achieving astonishing things when motivated by love, humility and solidarity. Indeed, it is the only means by which we have ever reached new, previously unimaginable horizons.

References

ABC (2010) Children of a Lesser God. *Australian Story*, ABC TV. See https://www.abc.net.au/austory/children-of-a-lesse + r-god/9172650 (accessed 10 October 2021).

AVI (2019) Living and accommodation allowances. See https://www.australianvolunteers.com/assets/Uploads/ResourceFiles/79357e925f/Living-and-Accommodation-allowances-Australian-Volunteers-Program-2019.pdf (accessed 25 January 2022).

CCT (2019) Lapo's story. See https://www.youtube.com/watch?v = _stMjm4mZRE (accessed 24 January 2022).

Edward, P. (2006) The Ethical Poverty Line: A Moral Quantification of Absolute Poverty. *Third World Quarterly* 37 (2), 377–393, https://www.tandfonline.com/doi/abs/10.1080/01436590500432739.

FCF-REACT (2019) Investing in innovation: Transformation of the dominant care model. See https://www.fcf-react.org/download/investing-in-innovation_transformation-of-the-dominant-care-model/ (accessed 24 January 2022).

Freire, P. (1970) *Pedagogy of the Oppressed*. New York: Continuum.

Government of Cambodia and UNICEF (2011) With the Best Intentions…A study of attitudes towards residential care in Cambodia. See https://www.thinkchildsafe.org/thinkbeforevisiting/resources/Study_Attitudes_towards_RC.pdf (accessed 24 January 2022).

Hartley, M. and Walker, C. (2013) Cambodia's booming new industry: Orphanage tourism. See https://www.forbes.com/sites/morganhartley/2013/05/24/cambodias-booming-new-industry-orphanage-tourism/?sh = 4cbd5f99794a (accessed 24 January 2022).

Haynes, D.F. (2014) The celebritization of human trafficking. *The Annals of the American Academy of Political and Social Science* 653, 25–45.

Heynen, R. and van der Meulen, E. (2021) Anti-trafficking saviors: Celebrity, slavery, and branded activism. *Crime, Media, Culture*. https://doi.org/10.1177/17416590211007896.

Hickel, J. (2017) *The Divide: A Brief Guide to Global Inequality and its Solutions*. London: Penguin Random House.

Hickel, J. (2019) The imperative of redistribution in an age of ecological overshoot: Human rights and global inequality. *Humanity: An International Journal of Human Rights, Humanitarianism, and Development* 10 (3), 416–428.

Higgins-Desbiolles, F., Scheyvens, R. and Bhatia, B. (2022) Decolonising tourism and development: From orphanage tourism to community empowerment in Cambodia. *Journal of Sustainable Tourism*. https://doi.org/10.1080/09669582.2022.2039678.

Hitchens, C. (2012) *The Missionary Position: Mother Teresa in Theory and Practice*. London: Hachette.

King, M.L. (1968) Honoring Dr. DuBois. *Freedom Ways* Spring, 110–111.

Lediard, D.E. (2016) Host Community Narratives of Volunteer Tourism in Ghana: From developmentalism to social justice. Theses and Dissertations (Comprehensive). 1862. https://scholars.wlu.ca/etd/1862.

Lumos (2017a) Children in institutions: The global picture. See https://lumos.contentfiles.net/media/documents/document/2017/03/Global_Numbers.pdf (accessed 24 January 2022).

Lumos (2017b) Ending the institutionalisation of children globally – the time is now. See https://lumos.contentfiles.net/media/documents/document/2017/02/Lumos_-_The_Time_is_Now.pdf (accessed 24 January 2022).

Marks, S. (2014) Somaly Mam: The holy saint (and sinner) of sex trafficking. *Newsweek* (Online). See https://www.newsweek.com/2014/05/30/somaly-mam-holy-saint-and-sinner-sex-trafficking-251642.html (accessed 24 January 2022).

Michael, J. and Goldstein, M. (2010) Reviving the Whitehouse Conference on Children. See https://www.cwla.org/reviving-the-white-house-conference-on-children/ (accessed 28 January 2022).

O'Brien, E. (2018) Slavery and human trafficking campaigns by Hollywood celebrities can be misleading. ABC (Online). See https://www.abc.net.au/news/2018-07-07/beware-the-hollywood-hype-on-human-trafficking/9893330 (accessed 24 February 2022).

Shonkoff, J.P. (2018) Statement on separation of families. See https://developingchild.harvard.edu/about/press/shonkoff-statement-separating-families/ (accessed 24 January 2022).

Signs of Safety (n.d.) What is Signs of Safety? See https://www.signsofsafety.net/what-is-sofs/ (accessed 12 February 2022).

Sovuthy, K. (2021, 28 September) Garment industry minimum wage raised $2 for 2022 amidst criticism from unions and workers. See https://cambojanews.com/garment-industry-minimum-wage-raised-2-for-2022-amidst-criticism-from-unions-and-workers/ (accessed 26 January 2022).

UN (n.d.) Goal 1: End poverty in all of its form. See https://www.un.org/sustainabledevelopment/poverty/ (accessed 18 January 2022).

UN (2021) The Sustainable Development Goals Report 2021. See https://unstats.un.org/sdgs/report/2021/The-Sustainable-Development-Goals-Report-2021.pdf (accessed 4 March 2022).

UNCTAD (2014) Developing countries face $2.5 trillion annual investment gap in key sustainable development sectors, UNCTAD report estimates. See https://unctad.org/press-material/developing-countries-face-25-trillion-annual-investment-gap-key-sustainable (accessed 24 January 2022).

Case Study in Practice of Working with Communities

Freire's 'Pedagogy of the Oppressed' and Working in Mutuality: Pathways for Communities to Take Control of Their Futures

Adapted From: Higgins-Desbiolles, F., Scheyvens, R. and Bhatia, B. (2022) Decolonising tourism and development: From orphanage tourism to community empowerment in Cambodia. *Journal of Sustainable Tourism*. An Open Access article distributed under the terms of the Creative Commons Attribution License.

Certain voluntourism analyses have considered how the relations between communities, mediating organisations and volunteers can be reconfigured for justice (e.g., Guttentag, 2011). Simpson insisted that a 'pedagogy of social justice' (PSJ) (2004: 690) must underpin voluntourism. Henry (2019) advanced these discussions by outlining three steps to support a PSJ: researcher reflection, action research and continued engagement with returned volunteers to support their ongoing radical consciousness and action. Henry's proposed agenda was intended to foster an approach based on social justice 'embedded within and continuing beyond voluntourism…' (2019: 563).

This article employs Freire's (1970) analysis of the 'pedagogy of the oppressed' and his articulation of mutuality as a way of answering the research question of how we might move beyond 'doing good' in voluntourism and development. Here, Freire addresses how those in positions of power and privilege may want to support those who suffer oppression. He outlines the nature of oppression, explaining that the humanity of both the oppressed and the oppressor are undermined. Some in the oppressor category may develop a guilty awareness of their role in oppression and seek to mitigate it; but if these efforts fail to address the structural roots of oppression, they only offer 'false generosity', thus maintaining the oppressed in binds of dependency (1970: 49). For Freire, 'oppression is domesticating' (1970: 51) and so members of both the oppressor and oppressed classes must continually struggle for their humanisation and freedom. Critical to this is conscientisation (*conscientização*), which refers to mutual, critical awareness work that the oppressed do in understanding their conditions leading to a determination that 'the concrete situation that begets oppression must be transformed' (1970: 50). The fundamental basis of this work is the recognition of the oppressed's own knowing (1970: 68) and abilities to transform their situation.

The aim of this mutual work is to become, both oppressed and oppressor, 'human in the process of achieving freedom' (1970: 49). This work is achieved through 'praxis: reflection and action upon the world in order to transform it' (1970: 51) in an effort to gain full humanity and carry out 'revolutionary' change (1970: 66). Allies work in solidarity with the oppressed but they are cautioned to not fall into the trap of trying to lead and thereby return to oppressing. The method for this work is dialogical action where allies (revolutionary leadership in Freire's terms) support the oppressed in conscientisation and their praxis towards liberation (1970: 67). This transformative work is not something the oppressed can do alone nor something others can do for them; it is derived from working *with* and not *for* (1970: 67). 'Founding itself upon love, humility, and faith, dialogue becomes a horizontal relationship of which mutual trust between the dialoguers is the logical consequence' (Freire, 1970: 91).

This thinking from Freire offers a framework for transforming both tourism and development for working in mutuality rather than in the 'false generosity' of white saviourism. It is built on critical consciousness developed through dialogue, co-development of a praxis for changing the material conditions of the community, what we would call 'capacity sharing' and trust in the capacities of the people. Instead of the traditional 'capacity building' approaches to development, this Freirian practice is best described as 'capacity sharing'. It is an ongoing process in which the people and their allies come to understand their context and how it might be transformed for the better and co-create a shared humanity in this mutual praxis. Instead of the traditional 'capacity building' approaches to development, this Freirian practice is best described as 'capacity sharing', describing when parties in the development process share experiences and knowledge to strengthen decision making under an agreed development trajectory. Freire provides a useful way of understanding this transformation:

> Authentic help means that all who are involved help each other mutually, growing together in the common effort to understand the reality which they seek to transform. Only through such praxis – in which those who help and those who are being helped help each other simultaneously – can the act of helping become free from the distortion in which the helper dominates the helped. (cited in hooks, 1993: 150)

Instead of the 'false generosity' of a dependent path to development, Freirian approaches work with and through the receiving community: 'True generosity lies in striving so that these hands – whether of individuals or entire peoples – need be extended less and less in supplication, so that more and more they become human hands which work and, working, transform the world' (Freire, 1970: 46).

Linking together the concepts set out in the literature review, a framework is revealed pertaining to conventional international development which is based on an ideological worldview that recipients are inferior, supporting practices that may be best described as paternalistic. Integrated relationships feature between voluntourism, development, development aid NGOs and countries of the Global South which perpetuate a cycle of dependency. Yet, as Lediard demonstrated if justice is placed to the fore, volunteer tourism relations can be transformed 'from power-over to power-with' (2016: 100). This is the promise of the Freirian conceptualisation of mutuality in praxis.

Winkler's narrative of the Cambodian Children's Trust offers a case study of a non-governmental organisation and its co-founder that illustrates the development of a Freirian praxis of mutuality to collaboratively move from dependency to liberation and self-determination (see Chapter 9).

References

Freire, P. (1970) *Pedagogy of the Oppressed*. New York: Continuum.
Guttentag, D. (2011) Volunteer tourism: As good as it seems? *Tourism Recreation Research* 36 (1), 69–74. https://doi.org/10.1080/02508281.2011.11081661.
Henry, J. (2019) Directions for volunteer tourism and radical pedagogy. *Tourism Geographies* 21 (4), 561–564. https://doi.org/10.1080/14616688.2019.1567577.
hooks, b. (1993) bell hooks speaking about Paolo Freire. In P. McLaren and P. Leonard (eds) *Paolo Freire: A Critical Encounter* (pp. 145–152). London: Routledge.
Lediard, D.E. (2016) Host Community Narratives of Volunteer Tourism in Ghana: From developmentalism to social justice. Theses and Dissertations (Comprehensive). 1862. https://scholars.wlu.ca/etd/1862.
Simpson, K. (2004) 'Doing development': The gap year, volunteer-tourists and a popular practice of development. *Journal of International Development* 16, 681–692. https://doi.org/10.1002/jid.1120.

10 The Neighbourhood Where History, Community, Tourism and Truth-Telling Meet: A Tourism Practitioner Case Study from the Greenwood Cultural Center of Tulsa, Oklahoma

Bobbie Chew Bigby and Michelle Brown-Burdex

Introduction

In the year 1921, a devastating massacre occurred in the historically Black Greenwood District of Tulsa, Oklahoma. One hundred years later in 2021 this event was commemorated and remembered at its centennial anniversary with numerous local activities, widespread media coverage and the arrival of visitors from across the country and globe, including United States (US) President Joseph Biden himself. The importance of the Greenwood story is in part because Greenwood was one of the most successful and prosperous Black neighbourhoods of the United States in the 20th century, having been dubbed 'Black Wall Street'. Further, this story's central place in the experience of the US, Oklahoma and Black America is linked to the fact that this 1921 massacre represented the worst act of racial violence and domestic terrorism in US history (known as the 'Tulsa Race Riot'). The documented testimonials of many survivors recalled the seemingly unthinkable, from witnessing airplanes that were used to drop incendiaries from above, to the active deputisation of White mob members by the Tulsa Police Department in the midst of the violence (Ellsworth, 1982). The significance of the Greenwood story is also tied to the resilience of the Black Americans who survived the massacre, rebuilt, faced continued adversity and moved forward, with their descendants still carrying forward this spirit and work into the present day.

But beyond these critical and painful facts of what happened to a thriving Black American community, the Greenwood story is particularly remarkable because for so long it was not told and had been actively suppressed by the White community in Tulsa. In the official Report by the Oklahoma Commission to study the 'Tulsa Race Riot of 1921', Oklahoma historian Danny Goble writes:

> The Tulsa disaster went largely unacknowledged for a half-century or more. After a while, it was largely forgotten. Eventually it became largely unknown. So hushed was mention of the subject that many pronounced it the final victim of a conspiracy, this a conspiracy of silence. (Oklahoma Commission, 2001: 4)

While no mention of the massacre was made among White Tulsans, the memories of what happened in Greenwood lived on among Black Tulsans, with efforts made by many community leaders to keep the story alive and seek some form of justice. In 1995, the Greenwood Cultural Centre (GCC) was built and dedicated to passing along this story of massacre and trauma, but more importantly the story of resilience and rebuilding for those survivors of the massacre and descendants of the Greenwood community.

For the past 25 years, Michelle Brown-Burdex has served as the Program Coordinator and Tour Guide at the GCC, a central place and vehicle for sharing the story of the district's legacy and the events of 1921. In the period since its founding, the GCC has functioned both as a centre of learning, tourism and truth-telling, as well as a hub for the local Black community in their efforts to offer cultural, educational, historical and community-based programming. This practitioner case study delves deeper into the role that the GCC has played as it sits at the crossroads of the local Black Tulsa-Greenwood community, tourism and the process of truth telling. The chapter first sketches the story of the Greenwood community and massacre, along with the GCC's establishment and the events that took place at the Centennial Commemoration of the massacre in 2021. The chapter then turns to a dialogue and reflection on these themes with Michelle Brown-Burdex who has been at the forefront of navigating tourism and relationships with the local Greenwood community for the past decades. This writing project has come into being through ongoing dialogues and a writing partnership between co-authors Michelle Brown-Burdex and Bobbie Chew Bigby, both natives of Tulsa whose lives have intersected with the Greenwood story.

The Story of Greenwood

The Black Americans that called Greenwood home in the early 20th century came from two different sets of backgrounds in the Black American experience. Some of these early Greenwood inhabitants were descendants of 'Freedmen', the term for slaves that had been enslaved by or intermarried with some members of the Five Tribes forcibly relocated to Oklahoma from the southeast from the 1830s onward (i.e. Cherokee, Creek, Choctaw, Chickasaw and Seminole). Others were the descendants of freed Black people who had migrated from other states in the US north and south post-Civil War and Reconstruction (Johnson, 2014). While the end of the US Civil War and the ensuing Reconstruction period brought emancipation to Black Americans held as slaves, inequality and denial of rights for Black communities would quickly begin to be reimposed across large parts of the US, particularly throughout the southern states where racial segregation would be enforced through the adoption of Jim Crow Laws (Franklin, 1980). Despite the great adversity that these different generations of Black Americans had encountered, this neighbourhood of Greenwood stood as a testament to the resilience and hard-earned opportunities that Black Tulsans were creating for themselves during a period of extreme injustices perpetrated against nearly all non-White populations in the US.

By the early part of the 20th century, Greenwood was a large neighbourhood covering approximately 35 square blocks to the immediate north of downtown Tulsa where Black-owned enterprises of all sorts boomed and grew, thus earning the popular moniker of 'Black Wall Street'. Historian and Tulsa-born Scott Ellsworth's impactful book *Death in a Promised Land: The Tulsa Race Riot of 1921* was one of the first sources to document the wealth, success and diversity that was contained in this Black American community of nearly 10,000 people. Ellsworth notes that within the Greenwood District, there were 38 grocery stores, 30 restaurants, 12 churches, over 12 Black doctors' and lawyers' offices, one hospital, two schools, two theatres, four hotels and two newspapers that were owned and run by Black Tulsans, for local Black Tulsans (1982: 26–27). Created out of the harsh reality that Tulsa, Oklahoma was a deeply segregated city within a Jim Crow state where Black people needed to find their own ways to shop and access services, Greenwood was much more than a neighbourhood and thriving centre of commerce for Black Tulsans; it stood and still stands as an extraordinary example of the potential of Black Americans in building a thriving and prosperous community.

The era in which the Greenwood neighbourhood grew was also a time of deep racism, jealousy, forced segregation and high crime rates on the part of non-Black Tulsans, as well as regular racial violence against Blacks taking place in other parts of Oklahoma (Ellsworth, 1982; Oklahoma

Commission, 2001). This deeper racial hatred served as the backdrop for the events that would take place in 1921. On 30 May 1921, a Black shoe shiner named Dick Rowland entered an elevator operated by a White woman named Sarah Page and stepped on her foot, causing her to scream. The news of this incident spread quickly and was immediately misconstrued – particularly by the local media – as an attempt by Rowland to assault and rape Page, leading to outrage and mobilisation of an angry mob of White Tulsans looking for revenge. The publicising of untrue, inflammatory information in the White-run Tulsa newspaper further fuelled the flames of anger with open talk of lynching that quickly turned into violent action by the next day.

With the White mob unable to get a hold of Rowland to lynch him and a contingent of Black World War I veterans that stood armed and ready to offer security to the Tulsa Court House, a scuffle ensued as a White man tried to disarm a Black veteran. When a shot rang out, numerous White Tulsans descended upon Greenwood in armed mobs and violence broke out. Throughout the night, Black businesses and homes were immediately looted and ransacked before being burned to the ground, with Black Tulsans trying to defend themselves and their property. The Tulsa Police Department actively deputised members of the White mob allowing them to take violence into their own hands. The pause in shooting at around 2 o'clock in the morning was to be short-lived as by early dawn a large, organised mob of White Tulsans had amassed at Greenwood's southern border and the systematic invasion and destruction of the neighbourhood began, with constant shooting, burning and destruction. Figure 10.1 below depicts this systematic burning and destruction of the neighbourhood, while Figure 10.2 shows Black Tulsans being rounded up and moved in the street. The calls by the Oklahoma Governor for State Troops to restore order had come too late and by the middle of the day on the first of June, the Greenwood neighbourhood had been obliterated and the surviving Black residents of Greenwood placed in internment camps, as seen from Figures 10.3 and 10.4. The number of deaths has continued to remain unclear to the present day, with estimates ranging from the 70s to the 300s (Oklahoma Commission, 2001). Numerous survivors also recounted witnessing the mass disposal and burial of bodies, some thrown into the nearby Arkansas River with others mass buried in downtown graveyards, leading to the ongoing work of present-day task force commissions attempting to locate and identify the mass graves.

Concerned with the horrific image that this would place upon Tulsa and not adhering to commitments to help Greenwood rebuild, the White powerholders of Tulsa immediately went about destroying any evidence or mention of the massacre, with photos and newspaper clippings systematically removed from archives and destroyed (Ellsworth, 1982). Not one White Tulsan ever served time for the murders and destruction that took place and as the decades ensued, taboos on speaking publicly about the

198   Part 3: Practitioners' Views and Insights

Figure 10.1 The destruction and burning of the Greenwood neighbourhood. In the foreground is the building containing the Village Blacksmith Shop. The printed caption reads, 'Runing [sic] the negro out of Tulsa, June 1, 1921'. Credit: Tulsa Historical Society & Museum (used with permission)

Figure 10.2 A group being led to the Convention Hall during the 1921 Tulsa Race Massacre. Credit: Tulsa Historical Society & Museum (used with permission)

Figure 10.3 Blocks of destroyed homes following the Tulsa Race Massacre. Credit: Tulsa Historical Society & Museum (used with permission)

Figure 10.4 Black Tulsans detained during the 1921 Tulsa Race Massacre. Credit: Tulsa Historical Society & Museum (used with permission)

massacre grew among both Whites and Blacks. For many Black Tulsans, the trauma and reluctance to share these stories among younger generations was compounded by the fact that life for survivors had to go on, which meant that Blacks had to depend on White employers for earning livelihoods.

This wall of silence began to be slowly chipped away by members of the Greenwood and Tulsa community beginning in 1968 with the writing of Greenwood native and then journalist for the local Black newspaper, Don Ross. By 1995, while serving as an Oklahoma State Representative, Ross drew attention to the story of the Greenwood massacre in light of the Oklahoma City Murrah Building bombing, which many had cited as one of the worst acts of domestic terror on US soil. Thanks to efforts of Ross, along with numerous other Greenwood survivors, community members, educators and advocates, the construction of the GCC got underway. Additionally, their efforts also led to the launch of a Commission in 1997 to investigate and report on what was referred to as the 'Tulsa Race Riot'. Ever since the events of 1921 have been brought to wider public attention in the last decades, there has remained disagreement, debate and inconsistency until recently on the correct word to describe what happened, with many White Tulsans using the word 'riot' while most Black Tulsans have used the term 'massacre'. As noted by historian Scott Ellsworth, however, these tendencies on word usage have not always been clear cut along racial community lines, particularly for some Greenwood survivors who had felt that the word 'massacre' denoted a lack of agency on the part of Black Tulsans and did not reflect the reality that they had experienced in defending themselves from violence (Ellsworth, 2022). However, in light of the Centennial Commemoration and widespread media coverage that Greenwood received, nearly all references to the events of 1921 now refer to it as the 'Tulsa Race Massacre'.

Offering its final report to the public in 2001, the Commission provided thorough documentation of the events of the massacre, as well as recommendations for facilitating justice, healing and increased opportunities for Greenwood and the Black community of North Tulsa. Central to the Commission's recommendations was a call for reparations for the survivors and descendants of the Greenwood community, along with the establishment of a scholarship fund, an economic development zone in historic Greenwood and a search for mass graves and memorial for victims once reburied (Oklahoma Commission, 2001). To this day, however, certain recommendations – including those of reparations – have been ignored by the state of Oklahoma while state-backed funding for the GCC has come in cycles and has been highly unstable at times (Sulzberger, 2011). This disregard for these central recommendations regarding financial reparations has caused deep frustration and sadness among the Greenwood community survivors and descendants, particularly given the devastating financial and business losses that these families were dealt after the Massacre – and the impacts that were felt through the generations. In research conducted by *National Geographic* during the Centennial Commemoration in 2021, it was estimated that in today's dollar value, the material losses of the Massacre rampage would amount to over USD 26 million (see Brown, 2021). Yet when projecting the total accumulation of lost wealth that has failed to accrue over the generations for Greenwood families due to the destruction, the research further estimates that this value would be approximately USD 610,743,750 – an amount that lends perspective to conversations on racial inequity, wealth gaps and conversations about reparations and justice (Brown, 2021: 71).

The Greenwood Cultural Center and 2021 Centennial Commemoration

A product of the advocacy of numerous Greenwood and Tulsa community members – central among them, former State Legislator Don Ross – the GCC was first dedicated and opened in 1995 (see Figure 10.5). Former Representative Ross is quoted in the local Tulsa World newspaper as stating that: '[The center] has truly and wholly united our community. This was a grassroots project. The first $100,000 for the project was raised in the black community in seven months' (Latham, 1995). Located in the heart of the historic neighbourhood along Greenwood Avenue, the Center sits across from one of the two historic Greenwood churches that remain from 1921, the Vernon African Methodist Episcopal (AME) Church, as seen in Figure 10.6. At its east side, the Center also includes the Mable B. Little Heritage House that was built by the Mackey family massacre survivors and highlights the Greenwood legacy, along with the story of its namesake, Mabel B. Little. The mission of the GCC is: 'to preserve African-American heritage and promote positive images of the African-American community by providing educational and cultural

Figure 10.5 The front entrance of the Greenwood Cultural Center in Tulsa, Oklahoma. Credit: Co-author Bobbie Chew Bigby

Figure 10.6 A plaque commemorating the 1921 Tulsa Race Massacre. Behind the plaque stands the historic Vernon AME Church, one of only two churches to have survived the Massacre. The church stands facing the Greenwood Cultural Center. Credit: Co-author Bobbie Chew Bigby

experiences; promoting intercultural exchange; and encouraging cultural tourism' (GCC, n.d.) In its vision statement, the Center sees itself as: 'the keeper of the flame for the Black Wall Street era, the events known as the Tulsa Race Massacre of 1921, and the astounding resurgence of the Greenwood District in the months and years following the tragedy' (GCC, n.d.).

Some of the key components of the Center include: a pictorial exhibit, an art collection, a large event space/banquet hall and more recently, an office for the Terence Crutcher Foundation in memory of Terence Crutcher, an unarmed Black Tulsan shot dead by police in 2016. Until 2007, the GCC had also housed the Oklahoma Jazz Hall of Fame. Central to the GCC's programming are a range of educational and cultural activities that are delivered throughout the year, and particularly during the summertime during school holidays and around the African American Juneteenth Holiday. Some of these programmes range from arts workshops to community discussions on popular topics to film nights and events for encouraging young entrepreneurs. For decades now, the Center has also served as a central hub for cultural and historical tourism, offering tours to all age groups and visitors of all backgrounds. As will be discussed in more detail below, the GCC operates with a small number of staff and early on had relied on local community member volunteers in providing its tours. Due to uncertain and wavering commitments by the state of Oklahoma over the years, the GCC has been challenged by unstable funding resources. In 2011, all funding from the state was cut – comprising half of the GCC's budget – and the Center risked being forced to shut down (Sulzberger, 2011). In spite of these immense and recurring financial challenges, the GCC has remained open and has continued to lead the way in educating local community members as well as touring visitors alike about the legacy of Greenwood and the massacre. Additionally, the Center was recently allocated funds from the city of Tulsa for a USD 5.3 million renovation of its facility, allowing it to move forward with its mission and activities.

This past 2021 year marked a significant moment for Greenwood and Tulsa as the community faced the Centennial Commemoration of the horrific events that had happened 100 years earlier. The GCC was a central hub for numerous important activities going on throughout the months of May and June 2021 during the Commemoration. As depicted in Figure 10.7, a large mural showing 'Black Wall Street' was painted across from the Center and diverse artistic activities abounded during the Commemoration, ranging from visual art in public spaces to poetry readings and concerts. The Center hosted activities including a lecture series, film screenings, a special Kinsey African American art collection exhibition, panel talks, exhibits, a Sunday Brunch with notable entertainment stars, as well as the official delegation from the White House, including US President Joseph Biden. Biden took a tour of the GCC before delivering a nationally televised speech. Additionally, the GCC partnered with numerous other local non-profit organisations and businesses to host

Figure 10.7 A large mural of 'Black Wall St.' that is painted onto the side of the highway that intersects the historic Greenwood neighbourhood and faces the Greenwood Cultural Center. Credit: Co-author Bobbie Chew Bigby

several events connected to the Black Wall Street Legacy Festival that ran outdoors along Greenwood Avenue between 27–31 May in the lead up to the official Commemoration on 1 June.

Numerous other events also took place held by a variety of different organisations and actors telling the Greenwood story in the lead up to and during the Centennial Commemoration. The nearby John Hope Franklin Center for Reconciliation held an in-person and online symposium, as well as hosting various events at its Park a couple of blocks from the GCC. Various non-profit organisations, libraries, schools, universities and businesses throughout the city of Tulsa, across Oklahoma and even around the country held events that educated and encouraged discussion of the Greenwood story. On the television, numerous national stations, including the Oprah Winfrey Network (OWN), CNN, the History Channel, MSNBC, National Geographic and PBS, among others, created and aired their own documentaries that told the Greenwood story and were shown nationally.

But one of the most significant and controversial focuses of community attention has been on the construction of the Greenwood Rising Museum and History Center building backed by the Tulsa Race Massacre Centennial Commission. The Centennial Commission is a multi-faceted initiative launched in 2015 that aims to:

> … educate Oklahomans and Americans about the Race Massacre and its impact on the state and Nation; remember its victims and survivors; and create an environment conducive to fostering sustainable entrepreneurship and heritage tourism within the Greenwood District specifically, and North Tulsa generally. (Greenwood Rising, n.d.)

The Commission comprises six different committees including: arts and culture, economic development, tourism, education, reconciliation and marketing/public relations. Each of these different committees have taken charge of certain projects related to their particular focus. The central focus of the tourism committee and the most ambitious project of the Commission has been the development and opening of a brand-new museum and history centre named Greenwood Rising. Located a block down from the GCC along Greenwood Avenue, this USD 20 million centre that was opened a couple of months after the Centennial Commemoration considers itself as a central place where the Greenwood story is told.

Yet from the beginning, Greenwood Rising's presence has caused division and deep disagreement in the Greenwood community, as the original plans for the Commission involved fundraising and financing a renovation of the GCC. These plans were scrapped in 2019 when the different community stakeholders involved did not agree on the specific roles and responsibilities that these organisations would take, leading to a decision by the Commission, along with significant funding commitments from various Tulsa philanthropies, to build an entirely separate museum and history centre (Canfield, 2021a). Further confusion and frustration were generated in the immediate lead up to the Centennial events planned by the Commission, including in particular a large, televised concert event in Tulsa called 'Remember & Rise' that was due to feature notable Black speakers, politicians and entertainers such as Stacey Abrams and John Legend. Due to a disagreement over the terms by which the Commission was to provide funding to the three living survivors of the massacre, along with numerous other complaints regarding the lack of local input and involvement from Greenwood community members, public support for the event fell apart quickly and publicly, leading to its cancellation three days before it was planned to happen (Canfield, 2021b). In spite of the difficulties and disagreements that came to light in the lead-up and during the days of the Centennial Commemoration, most events held throughout Greenwood did proceed with large attendance and support, an important and meaningful feat not only given Greenwood's difficult history, but also the fact that 2021 was a year when the COVID-19 pandemic raged on throughout Oklahoma.

Discussion and Reflections with Michelle, a Tourism Practitioner and Storyteller in Greenwood

Bobbie: Where are you from originally and how did you first learn about the Greenwood story?

Michelle: I was born and raised here in Tulsa. I am a country girl at heart and love being able to sit in my backyard and be surrounded by beautiful trees, squirrels, rabbits, possums, birds and the occasional cat. Driving through North Tulsa where I live, it is not unusual to see deer, foxes and men riding horses. And yet, amidst the natural beauty of North Tulsa is the reality that most of North Tulsa is a food desert. There is a lack of affordable housing, access to mental health services, doctors, dentists, recreational facilities, entertainment venues and many assets that contribute to a happy and healthy lifestyle. But there are several significant initiatives to address these issues and I have hope for North Tulsa becoming a much more liveable community over the next five years.

I had heard about a riot, as it was called for many years, from my mother. As a child, I overheard my mother telling one of her in-laws that while she worked as a nurse aide who cared for an elderly White man, he had recounted a story of a riot that took place in Tulsa. The man spoke of fires, houses burning, gunshots, and hundreds of people being shot and killed. My mother, being new to Tulsa wondered what he could be remembering. She asked her in-law what he was referencing, and the in-law told her that: 'we don't talk about that 'round here, and don't go asking anybody about it'. My mother never mentioned it again. And while I grew up in Tulsa attending predominantly Black schools with Black educators and attending

Black History classes, I never heard anyone discuss it again, until my tour of the GCC several years later, just months after it had first opened in 1995.

Bobbie: My own experience of first learning about the massacre was tied to my time as a student at Booker T. Washington High School (an historically Black high school within the former Greenwood district boundaries that endured after the Massacre and became racially integrated in 1973). Most people in other schools dreaded and disliked 9th grade Oklahoma History, but I was really lucky to have been assigned to Dr Ronald Foore's history class as a high school freshman. Because Dr Foore said that history was his passion and calling, he devoted time and serious discussion to certain topics that other teachers apparently were not as focused on, including the 1921 Race Massacre as well as the murders of Osage Native people because of their oil wealth in the same 1920s period. On the first day that he talked about the Race Massacre and showed slides of the photographs taken from the time, I was overwhelmed by the stories and images – as well as by the fact that this was my first time to hear about this, having grown up in Tulsa my whole life. Dr Foore spent a week covering the Massacre in the classroom and our student discussions were very emotional and raw. Those days talking about Greenwood in class were my most memorable and important experience of high school. After I graduated and talked to other Oklahoma high school graduates from around the state who had never heard of the Massacre, I came to see just how unique the opportunity I was given to learn was… and how many other students haven't been taught our history.

Bobbie: What is the relationship like between the local Greenwood community (also understood as referring to Greenwood massacre survivors, descendants and Black residents of North Tulsa) and the GCC?

Michelle: The GCC has the distinction of being a community centre built by the community; from the initial concept to the securing of state funds through the efforts of North Tulsa's elected officials, the community has been involved throughout the planning and implementation of the Center.

The initial docents [at GCC] were part of a group of community members who helped build the GCC… these senior community members answered the call for people to lead guided tours. Those initial docents were retired school teachers, Sunday school teachers and grandparents – all North Tulsa residents – who had watched or participated in the development of GCC and were proud to represent the Center as docents.

We consider the land where GCC is located to be sacred ground.

Bobbie: How would you describe the relationship between the GCC and tourism?

Michelle: The GCC founders always envisioned guided tours of the centre. They did not anticipate a need for guided tours of the Greenwood District, but tourists are interested in learning about the history of Black Wall Street and then exploring the area.

In June 1996, when I was first hired by GCC as a full-time Office Assistant, the Center had a robust docent programme. There were at least a dozen docents, mostly retired educators and community members who would lead tours… they greeted tourists with such pride and enthusiasm… Eventually as the number of docents dwindled, I was asked to learn to give tours by following lead tour guides Carmen Pettie (North Tulsa Heritage Foundation) and Eddie Faye Gates (renowned educator, author and historian). When I began giving tours on my own, I was terrified… This history is so important to me and it was important that I not only represented GCC well, but also the stories of the survivors, some of whom I had met as they began to tell their stories, often for the first time. I also wanted to make my mentors proud, people that I had followed and listened to as they gave tours – most importantly Eddie Faye Gates. Ms Gates was responsible for interviewing dozens of massacre survivors, served on the 1921 Tulsa Race Riot Commission that studied the massacre and wrote numerous books. One of the first Greenwood tours that she gave was to Rosa Parks. She received what she referred to in interviews as hate mail and 'something just short of death threats,' and yet she refused to stop sharing the history, to stop giving tours and advocating for reparations for survivors and descendants. I spent a lot of time with Eddie doing tours, whether I was transporting people or sitting in the background listening. So, it became a heavy responsibility placed on my shoulders to tell these Greenwood survivors' stories in a way that was honest and truthful. I watched as each one of the survivors passed away, with now only three living today. People will continue to tell this Greenwood story, but they won't tell *their* story, the stories of the survivors. And I wanted to be a part of telling *their* story.

Giving tours has always been challenging, yet doing these tours and presentations became my passion and perhaps my greatest accomplishment. Over the past 25 years, I have given tours and presentations to thousands of students, tourists, businesses and even the current President of the United States, Joe Biden (see Figure 10.8).

After people have toured the Center, they often visit the administrative office to share their experiences, to express their feelings, often feelings of shame, guilt and disbelief. What has happened naturally has been allowing visitors to share their feelings, to listen to them as they process what they have experienced and to discuss together our hopes for restoration and reconciliation. But the challenging part for us is that it is as if the historically oppressed are forced to serve as the consoler, often while suppressing our own feelings of anger or hurt.

Bobbie: What have been some of the main challenges that the GCC has faced?

Michelle: Most challenges can be attributed to a lack of consistent and significant funding and financial support from city and local government and philanthropic institutions. That support is inherently political and

Figure 10.8 Michelle Brown-Burdex speaking with President Joseph Biden on his visit to the Greenwood Cultural Center during the Centennial Commemoration on 1 June 2021. Credit: Christopher Creese, Creeseworks (used with permission)

associated with the intent to disregard organisations whose mission is to recognise, amplify and celebrate Black history, unless it is told through a White lens. It is the desire to deny support for an organisation that seeks to recognise the history of the 1921 Tulsa Race Massacre, the nation's worst race massacre that happened more than 100 years ago, that many people are still unaware of. Since the GCC opened its doors nearly 30 years ago, its mission has included providing educational and cultural opportunities. The education that the GCC was most committed to sharing and acknowledging was the history of the 1921 Tulsa Race Massacre and Black Wall Street.

The response from some White Tulsans was that this was Tulsa's dirty little secret and should not be discussed, that Blacks only wanted to discuss the massacre because they sought some form of reparations and that remembering this history does nothing to unify our community, or to help the Black community to heal. It is as if people who are a part of the race of people that murdered hundreds of innocent Black people in their homes, in their own community, could begin to tell the survivors of the atrocity and their descendants how to heal.

Aside from funding, it has been challenging to find new ways to tell the same story. The history remains the same, but we are challenged with making the history interesting for each age group while maintaining the integrity of the history and paying respect to the victims, survivors and descendants.

Bobbie: From your perspective at GCC, is there tension between the local Greenwood community and tourism? If so, how would you describe this disconnect or tension and how has it played out in relation to national attention on Greenwood during the 2021 Centennial Commemoration?

Michelle: The GCC has been active in the Tulsa community for nearly 30 years and we've been at the forefront of educating students and the wider public long before it was encouraged or considered acceptable. In the lead up to the 2021 Centennial Commemoration, it was frustrating to hear our local politicians and leaders talk about how Tulsa has moved forward, embraces reconciliation and how the whole Greenwood story and neighbourhood can serve as a tourism destination and important example of improved race relations. It feels that some have capitalised on the Greenwood story, place and community to use this as a tourism attraction. This is upsetting and disheartening because before the Centennial, there had not been the active encouragement or recognition of Greenwood and what we are doing at the GCC the way there is now. But right now, the Centennial has been seen as an opportunity to create a tourism destination.

The activities and preparations in the lead up to the Centennial left our Black Tulsan community quite divided. On the one hand, there are those who believe that there cannot be real reconciliation without reparations and conversations are essential about what reparations can and should look like. For these community members, the question of reparations – along with other difficult, unresolved issues such as continuing the work of the mass graves investigation – are the most important. But the other group in the community believes that we in the Black community need to move forward and work together, as one Tulsa, and put some of these divisions behind us. This group thus prioritises transforming Greenwood into a tourism destination that can give North Tulsa an important economic boost and accepting any financial assistance that is offered to make Greenwood that tourism destination.

Another divisive issue has been the creation of the new museum just down the street, Greenwood Rising. This new museum has received generous funding from some of Tulsa's large philanthropies and has been able to build a brand new, state-of-the-art facility, when the initial vision was to renovate and improve our GCC. Many in the Greenwood community have felt that those who stand with the new museum and the Commission have wanted to take over the Greenwood story and control the narrative. Many people in the Black community in North Tulsa – including the descendants and families affected by the massacre – feel that they have not been heard. When someone sees that they've built a museum just down the block from us here at the GCC, when we've been here all along doing what we've done with tourism and education, then it shows that there is a problem. It should be us here in Greenwood telling our story and we want to share the stories that we have been told from the survivors over the decades.

Bobbie: Apart from some of the challenging dynamics identified above and on a concluding note, do you see tourism as having potential for empowering Greenwood specifically and Black Americans more generally?

Michelle: Absolutely. I think that there is a realisation that as families are planning their vacations and travels that they want to have a significant experience that is cultural in nature, and not just all about fun and excitement. We are encountering more and more families coming to visit. There are more families that are home-schooling and more families that are taking responsibility for educating their children about America's history and Black history. It's also important to recognise that Black families travel as well. This form of tourism empowers our community and our organisation because for one, it acknowledges our value. It acknowledges the importance of the story that we tell and it gives us a place in tourism – a place that we've not always had. We are wanting to promote our history, our community and our cultural organisations. Through tourism that is led by our community, we are empowered to have a voice in how our history is promoted, recognised and defined. We have a voice in shaping that narrative. So, it's empowering for us to have the opportunity to be involved in some of the decision-making process as it relates to how our story is told.

A by-product of this is that it highlights the need to invest money in this community and in these cultural organisations that are telling the history that serve as tourist destinations. We benefit from the economic investment in these organisations and historic sites and that type of investment is so deeply needed. Real investments are so long overdue for this community, the Greenwood district. It's incredible to think that Greenwood has been overlooked for so long and now it is becoming one of Tulsa's tourism destinations. The voices and stories of the Greenwood survivors, descendants and community members must be at the centre.

Conclusion

The global attention that the Greenwood story and community received last year in light of the Centennial Commemoration of the horrific events of 1921 should not and cannot be a fleeting media headline or used to promote it as a tourism destination with the latest hot travel trend. The Greenwood story must be told and understood as an indispensable part of understanding the place of Black Americans in Oklahoma and US history, focusing on their incredible resilience and strength in the face of horrific challenges. The GCC stands as an important site where these truths are told and where this intersection of community, history and engagements with tourism can be carefully and thoughtfully navigated, with the central purpose of uplifting the Greenwood community and assisting it to thrive into the future.

Acknowledgements

Michelle and Bobbie would like to give special thanks to Christopher Creese of Creeseworks for his kind sharing of the photo taken at Greenwood Cultural Center during President Biden's 2021 visit. We would also like to thank the Tulsa Historical Society & Museum for sharing some of their photos of the Tulsa Race Massacre.

References

Brown, D. (2021, June) Generations Lost. *National Geographic*.

Canfield, K. (2021a) 'A hot-button issue': Not all Black Tulsans are happy about Greenwood Rising. *Tulsa World*, 28 May 2021, updated on 2 June 2021 (accessed 15 January 2022). https://tulsaworld.com/news/local/racemassacre/a-hot-button-issue-not-all-black-tulsans-are-happy-about-greenwood-rising/article_83c5f376-bef0-11eb-901e-5b122f1dc1e8.html.

Canfield, K. (2021b) Watch Now: Remember & Rise event collapsed after frantic week of meetings, emails and talks about survivor payments. *Tulsa World*, 29 May 2021, updated on 30 May 2021. See https://tulsaworld.com/news/local/watch-now-remember-rise-event-collapsed-after-frantic-week-of-meetings-emails-and-talks-about/article_29d9ef7a-bfc7-11eb-94ea-0b3aa94ba85a.html (accessed 15 January 2022).

Ellsworth, S. (1982) *Death in a Promised Land: The Tulsa Race Riot of 1921*. Baton Rouge: Louisiana State University Press.

Ellsworth, S. (2022) The Tulsa Race Massacre: Causes, Cover-Up and the Ongoing Fight for Justice (Scott Ellsworth, University of Michigan). Online lecture as part of the DAAS History-EIHS MLK Day Symposium Event. See https://events.umich.edu/event/90099 (accessed 17 January 2022).

Franklin, J.L. (1980) *The Blacks in Oklahoma*. Norman: University of Oklahoma Press.

Greenwood Cultural Center (n.d.) About Us. See https://www.greenwoodculturalcenter.org/about-us (accessed 19 January 2022).

Greenwood Rising (n.d.) About. See https://www.greenwoodrising.org/about (accessed 19 January 2022).

Johnson, H.B. (2014) *Tulsa's Historic Greenwood District (Images of America)*. Charleston: Arcadia Publishing.

Latham, A. (1995) RUINS to Renaissance- Greenwood Cultural Center Symbol of History, Hope. *Tulsa World*, 15 October 1995. See https://tulsaworld.com/archive/ruins-to-renaissance-21-race-riot-devastated-greenwood/article_1fe3d6eb-9af3-5b14-94fb-fa0af1445339.html (accessed 19 January 2022).

Oklahoma Commission to Study the Tulsa Race Riot of 1921 (2001, 28 February) *Tulsa Race Riot: A Report by the Oklahoma Commission to Study the Tulsa Race Riot of 1921*. Oklahoma City, Oklahoma.

Sulzberger, A.G. (2011, 20 June) As survivors dwindle, Tulsa confronts past. *New York Times (Online)*. See https://web.archive.org/web/20151211093423/http://www.nytimes.com/2011/06/20/us/20tulsa.html?_r = 0 (accessed 15 January 2022).

Part 4
Imagining New Futures

11 Convivial Tourism in Proximity

Nora Müller, Robert Fletcher and Macià Blázquez-Salom

Introduction

The COVID-19 lockdown has demonstrated the importance of the mental and physical health benefits provided by undertaking outdoor activities in open spaces (Sandifer *et al.*, 2015). In the context of the pandemic, such spaces also function as a refuge from potential infection. In this chapter we explore the consequences of this dynamic for the tourist–recreational use of natural spaces and their conservation. First, we place the coronavirus pandemic in the structural context of the ecological crisis and discuss the pandemic's impact on proximity and ecotourism. Second, we describe the history of conservation policy and practice that has tended to promote alienation of humanity from non-human nature. We present the alternative of a 'convivial conservation' to confront both structural and conjunctural problems of the human–non-human relationship the pandemic has exacerbated. To ground our analysis, subsequently, we present and discuss the case of *La Trapa* as an example of the intersection of convivial conservation and proximity tourism. This case has been chosen due to both its location within a mass tourism destination, Mallorca, and its management as a public common by a local non-governmental organisation (NGO) with the aim to facilitate local residents' contact with non-human nature. Our research concerning this case has been qualitative, entailing interviews with key local stakeholders (a farmer and two volunteers) as well as direct involvement as insider activists and participation in conferences on proximity and ecotourism. We have also analysed grey literature on the case, including public debate in newspapers and webpages produced by the managing NGO. We conclude by advocating a convivial shift in tourism rooted in local collective management of natural spaces and their recreational use.

COVID-19 in the Context of the Ecological Crisis

According to Malm (2020), the transmission of pathogens by animal species (zoonosis) is in large part the result of the excessive pressure humans exert on ecosystems. In Malm's analysis, zoonosis appears primarily due to deforestation – especially in tropical forests – to extract wood and extend the industrial cultivation of products such as palm oil or soybeans. Likewise, livestock farming to meet the increased global demand for meat increases contact between human and animal species (Hall *et al.*, 2020). Processes like these contribute to a decrease in biodiversity that erodes the natural buffer between humanity and non-human spaces given that, in general, 'greater biodiversity' means 'less risk of zoonotic transmission' (Malm, 2020: 58).

The COVID-19 pandemic is thus yet another signal of the mounting global ecological crisis rooted in human exploitation of the non-human world that has caused the destruction of habitats and the consequent loss of species worldwide. According to the World Wildlife Fund's *Living Planet Report 2018*, global species populations diminished 60% between 1970 and 2014. Species decline is especially pronounced in the tropics, with a loss of 89% compared to 1970.

Tourism is one key contributor to the ecological crisis, in that the industry contributes to the urbanisation of previously open spaces and the consequent destruction of habitats. The initial spread of COVID-19 along important tourism routes as well as the dramatic efficacy of the subsequent global lockdown in slowing the virus' further spread demonstrates that global hypermobility to transport both people and commodities also plays an important role in zoonosis (Hall *et al.*, 2020). Tourism flows also exacerbate the energetic and climatic crises; the industry is responsible for 8% of global greenhouse gases (GHG) emissions (Lenzen *et al.*, 2018) and 75% of the industry's energy demand comes from transport (Gössling *et al.*, 2010).

Capitalism can be understood as a root cause of the ecological crisis as the culture of consumption on which it relies implies the exploitation and commodification of nature and promotes the long-distance movement of goods and people (Malm, 2020; Moore, 2015). At the same time, it also produces social crises in encouraging sociospatial segregation on the basis of social class (Castree, 2005; Smith, 2008). The COVID-19 pandemic has also exacerbated this crisis in allowing the wealthiest class to disproportionately enjoy use of safe spaces of the highest environmental quality.

Tourism Debates in the Pandemic Setting

With the end of the hard pandemic lockdown in 2021, European protected areas received an upsurge in visitor numbers (McGinlay *et al.*, 2020) as a response to the increased local demand for outdoor recreation in proximity while much international travel remained off limits. In this context, the provision of public recreational areas for leisure activities and relaxation gained importance, while at the same time 'proximity tourism' was offered as a potential alternative in conditions of restricted hypermobility (Blanco-Romero & Blázquez-Salom, 2020; Cañada, 2020).

As Díaz-Soria and Llurdes-Coit (2013) demonstrate, proximity tourism can be understood in various ways: spatially, in terms of geographical distance from one's place or origin; organisationally, in terms of separation between spaces of production and consumption; and conceptually, in terms of coming back to known places even though this implies travel over large geographical distance. In all of these senses, proximity is understood in opposition to the World Tourism Organisation's (UNWTO, 2001) characterisation of conventional tourism as 'staying in places outside the usual environment' (which of course raises further questions concerning how to define this 'usual environment' (see Diaz-Soria & Llurdés-Coit, 2013: 71 ff.). Here, we understand proximity tourism primarily in the spatial sense of physical distance, entailing the aim to cultivate appreciation of unfamiliar aspects of one's everyday environment.

The demand for proximity tourism rises when long-distance travelling is no longer possible (Diaz-Soria, 2017). During the COVID-19 pandemic, travel restrictions have impeded tourism activity, while the economic recession has also hindered visitation to faraway places. In this context, proximity tourism allows one to avoid high travel costs while also providing the possibility to get in touch with nature and to leave daily routines or 'usual environments' behind (Govers *et al.*, 2008). Proximity tourism is also commonly related to an increased environmental awareness as a response to the ecological impacts of long-distance travelling.

On Spain's Balearic Islands, the halt of tourism due to the pandemic entailed serious economic difficulties as the Islands' economy is highly dependent on tourism activity. Consequently, inter-island travelling was incentivised through travel vouchers (AETIB, 2021) and promotion of national tourism intensified. In this context, proximity tourism was encouraged as a way to (re)activate the economy dependent on tourist flows. Yet Hall *et al.* (2020) caution that restimulation of tourism based on nearby domestic destinations may be short lived and not a long-lasting transformation.

The pandemic has also been framed as an opportunity to implement ecotourism more widely and reactivate the tourism sector in the Balearics and Spain in relation to increased demand to visit open, natural spaces without crowding and in small, familiar groups. During the 5th National Meeting of Nature Observation Tourism (29–31 October 2020, Majorca) the implementation of 'authentic ecotourism' was presented as a way to 'deseasonalise' and 're-grow' tourism. The payment to enter (protected) natural areas in compensation for the negative impacts of ecotourism activity is proposed as an answer to the environmental crisis, and yet also reinforces previous patterns of commodification and privatisation of nature. With pandemic travel restrictions, critiques concerning the consequences of commodifying nature through ecotourism (e.g. West & Carrier, 2004) are underscored, particularly in established ecotourist destinations where the sudden lack of income results in reduced financing for protected areas and conservation activities that affects simultaneously workers and businesses as well as species and habitats. The pandemic has made clear that ecotourism cannot provide for a stable economy for local peoples while depending on external market forces, nor can it facilitate durable conservation if this is only based on the monetary income ecotourism generates (Refisch, 2020; Serhadli, 2020). Issues such as these have led to calls for the need to socialise tourism to enhance its contribution to the well-being and livelihood of local communities (Higgins-Desbiolles, 2020) and for alternative support for conservation beyond ecotourism markets (Fletcher *et al.*, 2020).

However, the increase in visitor numbers to public recreation areas (like protected areas, parks, picnic areas, etc.) also puts pressure on these sites and demonstrates the immanent contradiction entailed in efforts to demarcate areas of 'nature' with the aim to protect them from human intervention while simultaneously promoting their recreational-tourist uses. These have long been the basic aims of the creation of protected areas within Western mainstream conservation (Brockington *et al.*, 2010). The overcrowding experienced in certain natural sites after the lockdown impacted the environment, especially due to littering and issues related to parking and traffic increase, disturbance of fauna and harming of vegetation, as well as provoking conflicts between residents and visitors (McGinlay *et al.*, 2020; Muñoz-Navarro, 2020). In the face of such impacts, calls for respectful behaviour in natural sites were pronounced (Muñoz-Navarro, 2020). Others proposed to put limits on visitor numbers and stronger restrictions have been implemented in some cases (Bosch, 2020; Eldiario.es, 2020; Pareja, 2021). But such measures are problematic considering the contribution that contact with non-human natures can make to social well-being and the imperative to provide spaces for recreation without discrimination due to social class. They also hint at the risk of land enclosure in private reserves for the privileged. How, then, should outdoor recreation and conservation be managed, bearing in mind both the biophysical integrity of the sites and social concerns of equity and well-being?

The Proposal of Convivial Conservation in the Realm of Tourism

In response to issues with conventional conservation approaches such as those previously outlined, Büscher and Fletcher (2020) propose the 'convivial conservation' approach. This is presented as a post-capitalist alternative that seeks to overcome the human-nature dichotomy inherent in the conventional approaches. It also is based on principles of social and environmental justice rather than market logics.

Büscher and Fletcher (2020) describe five elements of the convivial conservation proposal. First, the establishment of 'promoted areas' with the aim to transcend the idea of protected areas that aim to save nature from humanity, but instead 'to promote nature for, to and by humans' (2020: 163). Second, the 'celebrating of human and non-human nature' (2020: 165), which entails the overcoming of human-nature dichotomies entangled in the idea to save (non-human) nature. Instead, convivial conservation aims to appreciate diversity and recognise the various needs, wishes and actions of both human and non-human natures. Third, 'engaged visitation' (2020: 168) is envisaged to contest the consumerist idea of conserved natures for tourism that implies the risk of becoming an elite privilege. Conviviality, by contrast, means fostering long-term visitations, close to where one lives. Fourth, to go beyond the 'spectacle' of nature – rooted in commodification and promoted in the frame of capitalist conservation – towards 'everyday environmentalism' (2020: 170). This implies the appreciation of the mundane everyday nature around us. Finally, the encouragement of 'common democratic engagement' is fundamental for conviviality, implying a democratic management of 'nature-as-commons' and 'nature-in-context'. This entails also the change of the value system from understanding 'nature as capital'. In other words, decisions concerning nature management should not be based on promoting continuous economic growth and monetary valuation but instead on the local context and non-capitalist needs of the affected actors.

What does all of this imply in the realm of tourism specifically? Proximity tourism, as discussed, can be linked to the convivial conservation proposal of engaged visitations and everyday environmentalism by appreciating the mundane nature close to where one lives. Consequently, proximity tourism has the potential to foster responsibility with respect to the local natural and socioeconomic environment (including local customs, biodiversity and traditions), to create the possibility of getting to know and connect with natures close to where one lives and to encourage recreation for local peoples as a space of rest, enjoyment and personal and collective development (Schenkel, 2021). Additionally, for conviviality to be realised in tourism, 'nature-as-commons' and 'nature-as-context' has to be internalised. This implies that tourism is directed to provide value that is grounded in the local context, appreciating local ecosystems and cultural forms (Higgins-Desbiolles & Bigby, 2022) and within common property regimes (Fletcher, 2019), thereby contesting the privatisation of nature and the sociospatial segregation this entails. As a practical example that illustrates the potential of these five proposals in tourism development, we present an analysis of a case study located in the mass tourism destination of Mallorca.

Conservation for the Elite

The issues previously highlighted resonate with the history of conservation activity more generally. Demarcated natural areas are based on the physical and social division between humans and non-humans, thereby creating separate areas of 'nature' to be enjoyed by the elite (Brockington *et al.*, 2010). Historically, private hunting reserves can be considered the first protected areas with restricted access for the aristocracy (Santamarina Campos, 2019). With the creation of Yellowstone National Park in the USA in 1872, the idea of protected areas spread globally and outdoor recreation became a fundamental value of Western conservationism. The widespread demarcation of protected areas is promoted in a Fordist and colonial context in which the state plays a central role in their creation and management (Büscher & Fletcher, 2020) and is globally established with the classification system of protected areas first developed by the International Union for Conservation of Nature (IUCN) in the 1970s. In this way, the environmental problems caused by the capitalist exploitation of the planet are addressed through the creation of protected areas as a compensatory means to conserve nature and promote its recreational use, while most of the rest of the planet continues to be exploited.

Critical analysis by Adams and Hutton (2007) reveals that the creation of protected areas – also of tourist areas – implies dispossession of local people that manifests power relations and reproduces class structures. Dispossession affects local communities as they may lose their lands, which are often the means of subsistence, thus impacting income, livelihood and cultural roots (Holmes & Cavanagh, 2016). The displacement of local residents also reinforces the division between urban and rural populations and between rich and poor. Protected areas are rooted in the idea of a landscape created by and for the elite; originally the aristocracy and sovereigns, today the capitalist class. Moreover, environmental degradation due to pollution or the spread of zoonosis does not stop at the boundaries of protected areas.

Demarcation of natural areas creates an imaginary of nature as a wild, pristine and idyllic place for recreation, which can only be found apart from society. In this way, protected areas become the basis for a niche in the tourist market focused on nature-based or ecotourism. In fact, all forms of tourism depend on nature as a foundation, whether in terms of beaches, mountains, coral reefs, etc. But a romanticised imaginary of nature as a space of tranquillity and wilderness is created that can be commercialised through tourist-recreational offers like ecotourism in particular (Igoe, 2010).

Ecotourism is commonly presented as a win-win strategy in which local communities and ecosystems both benefit from the income generated (Stronza *et al.*, 2019). However, if the hypermobility underpinning ecotourism involves long-distance journeys, often by airplane, this contradicts its fundamental objective of conserving nature (Carrier & Macleod, 2005). Ecotourism in a neoliberal context is characterised by a displacement of public actors in favour of promotion of local governance and private actors in conservation (Büscher & Fletcher, 2020). Ecotourism revenue finances conservation activities and exemplifies the commodification of nature enjoyment. It provides an economic incentive for conservation based on market logics that is presented as a path to development and modernity in impoverished countries (Duffy, 2006). In this perspective, private protected areas offer a business model based on the commodification of nature for leisure and recreation – like ecotourism products – and the privatisation of the territory with the rationale to protect nature (Artigues-Bonet & Blázquez-Salom, 2016; Borrie *et al.*, 2020). This business model is embedded in the logics of neoliberal conservation that stimulate the expansion of capitalism to all spheres of life (Fletcher, 2020) and that thereby incorporates nature into processes of privatisation, commodification and deregulation (Rose & Carr, 2018). McAfee (1999) questions this approach to conservation as 'selling nature to save it', which responds to the dissociation between capitalism and nature. Neoliberal conservation is critiqued for provoking negative impacts like restrictions to resources and displacement, reinforcing inequalities in economic, social and environmental terms (Holmes & Cavanagh, 2016).

In summary, the tourism–nature relationship is based, on the one hand, on the separation of humans and non-humans through the material and symbolic redefinition of one part of the environment as a distinct realm of 'nature' (Ruiz-Ballesteros, 2019). On the other hand, the value of this isolated nature is emphasised in economic terms by putting a price on it via commercialised tourist experiences. Nature's conservation, then, is necessary to be able to sell tourism products that in turn finance conservation activities (Corson *et al.*, 2014). This neoliberal approach to conservation is described by Büscher (2012: 29) as 'the paradoxical idea that capitalist markets are the answer to their own ecological contradictions'. But can market-based instruments solve environmental problems that they provoked in the first place?

The Case of La Trapa (Mallorca)

La Trapa is a mountain farmstead of 80 hectares in the municipality of Andratx in the southwest of Mallorca (Figure 11.1). It owes its name to the Trappist monks who inhabited the area at the beginning of the 19th century after fleeing the French revolution. The Trappists built a chapel and agricultural infrastructure, including a watering system, threshing floor and terraces for cultivation (Bover i Pujol, 1987). After 1824 the property holders changed various times and the agricultural infrastructure was enlarged.

Figure 11.1 La Trapa. Credit: Co-author Macià Blázquez-Salom

The farmstead was abandoned in the middle of the 20th century and at the end of the 1970s plans to subdivide and urbanise the territory of La Trapa were presented. It was in this moment that the Balearic Group of Ornithology and Nature Defence (GOB), a local NGO, initiated a campaign to prevent this urbanisation and to protect the territory and its cultural and natural values. Through a membership subscription, an art auction and charitable contributions from individuals, enterprises and organisations, sufficient money was collected to purchase the farmstead in 1980.

Nowadays it is owned and managed by the GOB for the purpose of conserving the natural and cultural values of La Trapa, protecting nesting areas and preserving agricultural facilities. Formally, La Trapa was protected through different designations: as a Particular Zone for the Protection of the Birds; a Zone of Special Conservation; as embedded in the Natural Site Serra de Tramuntana (*Paraje Natural*); and also in the World Heritage Serra de Tramuntana. GOB supports scientific research, the dissemination of knowledge about La Trapa and its public enjoyment.

The management of La Trapa encourages volunteering for the maintenance and conservation of the farmstead and its values. Therefore, on the first Sunday of every month the GOB invites volunteers to participate in farm work and other tasks. La Trapa is also a site for environmental education for local schools and the youth of the GOB. Free, public access for recreational purposes is facilitated through sign posting and without limitations on visitor numbers.

La Trapa is financed from different sources. Firstly, part of the membership fee is dedicated to La Trapa. Furthermore, there are public subsidies, for example for projects of land stewardship from the Consell de Mallorca (BOIB 2/9/2021, Nr. 119, Section III, 9341) and for agricultural activities. Additionally, donations from individuals and organisations support the financing of La Trapa. There is also a donation box when arriving at the houses of La Trapa for collection from the visitors. Finally, some products are given in return for donations, like honey or chamomile (Figure 11.1).

The decision-making process of La Trapa is organised collectively. The general assembly of the GOB is made up of members who volunteer; they discuss and decide on fundamental issues regarding the management of La Trapa. A commission is formed that executes these decisions and presents proposals to be discussed in the general assembly. La Trapa is only one small part of the GOB's activities; GOB's larger focus includes improving awareness of nature conservation, enhancing democratic decision-making and promoting socioecological sustainability.

La Trapa as a 'promoted area' in proximity

In the following section, we discuss La Trapa in the framework of the convivial conservation proposal. Convivial conservation aims to overcome human-nature dichotomies in a way that goes beyond capitalist logics and encourages active democratic participation in promoting *living with* nature.

The initial motive for the campaign to purchase La Trapa reflects the same logic as found in most protected areas; that is, the aim to save nature from humanity and in the form of a planned subdivision and urbanisation of the property. Still today this argument is prevalent among stakeholders. One interviewee claimed that it is critical that the GOB maintains La Trapa as a reserve to ensure its conservation (Interviewee #2, pers. comm., 15 September 2021) reflecting the predominant human-nature dichotomy in conservation planning.

On the other hand, foundational to the management aims of La Trapa is the conservation of both the ecosystem functioning and the cultural heritage. This corresponds to the convivial aim to respect all natures – human and non-human – without discrimination and accepting the interrelation between these. In this perspective human nature forms and is formed by non-human nature. One of the interviewees explained it as follows when describing how he came to an abandoned farmstead to undertake maintenance:

I arrived there. It was all abandoned. There was nothing. [...] Of course, this is what nature does, eating up everything. But, when you arrive and start to clear up and to water, if you want it or not, more birds are appearing, there is more life, more food, more animals and worms. Eagles start to come. I see more nature. (Interviewee #1, pers. comm., 22 July 2021)

Thus, the maintenance and reconstruction of the agricultural facilities and infrastructure forms part of the convivial management of La Trapa to establish an equilibrium between different species – human and non-human – and their shared habitat. In this form of conservation, knowledge of the environment – including its cultural and natural values – is fundamental. At La Trapa scientific research and the dissemination of the values and knowledge obtained from this are stimulated. For example, bird censuses are undertaken and a collaboration with the association *Varietats Locals* (Local Varieties) was established to investigate local vegetable species with the aim to preserve these species that are more adapted to the local ecosystem.

The environmental education programme at La Trapa further supports the dissemination of knowledge related to the nature and culture of La Trapa. Moreover, the regular volunteering promotes a process of learning by doing in conservation and farm tasks. Additionally, the volunteering programme creates the possibility to participate in natural spaces close to where one lives and to experience everyday natures through sustained, engaged visitation. Interviewees explained that it is an established group of constantly engaged volunteers, being always 'the same', who participate in the work (Interviewee #2, pers. comm., 15 September 2021). This may lead to the creation of a coherent community that is fundamental to effective management of a commons (de Angelis, 2017). However, it also presents the risk that 'outsiders' are not welcomed or that they may perceive it like this.

The opening of free and public access to La Trapa provides the opportunity for short-distance visitation as an alternative to long-distance travelling (Figure 11.2). Nevertheless, La Trapa is embedded in the broader tourist destination, the Serra de Tramuntana in particular, and Mallorca in general. Both destinations – the Serra de Tramuntana and Mallorca – are promoted by different agents including international tour operators, the local tourism agency and local tourism companies. Additionally, the visitors themselves engage in promotion through the use of social media, 'the 'word of mouth' propaganda of nowadays,' as one interviewee called it (Interviewee #2, pers. comm., 15 September 2021). In contrast, the GOB themselves do not undertake tourist advertising to promote visitation to La Trapa, but instead promote volunteering and engaged visitation.

La Trapa facilitates access by all social classes without discrimination, thereby preventing sociospatial segregation. The access from the coastal village Sant Elm is facilitated through signposting and the interviewees confirm that the visitor profile is diverse. There are 'professional hikers'

Figure 11.2 Visitors at the viewpoint of La Trapa. Credit: Co-author Macià Blázquez-Salom

walking routes like the Grand Route 221 (crossing the whole Serra de Tramuntana), and 'Sunday recreationists' (*domingueros*), families and friends coming to have a picnic, trail runners, unprepared walkers, and many more. 'It's normal people. Workers. Sometimes people with a lot of money,' as one interviewee explained (Interviewee #1, pers. comm., 22 July 2021,). All enjoy the easy and economical opportunity to spend their leisure time and leave behind their daily routines (Figure 11.2).

However, various conflicts and contradictions are related to the recreational-tourist use of La Trapa. First, there appear to be issues from overcrowding like erosion and disturbance of the fauna, littering and degradation of the farmsteads' facilities (Interviewee #2, pers. comm., 15 September 2021). These were especially problematic after the end of the lockdown – as in other places – highlighting also the need the pandemic stimulated to reconnect human and non-human natures. However, one interviewee explained, 'generally the people who visit La Trapa are respectful' (Interviewee #2, pers. comm., 15 September 2021). The closure or limitation of visitor numbers is not seen as a solution to overcrowding. While the commission has discussed limiting visitor numbers, the general assembly of the GOB strictly defends open and free access to La Trapa. 'The same right [to visit the mountains] that holds for me, holds for all the others,' one person asserted (Interviewee #2, pers. comm., 15 September 2021). What is missing, according to interviewees, is a culture of respect that can only be achieved through education and awareness-raising (Interviewee #2, pers. comm., 15 September 2021). Hence, the educational and dissemination work undertaken by the GOB presents an opportunity to reinforce a more convivial culture, at the same time as it provides a recreational offer for all publics.

The second contradiction arises with the property regime of La Trapa. The interviewees themselves are confused as to exactly how this functions. At certain points in the interviews, they all emphasised that La Trapa is a public commons and everyone has a right to access it: 'But actually, La Trapa is a good of everyone' (Interviewee #2, pers. comm., 15 September 2021). At other points, however, they emphasised the fact that La Trapa is the property of the GOB. For example: '[Some visitors said:] "the mountain belongs to everyone". No. The mountain does not. This is owned by the GOB' (Interviewee #2, pers. comm., 15 September 2021). Another stated: 'Lots of people came and I ask: "Do you know who owns La Trapa?" They answer: "Yes. It belongs to Mallorca". No! La Trapa is not of Mallorca, it is of the GOB!' (Interviewee #1, pers. comm., 22 July 2021).

La Trapa and its commons also face the structural threat of enclosure and privatisation of nature. On the one hand, the members of the commission and volunteers are discussing issues in restricting access to the farmstead. On the other hand, the ownership of La Trapa is emphasised as a means to defend the territory from unwelcomed visitors. It is worth noting that this question is particularly accentuated with respect to visitors from mainland Spain (Interviewee #2, pers. comm., 15 September 2021), demonstrating a widespread conflict between the Mallorcan people and those from the Spanish mainland. A cultural turn towards a more tolerant ethos of *living with* is needed to embody conviviality without discrimination.

The third contradiction relates to the embeddedness of La Trapa in the international tourism destination Mallorca, and more specifically, the Serra de Tramuntana that is promoted as a destination for nature-based, active and luxury tourism for visitors, mostly from Europe. The simple fact of being located on an island makes it difficult to encourage an alternative to long-distance travelling and underlines the need of a more general transformation in tourism and its promotion to pursue conviviality. Yet La Trapa still holds potential to offer a natural site in which to leave behind daily routines or the 'usual environment' as an alternative to long-distance travelling based on collective management by and for the local peoples.

The financial situation of La Trapa is critical and the interviewees complain that there is a general lack of finance (Interviewee #2, pers. comm., 15 September 2021; Interviewee #1, pers. comm., 22 July 2021). Basically, two strategies to address this situation are proposed: first, the selling of products and second, the offer of accommodation. Some products are already sold for donations, like herbs or honey (Figure 11.1); however, different ideas to diversify the agricultural production to sell wine or flour, for example, were discussed by one of the interviewees (Interviewee #1, pers. comm., 22 July 2021). However, the human and financial resources to begin such projects are limited. This interviewee insisted, anyhow, that 'La Trapa has a huge marketing potential' (Interviewee #1, pers. comm., 22 July 2021), putting emphasis on the structural risk of commodifying nature to protect it – in this case, to preserve the farmstead and its territory. Moreover, reconstruction of the houses for the provision of accommodation has been identified as a source of additional financial income. Interviewees named a diversity of possible target groups for the rebuilt houses: providing mountain shelter for hikers, scouts or youth groups, offering a site for yoga retreats, and other events or seminars with different topics, like ornithology or alternative agriculture techniques. The stimulation of activities of this kind at La Trapa is proposed as fundamental to gain revenue (Interviewee #2, pers. comm., 15 September 2021), yet this may provoke overcrowding and the related negative impacts of disturbance and degradation. Nevertheless, the main threat described here is the consumerist logic to exploit and commodify natural environments. From a convivial conservation perspective, it is therefore critical to explore new forms of redistributive financing that are not based in such market logics.

Finally, the collective process of decision-making as described by the interviewees facilitates the construction of meaningful long-term engagement with nature. This is further fostered through the organisation of in-situ volunteering, with volunteers that are long-term and committed. These aspects are fundamental for a cultural turn in tourism that is embedded in the local community.

In summary, La Trapa can be conceptualised as a 'promoted area' where 'people are considered welcome visitors, dwellers or travellers' and that encourages 'long-lasting, engaging and open-ended relationships' (Büscher & Fletcher, 2020: 164). At La Trapa, in particular, through volunteering and knowledge generation and dissemination, the possibility to participate in decision-making and open and free access without discrimination is available. That is fundamental for the dismantling of the human-nature dichotomy. Furthermore, these activities and management actions encourage appreciation of everyday mundane natures and enable sustained visitation and prolonged human-nature relationships.

Nevertheless, the demarcation of a concrete space and the private property regime also presents a threat to conviviality at La Trapa because in some situations visitors are not welcomed and the private property is defended. Thus, it is important to explore how the demarcation of a concrete site stresses the separation between different natures and what other property regimes may be more appropriate to overcome this separation between human and non-human natures, such as co-ownership or cooperative properties.

Currently, La Trapa is insufficiently financed through subsidies and donations. The risk of the implementation of market mechanisms such as tourism and recreational exploitation and commodification of the site has been proposed by some interviewees as a solution. Therefore, the exploration of non-market financial mechanisms is important for the cultural transformation towards conviviality, such as has been proposed with the idea of a Conservation Basic Income (CBI): an unconditional payment to community members residing inside and close to promoted areas with the aim to empower people to manage their resources democratically and independently from market logics (see Büscher & Fletcher, 2020: 292).

Conclusion

The COVID-19 pandemic revealed structural problems in current efforts aimed at the conservation of nature and its tourist-recreational use. The consequences of the pandemic for tourism have been disastrous and call for reflection on the various ways that the industry's embeddedness within capitalism have contributed to this. The convivial conservation proposal presents a post-capitalist approach that promotes social and environmental justice based on connection and respect. Part of this proposal entails collective management of natural areas that serve the tourist-recreational use rooted in the local community. Finally, it advocates for financial mechanisms based on local sources and subject to collective control to avoid the dependence on highly volatile global markets.

Transferring conviviality to tourism praxis therefore implies the following:

(1) the strengthening of proximity and long-duration visitation to minimise the environmental impacts of displacement and to promote long-lasting relationships of respect and care;
(2) the decommodification of leisure in natural areas and the provision of free public access to spaces without discrimination due to social class;
(3) active and broad-based participation in the process of decision-making, enhancing collective management of tourism for and by the local community;
(4) protection of all natures – ecosystems and cultures – based on collective knowledge and equality;
(5) financing based on redistribution, independent from global markets.

In relation to these principles, the case of La Trapa can be understood as a relatively successful example of the convivial management of nature and leisure activities that is grounded in a common property managed and controlled by and for the local community. Nevertheless, La Trapa is still embedded in global markets that promote long-distance travelling to the Balearics, Mallorca, and more specifically the Serra de Tramuntana. Thus, the transformation of tourism here implies also the transformation of more global trends to ensure social well-being and local livelihoods. Consequently, the exploration of new approaches to financing nature enjoyment and its sustainable management is needed that resolves the contradictions of capitalist touristic markets.

Future exploration of this potential would include analysing everyday environmentalism closer to people's homes in their cities of origin. As Higgins-Desbiolles (2018) proposes, this could make the places of residence more attractive and long-distance travelling less important for cultural learning and tolerance. In closing, our analysis here demonstrates that a convivial conservation approach can contribute to addressing the ecological crisis we described. In the realm of tourism, in particular, proximity tourism presents an alternative to hypermobility but only if it is able to establish a relationship of respect and care with the environment and its inhabitants based on common management and embeddedness in the local context.

References

Adams, W.M. and Hutton, J. (2007) People, parks and poverty: Political ecology and biodiversity conservation. *Conservation and Society* 5 (2), 147–183.

AETIB (Agència d'Estratègia Turística de les Illes Balears) (2021) Bonus turístic per a residents. See https://bonusturistic.illesbalears.travel/ca/index.html?privacy (accessed 3 November 2021).

Artigues-Bonet, A.-A. and Blázquez-Salom, M. (2016) Huídas al paraíso y la realización mercantil del sueño. In N. Benach, M. Zaar and M. Vasconcelos (eds) *Las Uropías y la Construcción de la Sociedad del Futuro* (Online). Symposium conducted at the meeting of Universitat de Barcelona, Universitat de Barcelona. See http://www.ub.edu/geocrit/xiv-coloquio/ArtiguesBlazquez.pdf (accessed 16 March 2022).

Blanco-Romero, A. and Blázquez-Salom, M. (2020) Domesticar el turismo: La proximidad en la desescalada. *Albasud*, blog post, 14 May. See www.albasud.org/blog/es/1216/domesticar-el-turismo-la-proximidad-en-la-desescalada (accessed 12 April 2021).

Borrie, W.T., Gale, T. and Bosak, K. (2020) Privately protected areas in increasingly turbulent social contexts: Strategic roles, extent, and governance. *Journal of Sustainable Tourism* 1–18. https://doi.org/10.1080/09669582.2020.1845709.

Bosch, R.M. (2020) El verano más inquieto del Montseny. *La Vanguardia*, 26 September. See https://www.lavanguardia.com/local/barcelona/20200926/483661042520/montseny-parque-natural-impacto-pandemia.html (accessed 22 April 2021).

Bover i Pujol, J. (1987) La Trapa de s'Arracó: Breu resum històric. *Revista Del Centre D'estudis Teològics De Mallorca* 45–46, 43–50.

Brockington, D., Duffy, R. and Igoe, J. (2010) *Nature Unbound: Conservation, Capitalism and the Future of Protected Areas*. London: Earthscan. https://doi.org/10.4324/9781849772075.

Büscher, B. (2012) Payments for ecosystem services as neoliberal conservation: (Reinterpreting) evidence from the Maloti-Drakensberg, South Africa. *Conservation and Society* 10 (1), 29–41. https://doi.org/10.4103/0972-4923.92190.

Büscher, B. and Fletcher, R. (2020) *The Conservation Revolution: Radical Ideas for Saving Nature Beyond the Anthropocene*. London: Verso.

Cañada, E. (2020) Turismos de proximidad, un plural en disputa. *Alba Sud*, blog post, 8 July. See http://www.albasud.org/blog/es/1236/turismos-de-proximidad-un-plural-en-disputa (accessed 22 April 2021).

Carrier, J.G. and Macleod, D.V. (2005) Bursting the bubble: The socio-cultural context of ecotourism. *Journal of the Royal Anthropological Institute* 11 (2), 315–334. https://doi.org/10.1111/j.1467-9655.2005.00238.x.

Castree, N. (2005) *Nature*. London: Routledge.

Corson, C., Gruby, R., Witter, R., Hagerman, S., Suarez, D., Greenberg, S. . . . Campbell, L. (2014) Everyone's solution? Defining and redefining protected areas at the convention on biological diversity. *Conservation and Society* 12 (2), 190–202.

De Angelis, M. (2017) *Omnia sunt communia: On the Commons and the Transformation to Postcapitalism*. London: Zed Books.

Diaz-Soria, I. and Llurdés-Coit, J.C. (2013) Reflexiones sobre el turismo de proximidad como una estrategia para el desarrollo local. *Cuadernos De Turismo* 32, 65–88.

Diaz-Soria, I. (2017) Being a tourist as a chosen experience in a proximity destination. *Tourism Geographies* 19 (1), 96–117. https://doi.org/10.1080/14616688.2016.1214976.

Duffy, R. (2006) The politics of ecotourism and the developing world. *Journal of Ecotourism* 5 (1–2), 1–6. https://doi.org/10.1080/14724040608668443.

Eldiario.es (2020) Catalunya restringirá el acceso a los parques naturales para evitar su saturación durante la pandemia. *El Diario*, 22 October. See https://www.eldiario.es/catalunya/catalunya-restringira-acceso-parques-naturales-evitar-saturacion-durante-pandemia_1_6311750.html (accessed 22 April 2021).

Fletcher, R. (2019) Neoliberalismo y turismo. In I. Murray Mas and E. Cañada (eds) *Turistificación global: Perspectivas críticas en turismo* (Vol. 493). (pp. 37–52). Barcelona: Icaria.

Fletcher, R. (2020) Neoliberal conservation. In Oxford University Press (ed.) *Oxford Research Encyclopedia of Anthropology*, encyclopaedia entry, 28 September. See https://oxfordre.com/anthropology/view/10.1093/acrefore/9780190854584.001.0001/acrefore-9780190854584-e-300?rskey=43lrkv (accessed 31 January 2022).

Fletcher, R., Büscher, B., Massarella, K. and Koot, S. (2020) 'Close the Tap!': COVID-19 and the need for convivial conservation. *Journal of Australian Political Economy* 85, 200–211.

Gössling, S., Hall, C.M., Peeters, P. and Scott, D. (2010) The future of tourism: Can tourism growth and climate policy be reconciled? A mitigation perspective. *Tourism Recreation Research* 35 (2), 119–130. https://doi.org/10.1080/02508281.2010.11081628.

Govers, R., van Hecke, E. and Cabus, P. (2008) Delineating tourism: Defining the usual environment. *Annals of Tourism Research* 35 (4), 1053–1073. https://doi.org/10.1016/j.annals.2008.09.001.

Hall, C.M., Scott, D. and Gössling, S. (2020) Pandemics, transformations and tourism: Be careful what you wish for. *Tourism Geographies* 22 (3), 577–598. https://doi.org/10.1080/14616688.2020.1759131.

Higgins-Desbiolles, F. (2018) Sustainable tourism: Sustaining tourism or something more? *Tourism Management Perspectives* 25, 157–160. https://doi.org/10.1016/j.tmp.2017.11.017.

Higgins-Desbiolles, F. (2020) Socialising tourism for social and ecological justice after COVID-19. *Tourism Geographies* 22 (3), 610–623. https://doi.org/10.1080/14616688.2020.1757748.

Higgins-Desbiolles, F. and Bigby, B.C. (2022) A local turn in tourism studies. *Annals of Tourism Research* 92, 103291. https://doi.org/10.1016/j.annals.2021.103291.

Holmes, G. and Cavanagh, C.J. (2016) A review of the social impacts of neoliberal conservation: Formations, inequalities, contestations. *Geoforum* 75, 199–209. https://doi.org/10.1016/j.geoforum.2016.07.014.

Igoe, J. (2010) The spectacle of nature in the global economy of appearances: Anthropological engagements with the spectacular mediations of transnational conservation. *Critique of Anthropology* 30 (4), 375–397. https://doi.org/10.1177/0308275X10372468.

Lenzen, M., Sun, Y.-Y., Faturay, F., Ting, Y.-P., Geschke, A. and Malik, A. (2018) The carbon footprint of global tourism. *Nature Climate Change* 8 (6), 522–528. https://doi.org/10.1038/s41558-018-0141-x.

Malm, A. (2020) *El Murciélago y el Capital: Coronavirs, Cambio Climático y Guerra Social. [Corona, Climate, Chronic Emergency: War Communism in the Twenty-First Century]*. Madrid: Errata naturae.

McAfee, K. (1999) Selling nature to save it? Biodiversity and green developmentalism. *Environment and Planning D: Society and Space* 17 (2), 133–154. https://doi.org/10.1068/d170133.

McGinlay, J., Gkoumas, V., Holtvoeth, J., Fuertes, R.F.A., Bazhenova, E., Benzoni, A. . . . Jones, N. (2020) The impact of COVID-19 on the management of European protected areas and policy implications. *Forests* 11 (11). https://doi.org/10.3390/f11111214.

Moore, J.W. (2015) *Capitalism in the Web of Life: Ecology and the Accumulation of Capital*. London: Verso.

Muñoz-Navarro, A. (2020) Gaudim del medi natural, però amb respecte a la biodiversitat, el paisatge i les persones. *GOB Mallorca*, blog post, 21 October. See https://www.gobmallorca.com/que-feim/biodiversitat/gaudim-del-medi-natural-pero-amb-respecte-a-la-biodiversitat-el-paisatge-i-les-persones (accessed 22 April 2021).

Pareja, P. (2021) El confinamiento municipal lleva al parque natural de Collserola al límite: 'Si seguimos abusando, se irá al garete'. *El Diario*, 6 February. See https://www.eldiario.es/catalunya/barcelona/confinamiento-municipal-lleva-parque-natural-collserola-limite-si-seguimos-abusando-ira-garete_1_7197399.html (accessed 8 April 2021).

Refisch, J. (2020) What COVID-19 means for ecotourism: Interview with Johannes Reisch, United Nations Great Apes Survival Partnership Programme. See https://www.unep.org/news-and-stories/story/what-covid-19-means-ecotourism (accessed 22 April 2021).

Rose, J. and Carr, A. (2018) Political ecologies of leisure: A critical approach to nature-society relations in leisure studies. *Annals of Leisure Research* 21 (3), 265–283. https://doi.org/10.1080/11745398.2018.1428110.

Ruiz-Ballesteros, E. (2019) De la naturalización al naturing: La emergencia del entorno como naturaleza. In J.A. Cortés-Vázquez and O. Beltran (eds) *Repensar la Conservación: Naturaleza, Mercado y Sociedad Civil. Reconsidering Conservation: Nature, Market and Civil Society* (Vol. 21). (pp. 107–123). Barcelona: Edicions de la Universitat de Barcelona.

Sandifer, P.A., Sutton-Grier, A.E. and Ward, B.P. (2015) Exploring connections among nature, biodiversity, ecosystem services, and human health and well-being: Opportunities to enhance health and biodiversity conservation. *Ecosystem Services* 12, 1–15. https://doi.org/10.1016/j.ecoser.2014.12.007.

Santamarina Campos, B. (2019) El inicio de la protección de la naturaleza en España: Orígenes y balance de la conservación. *Revista Española De Investigaciones Sociológicas* 168, 55–72.

Schenkel, É. (2021) El turismo en la agenda pública latinoamericana: ¿Cómo llegamos hasta aquí?. In E. Cañada and I. Murray Mas (eds) *#TourismPostCOVID19: Turistificación Confinada* (pp. 126–133). Barcelona: Alba Sud Editorial.

Serhadli, S. (2020) Market-based solutions cannot solely fund community-level conservation. *Mongaby*, commentary, 11 May. See https://news.mongabay.com/2020/05/market-based-solutions-cannot-solely-fund-community-level-conservation-commentary/⁣ (accessed 22 April 2021).

Smith, N. (2008) *Uneven Development: Nature, Capital, and the Production of Space* (3rd edn). Athens: Georgia Press.

Stronza, A.L., Hunt, C.A. and Fitzgerald, L.A. (2019) Ecotourism for conservation?. *Annual Review of Environment and Resources* 44, 229–253. https://doi.org/10.1146/annurev-environ-101718-033046.

UNWTO (2001) Tourism satellite account: Recommended methodological framework. See https://unstats.un.org/unsd/tradeserv/EGTS/updated%20TSA%20RMF%20v.1.pdf (accessed 20 December 2021).

West, P. and Carrier, J.G. (2004) Ecotourism and authenticity: Getting away from it all? *Current Anthropology* 45 (4), 483–498. https://doi.org/10.1086/422082.

World Wildlife Fund (2018) *Living Planet Report – 2018: Aiming Higher*. See https://c402277.ssl.cf1.rackcdn.com/publications/1187/files/original/LPR2018_Full_Report_Spreads.pdf (accessed 20 December 2021).

12 Towards a 'More-than-Tourism' Perspective for Localising Tourism

Phoebe Everingham and Sinéad Francis-Coan

The COVID-19 tourism pause has led to a plethora of analyses by tourism scholars examining the implications for tourism recovery around the world. Many scholars have used the pause to emphasise opportunities to rethink and transform unsustainable forms of tourism (see Ateljevic, 2020; Brouder, 2020; Cheer, 2020; Everingham & Chassagne, 2020; Lew *et al.*, 2020), with a push towards local tourism as part of this agenda (Higgins-Desbiolles & Bigby, 2022). While a localised agenda goes someway in rethinking tourism towards more equitable, sustainable and even regenerative forms of tourism, for Hall *et al.* (2020) global approaches and solutions should be prioritised; otherwise, there is a risk that the localisation approach to tourism recovery may well just be a form of 'selfish nationalism', where governments turn inwards, prioritising national interest rather than international solidarity.

We take an approach in this chapter that micro/macro, global/local binaries are not useful for rethinking tourism futures – the local and global are inextricably linked. Attunement to the multiplicities of tourism realities requires both local *and* global analyses. For example, rethinking our relationship to travel, tourism and well-being in the realms of the everyday is an important layer in this challenge. Power operates at multiple levels and working towards more equitable tourism requires analyses and solutions that are attentive to these multiplicities. For example, at the local level, place-based perspectives are key for regenerative tourism agendas, where local community should be empowered to contribute to decision making for inclusive tourism development (Bellato & Cheer, 2021).

The challenges for working towards a more sustainable and equitable planet are huge; however, there are many levels that need to be addressed for social change to occur, both at the global and the local levels. As Edensor notes (2007: 60) tourist behaviour is connected to the everyday through embodied habits and dispositions – even when we travel to far-flung destinations. The everyday is an important site within which to reframe travel and tourism more holistically, to understand the relationships to place, and to ensure that local communities are at the centre of inclusive forms of tourism and recreation.

In 2006 Higgins-Desbiolles posed an important intervention into tourism studies – that tourism is more than an industry. When tourism is harnessed for the wider public good, rather than a tool for unfettered capitalist growth, it can achieve many important goals for humans and the environment (Higgins-Desbiolles, 2006). We take this perspective as a starting point to consider a more-than-tourism agenda; a critical perspective that not only challenges the dominance of the discourse of tourism as a mechanism for boosting economic growth, justified by the neoliberal capitalist fallacy of the trickle-down effect, but also as a site for building more peaceful, globally minded and sustainable worlds. Mass models of tourism have exacerbated the hegemony of tourism as an industry connected to economic growth at the expense of the environment and the well-being of local communities in many tourism destinations (see analysis of overtourism by Milano *et al.*, 2019). Behind every tourism façade there is more going on than a simple economic transaction, including for example, considerations of power, politics, vested interests and contested notions of how local place is utilised for tourism and recreation purposes. When tourism is understood as 'more than an industry', analysis of tourism is expanded (Everingham & Francis-Coan, forthcoming). The COVID-19 crisis has reinvigorated the impetus to reclaim tourism as a powerful social force, with possibilities for connecting across cultures, the formation of global consciousness and even peace and justice (see Higgins-Desbiolles *et al.*, 2022; Higgins-Desbiolles, 2020).

Taking this approach of tourism as more than an industry further, we centre the importance of emotions and affect in relation to attachment to place, to further understand tourism as an 'experience economy' and the implications for thinking more broadly about tourism in the everyday local spaces of leisure and recreation. We put forward this more-than-tourism perspective to consider the extent we can reframe the experience economy towards slower, more localised tourism experiences. A more-than-tourism perspective puts greater emphasis on collective well-being, rather than just individual well-being, one that opens up analysis beyond dominant industry pro-growth narratives (Everingham, 2023). We also utilise more-than-representational insights and geographies of enchantment (Edensor & Millington, 2018; Woodyer & Geoghegan, 2012) to argue for a more-than-tourism agenda that centres embodiment and affect in the localised tourism agenda. We consider how more-than-tourism notions of well-being can (re)enchant our notion of being-in-local-place. The COVID-19 'tourism reset' allows us to reimagine our lives, habits and routines in ways that foreground social and ecological sustainability through rethinking tourism in relation to leisure time more holistically (Everingham & Francis-Coan, forthcoming).

A 'more-than-tourism' perspective highlights that the underlying motivations for travel – such as relaxation, joy, wonder and connection – can and should be implemented in our *everyday* lives, and that this is necessary for rethinking our relationships to each other and the natural world for more sustainable, equitable and regenerative futures (Everingham & Francis-Coan, forthcoming).

In this chapter we take an autoethnographic approach to explore a more-than-tourism perspective of rediscovering being-in-local-place from the geographical location of Mulubinba/ Newcastle, New South Wales (NSW) Australia. We discuss our autoethnography of a local tour, to reveal a more contextualised, nuanced account of what localised tourism actually looks like, and the possibilities that engaging in local space as a tourist creates in (re)enchanting our everyday lives and attachment to place. We also discuss the contested nature of local public space and recreation and consider how experiencing local place as a tourist can deepen our perspectives of these issues. We conclude that rethinking the everyday local spaces of tourism and recreation can contribute to well-being without having to utilise massive amounts of carbon to travel to far flung places, as well as enabling local communities to benefit directly from tourism.

An Autoethnographic Approach for a More-than-Tourism Localised Tourism Agenda

The authors of this chapter are friends with a keen academic and practical interest in leisure, tourism and recreation. We are both Novocastrians – that is we have both spent most of our lives living in Newcastle – on the unceded lands of Mulubinba, Awabakal and Worimi Country, NSW, Australia. We are both deeply embedded within our community with a strong attachment to place, both of us advocates for community-led decision making regarding any form of development that impacts our beautiful region. We both care deeply about community voices being heard and prioritised as key stakeholders – the local community acts as custodians of place (which is also not to say that local agendas are homogeneous or uncontested).

Taking an autoethnographic approach to being-in-place during the COVID-19 international border closure in Australia, we have both reflected and dialogued with one another regarding our needs, desires and emotions of living in a global pandemic in the context of unprecedented impacts of climate change. Indeed, many parts of Australia were significantly impacted by some of the worst bushfires the country has recorded in the summer of 2019 (Cook *et al.*, 2021), during which Sinéad's close family were required to evacuate. At the time of writing, eastern parts of Australia have been experiencing unprecedented levels of flooding, with Sinéad's family again forced to evacuate. Our anxieties around climate change, as well as the effects of COVID-19 on vulnerable communities around the world, have led us to reflect in depth about rethinking tourism futures within regenerative, localised models and the connections to our own embodiment of being-in-place under lockdown.

Autoethnography is a phenomenological approach to research that centres the 'experiencing person', documenting one's social reality (Coghlan, 2012). Autoethnography is an approach to writing and research that describes and analyses personal experience in order to understand cultural/ social experience (Ellis *et al.*, 2011). This approach challenges traditional ways of doing research and how sociocultural life and relationships are represented, treating research as 'a political, socially-just and socially-conscious act' (Ellis *et al.*, 2011: 273). Autoethnography allows the researcher to explore emotions and subjectivities (Ellingson & Ellis, 2008) and gives insights into how we embody and experience space.

A more-than-tourism perspective through autoethnography widens the scope of academic analysis to include a more-than-rational perspective of social life (Anderson, 2006), expanding the scope of how we understand engagement in the everyday and attachment to local place. We utilise this more-than-rational perspective to reflect on the role of embodiment, emotions and affect and the entanglements between human and non-human worlds in working towards regenerative and localised tourism futures. The more-than-rational components of affect mean that some aspects of our experiences cannot be codified into fixed models of understanding reality (Everingham, 2016, 2022). The significance of analysing the role of affect through autoethnography in relation to regenerative tourism futures is to centre the importance of the body, experiencing local place as an 'ever present layer, with material implications' (Anderson, 2006: 736). Through autoethnography, we critique the notion of 'rational actor models' (Barnes & Sheppard, 1992) in relation to tourism consumption. In rational actor models of consumer decisions, the significance of affect is often absent, negated within codified structures and models. Yet tourism, as embedded within the 'experience economy' demands a perspective that is attuned to the embodiments, habits and routines of the everyday and how this impacts how we *do* tourism (Edensor, 2007; Everingham, 2022).

Autoethnography is an approach that allows for a broadening of the parameters of research, to be 'endlessly expansive, inventive, and creative' (Gannon, 2018: 978). In relation to experiencing COVID-19 and being-in-place, affect is an important lens for how we have dialogued together around broadening our understanding of connection to local place. We have reflected on the importance of local custodianship of place for the well-being of host communities and local environments, in turn generating deeper and more meaningful tourism for everyone involved in generating and consuming tourism experiences (Everingham, 2022).

The primary component of this research was conducted through a number of different mechanisms. At certain points throughout the tour, Phoebe and Sinéad found the opportunity to share brief observations with each other, allowing them to calibrate and reflect on these in real time. Both researchers took brief notes through the tour in an informal field journal style and took photos to capture details of the experience. Following the tour, both Phoebe and Sinéad wrote detailed reflections outlining their observations, interpretations and emotional responses throughout the tour. These reflections were written using a shared online document allowing Sinéad and Phoebe to view each other's responses to the tour, which prompted further responses and a richer overall reflection. This process resembles the process known as duo or collaborative autoethnography described by Lapadat (2017) and applied to tourism studies by the likes of Mair and Frew (2018).

The following section provides an expanded and readapted account of an autoethnography written for another book chapter (Everingham & Francis-Coan, forthcoming). We describe a tour within our local government area and consider how we have both been (re)enchanted by local space in our everyday embodiment of being-in-place under lockdown between 2020–2021. We reflect on how we embody local space – and to what extent our bodies feel (re)enchanted by being-in-place. This has implications for regenerating local place, not only for the betterment of our everyday recreation for well-being, but also for considering the layers of how being-in-place is mediated by politics and decision-making related to recreational and public space.

Newcastle Afoot: The Politics of Public Space and Being (Re)Enchanted by Local Place

The COVID-19 lockdown led to Phoebe and Sinéad rethinking local place for tourism and recreation and we decided to start by taking some local tours. One of these tours was with a solo entrepreneur who had set up a successful walking tour business of our city – *Newcastle Afoot*.

We were joined by another four participants, all with masks on and ready to see a different perspective of a city that we had been brought up in. The small number of people on the tour allowed for some camaraderie between tour participants, but also allowed each person the opportunity for personal connection with the tour guide, Becky, and to ask questions. When travelling overseas – especially as solo women travellers – we are often interested in meeting and connecting with other tour participants. But in this case, it wasn't so much the interpersonal connections (that may lead to hanging out more or even traveling on to the next destination) as we felt more interested in the task of understanding what drove Becky to set up a such a tour and what angle she was going to take on articulating the wonders of our city. Would this capture the nuances we were familiar with? We were also interested in what we as locals would get out of the tour and how it fits into the wider picture of being (re)enchanted by local tourism.

Newcastle is a city that we know and love well, a city that has undergone many layers of change including increasing gentrification and urban planning decisions that have divided the community. The introductions involved being asked to contribute our own stories, reflections, ideas and knowledge of this city. This question brought up a lot of negative feelings for us about some of the urban planning decisions that have privileged property developers at the expense of the broader local community. These feelings were unlike if we were doing this tour in another part of the world, where we may not know the backstory of a place and have an emotional connection to it. We felt the need to balance out our own political beliefs and activist natures before contributing, trying to decipher the 'vibe' of the group. For example, the tour started at the old Newcastle train station which in late 2014 was truncated due to a decision by the NSW Government which involved running a light rail along Newcastle's main street to 'revitalise' the city by opening up the main street to the harbour (Virtue, 2016). This was a controversial decision for many living in the inner city – who frequently used the train to get to the main campus of the university – and onwards to the central coast and Sydney and for those around the Hunter region who use rail to commute to and from work in the Newcastle central business district (CBD). Locals were told that the rail line would be converted into public green space – not used for property development for private interest. Taking away the heavy rail was fraught with secrecy and misleading information presented to the public with no business case ever presented to support government decisions. While the opening up of the city and harbour has its benefits, the public 'green space corridor' has not eventuated at the scale that the original glossy brochures advertised. In fact, some of the rail line has already been sold off for private development.

Newcastle as a colonial city has a longstanding history of controversial planning and development decisions, particularly in relation to the preservation of heritage buildings. Aside from the rail line, the Palais Royale dance hall (Scanlon, 2015) was demolished and replaced by a KFC fast-food restaurant; 'The Store', a former community co-op with a historical façade was demolished despite community opposition (Perrett, 2021); and the redevelopment of the Newcastle Ocean Baths, a heritage ocean pool, has attracted community concern (Swinton, 2022); all roughly in the last decade. This was the context in which we found ourselves undertaking this local tour and we wondered what perspective the guide would take in relation to these issues.

If we were doing such a tour overseas, we wouldn't know the history and contested nature of such decisions. And we certainly wouldn't feel emotionally connected to the consequences of such decisions. We weren't sure what kind of political story or perspective of this city we were going to get from the tour leader, who disclosed she had only lived in the city for seven years, having grown up in Adelaide then subsequently living in Melbourne.

For Phoebe there was something a little bit grating about the tour guide capitalising on doing local tours as an 'outsider' – although this also raised questions as to at what point someone becomes 'a local'. While Phoebe is proud of the city she grew up in, part of her doesn't want people to know about this little gem of a city– she remembers the city as a rough place – but the roughness gave it edge and character. Not owning her own house or property, she is worried about the effects of gentrification which may lead her to be forced into the outer suburbs away from the harbour and the sea due to rising prices of housing. Was the tour leader going to have the same kind of nostalgia for this place that we had? What did she think about the decisions that had been made about the rail, for example? She admitted that she loved the fact it had opened up the city to the harbour; but while she mentioned that the decision was controversial, the politics of these decisions were glossed over. However, she did share many common values with us and in the end, we felt she could be a potential ally and collaborator in the local and regenerative forms of tourism that we want to see more of in our locality. In fact, as the tour went on, it became apparent that the dynamic of Becky herself as a *new local* was interesting. In that sense, the tours themselves became part of Becky learning about Newcastle, but also a mechanism for her to share this learning with others. Her experience seeing the city with new eyes, and her subsequent love and enthusiasm of the city, was contagious.

Despite our initial feeling of discomfort and unease about how our beloved city was being positioned within Becky's narrative of place, the information provided on the tour was interesting and there were layers of stories about our city of which we had no idea previously. As she unravelled more of these stories, we found ourselves being (re)enchanted with parts of the city. For example, part of the tour was to look at the local graffiti. The first graffiti that was shown depicted an image of the Prime Minister among fire and ashes. We immediately felt affected by this image – reflecting on lives and livelihoods lost and the effects of bushfires on regional tourism in Australia, as well as Sinéad's immediate family. While much of the east coast of NSW was impacted by the 'Black Summer' bushfires of 2019–2020 (Cook *et al.*, 2021), towns reliant on tourism income such as Mallacoota on the NSW south coast became some of the worst-affected areas (McGuire & Butt, 2020) and are yet to fully recover at the time of writing. Much public ire was directed towards the Australian Government, in particular Prime Minister Scott Morrison, in relation to perceived failure to both prepare for and respond to such disasters (Remeikis, 2020). In fact, Scott Morrison had jetted off on his own holiday to Hawai'i as the bushfires raged throughout the country, causing national outrage. The image of the Prime Minister among the ashes was instantly recognisable to Sinéad and Phoebe as symbolic of this critique.

Part of the tour involved going through a little arcade, with a bunch of shops run by small businesses showcasing their artistic wares (Figure 12.1). Sinéad was surprised to see that a friend of hers had a shop on the tour. We passed by and said hello – if we were overseas, we wouldn't know any of the vendors – but this was someone that Sinéad knew over a long period of time – yet didn't know she has relocated her business to this new place. Phoebe also had a backstory of doing a jewellery workshop with her many years earlier. We were delighted that the tour involved showcasing such local talent and supporting these local businesses. Everyone oohed and ahhed over the locally designed products – handmade jewellery and hand-designed tights with unique patterns. We both made a vow to go back shopping there later to buy some goods. The shop vendors all smiled and waved as we walked by. It was clear that they were benefitting from this local form of tourism, where tourists were guided towards their local wares. A localised tourism agenda should ensure that local businesses reap the benefits, and in this case the uniqueness of the artistic quality of items in these shops gives Newcastle a unique edge. In such an example, the local artisans benefit from the sales of their wares and an opportunity to directly convey the uniqueness, significance and value of their work. At the same time, the tourists here benefit from more direct and less mediated interactions adding value and authenticity to their experience.

Figure 12.1 The city arcade as it looks today. Credit: Author

Figure 12.2 Homage to the secret baths by local graffiti artist painted on power box outside the arcade. Credit: Author

The arcade itself has a unique history which neither Phoebe nor Sinéad knew about. It used to contain a bathhouse! (Figures 12.2 and 12.3). Becky showed us a photo of the baths, recounting the story of 'Newcastle's Secret baths'. In 1888, Newcastle Corporation Baths were built at a cost of £4500, measuring 27.5 by 10.5 m long. The baths were for men only (until the ladies-only day from 6–4), could be filled in six hours and were cleaned once a week. However, the baths were said to get quite dirty with dirty coal miners plunging straight in after work, leaving coal scum around the edge of the pool. The baths were popular with swimming carnivals and races; however, once ocean swimming and surfing became more popular, the baths became less utilised.

The story was astonishing, and we were awestruck. It gave us a both a glimpse into the lives of the early settlers in Newcastle, again affecting us on an emotional level to connect with the place with new and old layers of stories, imagining what life was like in the city we love back in those days. Moreover, we had no idea that the swimming culture of Newcastle had shifted so much over time, with ocean swimming only becoming popular in the 20th century. These days, ocean swimming, as well as swimming in the ocean baths (heritage-built coastal pools of ocean water) is one of Newcastle's most popular drawcards for visitors.

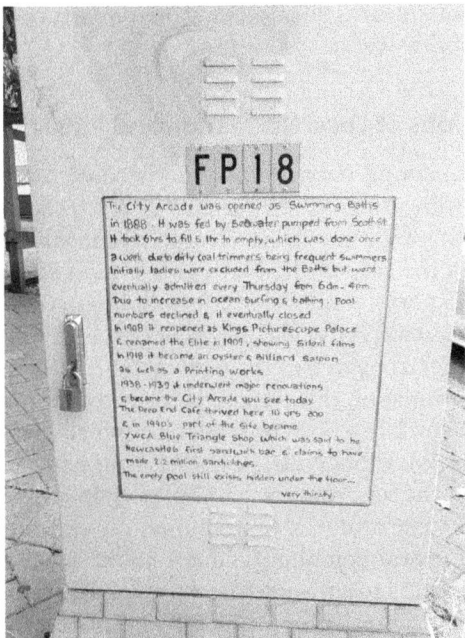

Figure 12.3 Detail of the secret baths' history presented in public graffiti art. Credit: Author

Next Becky took us to see a new development that had recently been constructed. Neither of us having been there yet; we both expected to be disgusted by it as we so often had been with other new developments in the city. However, we got such a surprise with how nice it was – a little nook in our city that we hadn't had a chance to get to know yet.

While initially we were a little disconcerted with the tour guide's legitimacy as an 'outsider' running the tour, we were impressed to learn that she had worked on a major street art project bringing many nationally recognised graffiti artists to the city to put murals on some of the buildings. Of course, we had noticed some of these murals ourselves – in fact a few of the murals are on walls in the inner-city university campus where Phoebe works and she has at times been able to see one of these murals from the window of her classroom. While Phoebe was unaware of how they got there, Sinéad had actively participated in the 'Big Picture Fest' and followed the self-guided tour of the new murals around the CBD which were close to where Sinéad was living at the time.

While the tour skirted on interesting political discussion, it did leave a small amount of tension unresolved. It made us wonder about the politics and positionality of telling such stories about place – what is included and what is excluded. While being intellectually aware of the politics of such storytelling the fact that we were experiencing this from our embodied positionalities as 'locals' being on tour attuned our attention to this politics in bodily ways.

Contested Notions of Local Recreation and Public Space

Decisions impacting public space can become controversial on the basis of the decisions themselves but also in many cases on the basis of how these decisions are communicated and the consultation that is undertaken (or not). In the case of the Newcastle Ocean Baths, a unique pool built into the rockpools, a drawcard for both international and domestic tourists – controversy arose when images of concept designs for the restoration of the facility appeared in the *Newcastle Herald* before stakeholders had been consulted on this design, before any expressions of interest to undertake the work had been received and before a tender for the work had been decided (McKinney, 2021). In fact, concerns were even raised in the community at the time that the baths may no longer remain free to the public. Although some feedback from community was subsequently incorporated into the restoration plans and the public was reassured that the council was committed to ensuring that access to the facility would remain free to the public, the seeds for suspicion and distrust were sewn. This dynamic has created further complications in communicating complex engineering decisions related to raising the external walls of the ocean pool and adding concrete to the pool floor to account for predicted sea level rise (Swinton, 2022). These contestations are ongoing, with recent protests from local baths users about the removal of sand from the bottom and concreting over the natural rock shelf. A localised tourism approach must centre the local communities in the design and decision making. For tourism to be meaningful for local communities, access to the local space and alignment with community and social values is necessary (Melubo & Doering, 2022).

Reclaiming Local Space, Centring Local Community as Custodians

Typically, tourism as an experience economy is driven by 'restlessness, boredom and new ways to escape reality', perpetuating 'the growth fetish of tourism and underpinning the ability of tourism corporates to sustain profit agendas from tourism growth' (Higgins-Desbiolles *et al.*, 2019: 1927). At the same time, travel and tourism contain possibilities for hope if goals are centred that enable a global consciousness that is sensitive to planet and people (Lew, 2018). In reflecting on the tourism experience in these ways, it strikes us that a local and regenerative tourism agenda needs to also address people's desire for exploration, awe, wonder and joy that travel brings. The role of awe in tourism experiences has been described as fostering stronger connections with 'the surrounding world' and 'sense of place' (Coghlan *et al.*, 2012: 1711). 'Awe is elicited through a sense of perceptual vastness… [requiring] cognitive accommodation to fit novel, unfamiliar information' (Coghlan *et al.*, 2012: 1711). Yet awe and its effects have largely been neglected in tourism research despite its potentially transformative and schema-changing effects (Coghlan *et al.*, 2012: 1711). The role of awe and wonder has a huge role to play in being (re) enchanted by local place – and is partly what makes a localised tourism agenda feasible.

The embodied and emotional aspects of travel and tourism, in terms of attachment to place, are an important layer to address in putting forward localised and regenerative tourism futures, to attune our mental schema towards custodianship, care and responsibility towards place. A regenerative tourism economic model is about adding value to places for hosts, tourists and the local environment. From this perspective the role of tourism is to 'provide hospitality and healing for the benefit of all stakeholders', where local/host 'tourism operators are regarded as crucial catalysts of change and stewards of regenerative tourism systems' (Bellato *et al.*, 2022: 8). Making our own local spaces more joyful, awe inspiring and wonderous by ensuring the regenerative value of a place is also a way of alleviating the boredom and restlessness that can come from the negative aspects of living our everyday lives.

When tourism is understood as 'more than an industry', an analysis of tourism is expanded away from economic growth models and towards the well-being of local communities and places. A 'more-than-tourism' perspective opens up possibilities for rethinking well-being more broadly, centring the importance of emotions and affect in the experience economy. A more-than-tourism perspective considers the extent we can reframe the experience economy towards slower, more localised tourism experiences, putting greater emphasis on collective well-being, rather than just individual well-being. This perspective opens up analysis beyond dominant industry pro-growth narratives and takes seriously the importance of collective regenerative futures and how we can also embed these experiences within our everyday lives.

Attunement to the everyday wonders of awe, enchantment and joy are key for enjoying local spaces (Edensor & Mundell, 2021). Collective regenerative futures involve making local space more liveable and enjoyable, resulting in less need to travel to far flung places with excessive carbon mileage; at the same time, these futures enable locals, especially those without the economic means to travel, to enjoy their own places as tourism spaces. It would also mean more local communities could enjoy their own local experiences. However, attention to the contestation of local space is also required. Local tourism and recreation agendas will not appease everyone. The Newcastle 500 supercars race, for example, has been a controversial event for the city, with local residents feeling disempowered in the decision-making processes, while fans in the region deploring local residents as being 'whingers' and trying to keep their place (which is right next to the beach and other inner-city attractions) contained for themselves (see Johnson *et al.*, 2020).

While local community is not homogenous, we argue that localisation is an important layer in building regenerative futures. The local turn in tourism is indicative of a necessary shift in thinking towards addressing structural inequality resulting from hyper neoliberal capitalism (where Global South and Indigenous communities can get locked into tourism-dependent economies) and for addressing the challenges of climate change. Travelling less to far flung places, travelling more slowly in international environments and contributing to local supply chains contributes to a sustainable climate change agenda and goes some way in addressing contemporary power imbalances and injustices in tourism by focusing on and empowering local communities from a global perspective (Higgins-Desbiolles & Bigby, 2022). Defining tourism by the needs of the local community is a way to make tourism more accountable within the society it occurs– to ensure that *both* tourists and hosts receive the benefits of tourism.

In relation to the *Newcastle Afoot* tour, it was clear to see that the focus on local businesses, artistry and graffiti is what gives Newcastle its uniqueness. Newcastle is well placed for the neolocal trend in tourism, where people are searching for more immersive and meaningful experiences in the places they visit. Encouraging local consumption patterns and redefining understandings of place and identity is important in the neolocal shift (Mody & Koslowsky, 2019). This agenda is increasingly necessary in relation to reducing emissions, not only in relation to where we travel, but also the 'role of local production, distribution and consumption in building networks of well-being' which can 'link people to their environment and contribute to deeper understanding to support actions at personal and political levels to address climate change' (Higgins-Desbiolles *et al.*, 2019: 1940). Emphasising the importance of local supply chains is a key element for rethinking economies more broadly in sustainable and regenerative ways – where benefits can be distributed more equitably.

Regenerative models including doughnut and circular economic models are gaining traction around the world and are increasingly being recognised within tourism (see Caterina-Knorr, 2021; Everingham, 2020; Sciacca, 2020). These models challenge traditional linear economic models where resources are merely consumed and disposed of, perpetuating negative social and environmental impacts. Acknowledging limitations to growth does not mean taking away from job growth or well-being. On the contrary, these regenerative models are forging innovative agendas for growing economies in more diverse ways. Doughnut economic models focus on finding the 'sweet spot' where everyone has what they need to live a 'good life' in line with the 'environmental ceiling' (Raworth, 2017) (see also Chapter 6). By and large those living in rich countries are living above this ceiling, while those in poorer countries fall below the social foundations needed for living a 'good life' (Nugent, 2021). Rethinking economies along regenerative lines can actually grow well-being by creating different kinds of jobs in non-extractive sectors such as regenerative tourism.

Conclusion: (Re)Enchanting the Local for Living in More Inclusive Regenerative Futures

In this chapter we have discussed how expanding the scope of leisure and recreation as 'more-than-tourism', can (re)enchant our experiences of being-in-place. Experiencing the COVID-19 crisis from the geographical location of Mulubinba/ Newcastle, Australia has given the authors novel ways to engage in local place. Rethinking tourism from a more-than-tourism perspective involves an expanded notion of tourism and placing 'tourism within its appropriate context of global mobilities, human well-being and sustainable futures' (Higgins-Desbiolles & Bigby, 2022: 1). These embodied experiences and reflections have enabled us to think about the implications of tourism and hospitality to be (re)enchanted and reconfigured towards the local, offering 'potentially positive transformation[s] in terms of activation of local relations, networks, connections, and multiplicities' (Tomassini & Cavagnaro, 2020: 714). Rethinking tourism mobility in such ways opens up 'such space to multiple novel functions designed not just for tourists and travellers but also for citizens' (Tomassini & Cavagnaro, 2020: 716). Taking a more-than-tourism perspective gives impetus for rethinking and redesigning infrastructure, parks and recreational facilities for the well-being of everyone who interacts within tourism destinations, including locals (Caterina-Knorr, 2021). This also involves taking care of the surrounding natural environment. Caterina-Knorr and Everingham (2021) argue that that there is not only a moral imperative for preserving natural communal resources, but also a business one – as without the natural wonders that visitors come to see, there will be a forced shift in labour markets, leaving many unemployed. Already too many tourism destinations across the globe have shut down for indefinite periods of time due to the destruction of local ecosystems.

Taking a more-than-tourism perspective allows us to think beyond tourism as an industry. Expanded notions of tourism as more than economic growth allow an analysis of the broader sociocultural and political relations in which tourism development takes place. This reveals not only the structural inequalities of tourism mobilities but also the importance of being attentive to our everyday lives, habits and routines. We suggest that (re)enchanting being-in-place is connected to a more-than-tourism agenda through rethinking how we engage in local spaces. Rethinking space in such ways opens it up for the local community to enjoy – all year long and not just in tourism seasons (Tomassini & Cavagnaro, 2020).

Discussion becomes about equitable use of resources – a tourism commons that the majority has access to rather than just a minority (Chassagne & Everingham, 2021). Public urban spaces, services and infrastructure should be 'conceived and designed to favour human encounters, connections, and spontaneous interactions, together with the coexistence of a diversity of inhabitants and visitors' (Tomassini & Cavagnaro, 2020: 717). In the context of the urban, sustainable tourism not only makes cities more liveable, but it also allows everyone to inhabit the cities in more equitable ways (Tomassini & Cavagnaro, 2020).

Finally, we argue that while attention to the need for structural change is important, falling into either/or, global/local, structural/micro analysis of rethinking tourism is not necessarily helpful, and negates the importance of the embodied, emotional and affective aspects of social change that happen within the realms of the everyday. Critical perspectives are too often entrenched in binary thinking that limits the potential for knowledge to translate into reflexive practise (Fullagar & Wilson, 2012). As Pritchard *et al.* (2007: 9) remind us, '[a]s tourism academics we have an obligation to challenge injustices and inequalities whether in tourism's material or symbolic domains'. Building more equitable futures involves multiple pathways forward which Gibson argues '…will be messy, contested, lived, and experienced emotionally. The non-human, political–economic and emotional are inextricably entwined in the fabric of tourism places' (2021: 84). We have proposed that (re)enchanting the local is one out of many pathways towards living in more inclusive, regenerative more-than-human worlds.

Acknowledgement

This chapter is a revised and expanded version of a conference paper presented at the CAUTHE 2022 Shaping the Next Normal in Tourism, Hospitality and Events conference, convened online 7–9 February by Griffith University, Brisbane, Australia.

References

Anderson, B. (2006) Transcending without transcendence: Utopianism and an ethos of hope. *Antipode* 38 (4), 691–710.
Ateljevic, I. (2020) Transforming the (tourism) world for good and (re)generating the potential 'new normal'. *Tourism Geographies* 22 (3), 467–475.
Barnes, T.J. and Sheppard, E. (1992) Is there a place for the rational actor? A geographical critique of the rational choice paradigm. *Economic Geography* 68 (1), 1–21.
Bellato, L., Frantzeskaki, N. and Nygaard, C. (2022) Regenerative tourism: A conceptual framework leveraging theory and practice. *Tourism Geographies*. https://doi.org/10.1080/14616688.2022.2044376.
Bellato, L. and Cheer, J.M. (2021) Inclusive and regenerative urban tourism: Capacity development perspectives. *International Journal of Tourism Cities* 7 (9), 943–961. https://doi.org/10.1108/IJTC-08-2020-0167.

Brouder, P. (2020) Reset redux: Possible evolutionary pathways towards the transformation of tourism in a COVID-19 world. *Tourism Geographies* 22 (3), 484–490. https://doi.org/10.1080/14616688.2020.1760928.

Brymer, E., Crabtree J. and King, R. (2021) Exploring perceptions of how nature recreation benefits mental wellbeing: A qualitative enquiry. *Annals of Leisure Research* 24 (3), 394–413. https://doi.org/10.1080/11745398.2020.1778494.

Caterina-Knorr, T.C. (2021) Tourism infrastructure, well-being, & how to 'build back better' for all. *Good Tourism Blog*. See https://goodtourismblog.com/2021/01/tourism-infrastructure-well-being-how-to-build-back-better-for-all/ (accessed 6 March 2022).

Caterina-Knorr, T.C. and Everingham, P. (2021) Climate change, COVID-19, and global systemic change. *Journal of Responsible Tourism Management*. DOI: 10.47263/JRTM.01-02-05.

Chambers, D. and Buzinde, C. (2015) Tourism and decolonisation: Locating research and self. *Annals of Tourism Research* 51 (1), 1–16.

Chassagne, N. and Everingham, P. (2021) Buen Vivir: A guide for socialising the tourism commons in a post-COVID era. In F. Higgins-Desbiolles, A. Doering and B.C. Bigby (eds) *Socialising Tourism: Rethinking Tourism for Social and Ecological Justice* (pp. 214–229). Oxon: Routledge.

Cheer, J.M. (2020) Human flourishing, tourism transformation and COVID-19: A conceptual touchstone. *Tourism Geographies* 22 (3), 514–524.

Coghlan, A. (2012) An autoethnographic account of a cycling charity challenge event: Exploring manifest and latent aspects of the experience. *Journal of Sport and Tourism* 17 (2), 105–124.

Coghlan, A., Buckley, R. and Weaver, D. (2012) A framework for analysing awe in tourism experiences. *Annals of Tourism Research* 39 (3), 1710–1714.

Cook, G., Dowdy, A., Knauer, J., Meyer, M., Canadell, P. and Briggs, P. (2021, November 29) Australia's Black Summer of fire was not normal – and we can prove it. *CSIROscope*. See https://blog.csiro.au/bushfires-linked-climate-change/ (accessed 6 March 2022).

Edensor, T. (2007) Mundane mobilities, performances and spaces of tourism. *Social and Cultural Geography* 8 (2), 199–215.

Edensor, T. and Millington, S. (2018) Learning from Blackpool Promenade: Re-enchanting sterile streets. *The Sociological Review* 66 (5), 1017–1035.

Edensor, T. and Mundell, M. (2021) Enigmatic objects and playful provocations: The mysterious case of Golden Head. *Social & Cultural Geography*. https://doi.org/10.1080/14649365.2021.1977994.

Ellingson, L.L. and Ellis, C. (2008) Autoethnography as constructionist project. In J.A. Holstein and J.F. Gubrium (eds) *Handbook of Constructionist Research* (pp. 446–467). New York: The Guilford Press.

Ellis, C., Adams, T.E. and Bochner, A.P. (2011) Autoethnography: An overview. *Historical Social Research* 36 (4), 273–290. https://doi.org/10.12759/hsr.36.2011.4.273-290.

Everingham, P. (2016) Hopeful possibilities in spaces of 'the not-yet-become': Relational encounters in volunteer tourism. *Tourism Geographies* 18 (5), 520–538.

Everingham, P. (2022) Building regenerative futures: A more-than-tourism agenda for post Covid-19 recovery. In B. Moyle and A. Krail (eds) *CAUTHE 2022. Shaping the Next Normal in Tourism, Hospitality & Events*, online conference 7–9 February, Griffith University, Brisbane, Australia.

Everingham, P. (2023) Rethinking tourism for the long term: Covid-19 and the paradoxes of tourism recovery in Australia. In E. Çakmak, R.K. Isaac and R. Butler (eds) *Changing Practices of Tourism Stakeholders in Covid-19 Affected Destinations*. Bristol: Channel View Publications.

Everingham, P. and Chassagne, N. (2020) Post-COVID-19 ecological and social reset: Moving away from capitalist growth models towards tourism as Buen Vivir. *Tourism Geographies* 22 (3), 555–566. https://doi.org/10.1080/14616688.2020.1762119.

Everingham, P. and Francis-Coan, S.E (forthcoming) Rethinking tourism post COVID-19: Towards a 'more-than-tourism' perspective. In P. Mohanty, J. Nivas, A. Sharma and J. Kennell (eds) *The Emerald Handbook of Destination Recovery in Tourism and Hospitality*. Bingley: Emerald Publishing.

Fullagar, S. and Wilson, E. (2012) Critical pedagogies: A reflexive approach to knowledge creation in tourism and hospitality studies. *Journal of Hospitality and Tourism Management* 19 (1), 1–6.

Gannon, S. (2018) Troubling autoethnography: Critical, creative, and deconstructive approaches to writing. In S. Holman Jones and M. Pruyn (eds) *Creative Selves/ Creative Cultures. Creativity, Education and the Arts*. Cham: Palgrave Macmillan.

Gibson, C. (2021) Theorising tourism in crisis: Writing and relating in place. *Tourist Studies* 21 (1), 84–95. https://doi.org/10.1177/1468797621989218.

Hall, C.M., Scott, D. and Gössling, S. (2020) Pandemics, transformations and tourism: Be careful what you wish for. *Tourism Geographies* 22 (3), 577–599.

Higgins-Desbiolles, F. (2006) More than an 'industry': The forgotten power of tourism as a social force. *Tourism Management* 27 (6), 1192–1208.

Higgins-Desbiolles, F. (2020) Socialising tourism for social and ecological justice after COVID-19. *Tourism Geographies* 22 (3), 610–623. https://doi.org/10.1080/14616688.2020.1757748.

Higgins-Desbiolles, F. and Bigby, B.C. (2022) A local turn in tourism studies. *Annals of Tourism Research*. https://doi.org/10.1016/j.annals.2021.103291.

Higgins-Desbiolles, F., Blanchard, L.-A. and Urbain, Y. (2022) Peace through tourism: Critical reflections on the intersections between peace, justice, sustainable development and tourism. *Journal of Sustainable Tourism* 30 (2–3), 335–351. https://doi.org/10.1080/09669582.2021.1952420.

Higgins-Desbiolles, F., Carnicelli, S., Krolikowski, C., Wijesinghe, G. and Boluk, K. (2019) Degrowing Tourism: Rethinking tourism. *Journal of Sustainable Tourism* 27 (12), 1926–1944. https://doi.org/10.1080/09669582.2019.1601732.

Higgins-Desbiolles, F., Doering, A. and Bigby, B.C. (2022) Introduction: Socialising tourism: Reimagining tourism's purpose. In F. Higgins-Desbiolles, A. Doering and B.C. Bigby (eds) *Socialising Tourism: Rethinking Tourism for Social and Ecological Justice* (pp. 1–23). Abingdon: Routledge.

Johnson, P.C., Everingham, C. and Everingham, P. (2020) The juggernaut effect: Community resistance and the politics of urban motor-racing events. *Annals of Leisure Research* 25 (1), 93–115. https://doi.org/10.1080/11745398.2020.1818590.

Lapadat, J.C. (2017) Ethics in autoethnography and collaborative autoethnography. *Qualitative Inquiry* 23 (8), 589–603.

Lew, A.A. (2018) Why travel? Travel, tourism and global consciousness. *Tourism Geographies* 20(4), 742–749. https://doi.org/10.1080/14616688.2018.1490343.

Lew, A.A., Cheer, J.M., Haywood, M., Brouder, P. and Salazar, N.B. (2020) Visions of travel and tourism after the global COVID-19 transformation of 2020. *Tourism Geographies* 22 (3), 455–466.

Mair, J. and Frew, E. (2018) Academic conferences: A female duo-ethnography. *Current Issues in Tourism* 21 (18), 2152–2172.

McGuire, A. and Butt, C. (2020, January 19) Cut off: How the crisis at Mallacoota unfolded. *The Age*. See https://www.theage.com.au/national/victoria/cut-off-how-the-crisis-at-mallacoota-unfolded-20200117-p53sdn.html (accessed 6 March 2022).

Melubo, K. and Doering, A. (2022) Local participation as tourists: Understanding the constraints to community involvement in Tanzanian tourism. In F. Higgins-Desbiolles, A. Doering and B.C. Bigby (eds) *Socialising Tourism: Rethinking Tourism for Social and Ecological Justice* (pp. 72–90). Abingdon: Routledge.

McKinney, M. (2021, May 19). Revised designs for Newcastle Ocean Baths upgrade. *The Newcastle Herald*. See https://www.newcastleherald.com.au/story/7258265/revised-designs-for-newcastle-ocean-baths-upgrade/ (accessed 6 March 2022).

Milano, C., Novelli, M. and Cheer, J.M. (2019) Overtourism and degrowth: A social movements perspective. *Journal of Sustainable Tourism* 27 (12), 1857–1875.

Mody, M. and Koslowsky, K. (2019) Panacea or peril? The implications of Neolocalism as a more intrusive form of tourism. *Boston Hospitality Review* 7 (1), 1–10.

Nugent, C. (2021) Amsterdam Is embracing a radical new economic theory to help save the environment. Could it also replace capitalism? *Time2030*. See https://time.com/5930093/amsterdam-doughnut-economics/ (accessed 12 October 2020).

Perrett, B. (2021, September 12) Lower the bar for saving buildings; and rebuild The Store façade. *The Newcastle Herald*. See https://www.newcastleherald.com.au/story/7421547/lower-the-bar-for-saving-buildings-and-rebuild-the-store-facade/ (accessed 6 March 2022).

Pritchard, A., Morgan, N. and Ateljevic, I. (2011) Hopeful tourism: A new transformative perspective. *Annals of Tourism Research* 30 (3), 941–963.

Pritchard, A., Morgan, N., Ateljevic., I. and Harris, C. (2007) Editors' introduction: Tourism, gender, embodiment and experience. In A. Pritchard, N. Morgan, I. Ateljevic and C. Harris (eds) *Tourism and Gender: Embodiment, Sensuality and Experience* (pp. 1–13). Wallingford: CABI.

Raworth, K. (2017) *Doughnut Economics: Seven Ways to Think Like a 21st Century Economist*. White River Junction, VT: Chelsea Green Publishing.

Remeikis, A. (2020, January 3). Where the bloody hell was he? How Scott Morrison spent the past week of the bushfire crisis. *The Guardian Australia*. See https://www.theguardian.com/australia-news/2020/jan/03/where-the-bloody-hell-was-he-how-scott-morrison-spent-the-past-week-of-the-bushfire-crisis (accessed 6 March 2022).

Romagosa, F. (2020) The COVID-19 Crisis: Opportunities for sustainable and proximity tourism. *Tourism Geographies* 22 (3), 690–694.

Scanlon, M. (2015, May 5) HISTORY: Palais Royale – king of venues. *The Newcastle Herald*. See https://www.newcastleherald.com.au/story/3048136/history-palais-royale-king-of-venues/ (accessed 6 March 2022).

Sciacca, A. (2020) From linear to circular: How to build resilience in small island tourism destinations. *Good tourism Blog*. See https://goodtourismblog.com/2021/02/from-linear-to-circular-how-to-build-resilience-in-small-island-tourism-destinations/ (accessed 1 March 2022).

Swinton, S. (2022, February 19). *The Newcastle Herald*. See https://www.newcastleherald.com.au/story/7625786/year-long-major-upgrade-of-newcastle-ocean-baths-set-to-begin-next-month/ (accessed 6 March 2022).

Tomassini, L. and Cavagnaro, E. (2020) The novel spaces and power geometries in tourism and hospitality after 2020 will belong to the 'local'. *Tourism Geographies* 22 (3), 713–719. https://doi.org/10.1080/14616688.2020.1757747.

Virtue, R. (2016, August 5) Newcastle rail line truncated. *ABC News*. See https://www.abc.net.au/news/2016-08-05/newcastle-rail-line-truncated/7689352 (accessed 6 March 2022).

Woodyer, T. and Geoghegan, H. (2012) (Re)enchanting geography? The nature of being critical and the character of critique in human geography. *Progress in Human Geography* 37 (2), 1–20.

13 Reclaiming the City: Social Movements and the Local Impacts of the Global Tourism Industry

Alexander Araya López

Introduction: The Worst Year of the Industry

Before the pandemic, there was a complex public debate regarding the 'right to the city' and the impact of global mass tourism in several cities and communities all over the world. In the many acts of dissent that took place in European cities at the time, as well as in digital spaces, activists frequently defended their neighbourhoods, warning about ongoing processes of exclusion, 'touristification' and gentrification. The commonalities of the struggles were evident. The popularity of Airbnb and similar tourist rental platforms was considered a threat to affordable, social housing and the unruly behaviour of tourists affected the quality of life of those who were able to reside in trendy 'touristic' areas. Local social movements also reported the overall impact of mass tourism on the environment, with campaigns advocating for reducing the unnecessary emissions needed for mass transport or concerns about water overconsumption and excessive waste. In 2019, global tourism was indeed a trillion-dollar industry (World Travel and Tourism Council, 2021) that represented nearly 10.4% of global gross domestic product, and the number of domestic and international travellers was expected to grow uninterruptedly, with the industry being accountable for about 8% of global greenhouse gas emissions (Lenzen *et al.*, 2018). 'Overtourism' and 'tourismphobia' (particularly in its Spanish version *'turismofobia'*) became buzzwords in local and global media in the pre-pandemic times (Araya López, 2021a), while some media texts explored the idea of 'undertourism' both before and during the pandemic (Lindblad, 2020), perceiving it as a more critical issue for 'destinations' than 'overtourism'. In early 2020, when the world was facing the inevitability of lockdowns and social distancing, and while rising COVID-19 cases kept airports closed and cruise ships unable to dock, social movements continued to denounce the risks of depending exclusively on tourism revenues (i.e. the monoculture of tourism). Since these restrictions took place, the hope for an effective vaccine implied the possibility of recovering at least some of the previous tourism flows, but uncertainty was looming and the future of the industry was unknown.

This chapter will discuss the political dissent against global mass tourism in the pre- and 'post'-pandemic scenarios, with a particular emphasis in the social campaigns against this global industry in Europe. Firstly, the notion of the 'right to the city' will be explored, highlighting how this 'right' is linked to acts of dissent in both physical and digital spaces. Secondly, the complex 'externalities' of global mass tourism will be examined for Barcelona, while drawing brief comparisons with three other European cities, namely Venice, Lisbon and Amsterdam. Finally, the chapter concludes with an assessment of the potential future for global tourism, its challenges and recommendations for new avenues for research.

The Political Uses of the 'Right to the City': Challenging the Narratives of Global Tourism

There is an extensive debate about the meaning, limits and applicability of the notion of the 'right to the city'. Since its original formulation by Henri Lefebvre (1996), this right has captured the interest of a wide diversity of social movements, including movements campaigning for gender and racial equity or those denouncing exclusionary decision-making process apropos city budgets. The malleability of this right has also given rise to some (ab)uses of the term, and scholars have denounced how the idea of the right to the city has been 'co-opted' by international organisations, private businesses and local authorities (Kuymulu, 2013; Mayer, 2009).

'The right to the city' (Lefebvre, 1996), written as an essay in 1967, addresses the utopian possibility of overcoming the generalised segregation resulting from the current organisation of urban society. Lefebvre explains how this right could be understood as a democratisation of (urban) space:

> The right to the city, complemented by the right to difference and the right to information, should modify, concretize and make more practical the rights of the citizen as an urban dweller *(citadin)* and user of multiple services. It would affirm, on the one hand, the right of users to make known their ideas on the space and time of their activities in the urban area; it would also cover the right to the use of the centre, a privileged place, instead of being dispersed and stuck into ghettos (for workers, immigrants, the 'marginal' and even for the 'privileged'). (Lefebvre, 1996: 34)

In this sense, the right to the city reclaims these urban spaces not only for those who can afford them but also for the 'inhabitants' as a whole. This right is perceived as a way to counteract the ongoing exclusion from production (unemployment) and consumption (poverty), as well as the more subtle physical and symbolic exclusion taking place in urban environments. These overt and subtle forms of exclusion may be visible, for example, in the location of public parks or the establishment of schedules to use them. It expresses itself in the availability, cost and quality of public transport, including commuting time or traffic noise pollution. The exclusion could be visible in the access to education and health, which become highly localised and even racially segregated. Also problematic is the access to places that may be defined as semi-public (i.e. restaurants, sport facilities, shopping malls, airports) given that these are 'private' spaces relatively open to the public.

In his article, 'What *kind* of right is the right to the city?', Attoh (2011: 670) emphasises that the deliberation about 'the right to the city' needs to consider both the limits of this specific right and the meaning of its violation. 'The right to the city', he adds, 'may allow us to see rights to housing, rights against police abuse, rights to public participation in urban design, rights against established property laws, or rights to a communal good like aesthetics, as necessarily connected' (Attoh, 2011: 674). Similarly, Purcell (2002) has explored the practical propositions of the right to the city, remarking the impossibility to delimitate its scope. Considering scalar politics, Purcell questions what this right to the city means when an individual belongs to two different cities or what happens when the rights of the inhabitant of one particular city, in highly globalised and networked societies, become a risk to secure the right to the city of the inhabitants of another city (for example, in his reflection about conflicting rights between inhabitants in Los Angeles and Oaxaca (Purcell, 2002: 104)).

For our discussion in this chapter, the idea of the right to the city is utopian in two senses: on the one hand, it reflects Lefebvre's optimism in the search for 'the possible'; and on the other hand, the concept is extremely elusive, giving hints of the 'impossibility' of its realisation (Coleman, 2013). In Harvey's view, the right to the city is the result of the collective creation of the city as a product (or as an *ouvre*, in Lefebvre's term). If citizens/inhabitants contribute with their existence to the creation of this environment, they are entitled to take part in the opinion- and will-formation and decision-making processes (Harvey, 2012).

Following Harvey's interpretation of urban sociologist Robert Park, every city could be interpreted as a man-made (sic) environment (Harvey, 2008: 23), created after her/his/their heart's desire. The right to the city (a transformed and renewed right to the urban life) as proposed by Lefebvre (1996: 158) refers not only to the right to access the existing urban infrastructure, Harvey explains, but also entails the possibility of transformation of this environment (Harvey, 2012: 3–4). Lefebvre valued this utopian nature of cities as places of spontaneity and historical significance (Lefebvre, 1996: 176).

In relation to global mass tourism, cities have become contested spaces in which the needs of the local population frequently clash with the ongoing processes of branding and commodifying the 'city' for tourists. Colomb and Novy (2017) propose to understand these urban politics as resulting from the competition between cities in their efforts to attract investment and secure economic growth. Their edited book offers a panoramic view of the emergence of protest movements in various cities facing touristification all over the world, including Barcelona, Singapore, Hamburg, Belfast and Buenos Aires. In the introduction, Colomb and Novy (2017) identify four major sources of conflicts linked to global mass tourism: economic, physical, social and sociocultural, and psychological (which are frequently intertwined). This chapter contributes to the ongoing analysis of the impacts of global mass tourism in Barcelona, referencing several spatial struggles in the city that have the notion of the right to the city at their core.

Methods

The data included in this chapter have been collected as part of a larger study on the emergence of social movements campaigning against mass tourism in Venice, Amsterdam and Barcelona. An analysis of media texts between 2014 and 2017 (Araya López, 2021c) has been used to track the main conflicts both directly and indirectly linked to global mass tourism. Between 2018 and 2020, I undertook several research fieldtrips and participant observation of protest acts. Additionally, I collected and analysed physical manifestations of dissent (mainly through photographs). Specifically for Barcelona, I recorded a total of eight interviews with key activists from various action groups including Arran, Barris Per Viure, Las Kellys, Association of Neighborhoods for Tourism Degrowth (ABDT), Zeroport and Defensem Park Güell. These in-depth, semi-structured interviews followed strict ethical guidelines, including not only the mandatory informed consent protocols but also proactive strategies to secure the privacy rights of the research informants. In a few supplementary cases, I held informal, unrecorded conversations with activists who were not able to consent to a recorded interview (documented in written field notes), following the advice of Swain and Spire (2020) on their empirical value. The data collected through these informal conversations have been used here for contextual purposes.

Barcelona: Mapping the Conflicts Related to Global Mass Tourism

This section offers an analysis of conflicts related to global mass tourism in Barcelona, Spain. As discussed above apropos the right to the city, local dissent has pointed out processes of spatial segregation and exclusion, limited access to resources and the impact of the industry in terms of quality of life and environmental degradation. The 'touristification' of Barcelona, as several scholars and activists have pointed out (Bruttomesso, 2018; Milano & Mansilla, 2018; Nofre *et al.*, 2018) can be tracked back to the promotion of the city as the host for the 1992 Summer Olympics, and it could be argued that this 'branding' has continued until today (even with the forced impasse caused by pandemic).

In 2014, protests broke out at La Barceloneta neighbourhood, after the photographer Vicens Forner captured three naked Italian tourists attempting to enter a supermarket, undisturbed by the shocked reactions of the witnesses. The local population took to the streets in a series of acts of dissent, campaigning not only against the 'anti-social' behaviour of visitors, but specifically targeting the uncontrolled emergence of tourist rentals in the area, particularly those rented through Airbnb. In their own reports, Airbnb reported the economic contribution that the corporation provides for local hosts, describing how it helps them to pay their mortgages and to have an extra income for their private consumption (the company has also argued the same for other destinations such as Amsterdam, Venice, Bangkok and Kyoto) (Airbnb, 2018).

Although these protests of 2014 captured international media attention, there were other conflicts linked to the global tourism industry in Barcelona at the time. Since 2006, a group of local neighbours has been campaigning against the enclosure and the later introduction of an entrance fee to the renowned Park Güell (Arias-Sans & Russo, 2017). For the activists, the entrance fee represented an appropriation of shared public spaces, although local authorities anticipated some strategies to grant free access to the park to those living in its proximity. Park Güell, which attracted large numbers of tourists from all over the world, was experiencing not only overcrowding and 'unruly' behaviour, but there were also concerns about the preservation of the so-called Monumental Zone. A report published by the local government (Ajuntament de Barcelona, 2017) summarised the improvements in terms of heritage protection and reduction of vandalism resulting from this enclosure, while describing the strategic plan for re-connecting with the local community (for example, through the revival of 'lost' traditions). In 2018, the local youth group Arran performed a protest at the Monumental Zone in Park Güell, with two activists chaining themselves to the sculpture of a dragon, with banners in the background stating 'Prou turisme massiu!' (Stop mass tourism!) (Andrés, 2018). The group Defensem Park Güell (Araya López, 2021d) has actively campaigned against this enclosure, for example by offering flyers to tourists explaining the segregation and by documenting the physical alterations of the infrastructure of the park.

More recently, another touristic hotspot has emerged in Barcelona, the Bunkers of Carmel (Turó de la Rovira). With an altitude of 262 metres above sea level, the bunkers offer a panoramic view of Barcelona and they have become the meeting point of young people from all over the world. In informal conversations with neighbours, they reported fear of increased crime and the impact of a growing number of visitors on their quality of life. The local government of Barcelona has also noticed the damage to the infrastructure in the area – ranging from vandalism to the placing of love locks. A proposal for creating an enclosure for the affected areas has sparked protests, with locals fearing a new privatisation of public space (García Prat & Segura, 2021) similar to the abovementioned case in Park Güell. In 2019, neighbours in the area placed banners on their houses to denounce the planned evictions (Figure 13.1), specifically targeting Mayor Ada Colau (Barcelona En Comú) and the political party Esquerra Republicana de Catalunya (ERC, Republican Left of Catalonia).

Evictions are also an ongoing threat for the neighbours of La Sagrada Familia in the Eixample district, with the unfinished temple capturing headlines for a variety of reasons. News articles reported that the construction works have been 'irregular', given that the Catholic Church did not have the required permits for some 137 years (Casanova, 2019). Later during the pandemic, the Junta Constructora (a private foundation in charge of both building and restoring La Sagrada Familia) requested a delay for the compensation payments for the year 2020, allegedly due to economic hardship resulting from reduced numbers of tourists. These payments are part of an agreement with the Ajuntament of Barcelona as a compensation for its impact on the city and the surrounding environment (Cia, 2020). The church is also perceived as a threat for those who reside in Carrer de Mallorca, Carrer de Valencia and neighbouring streets,

Figure 13.1 Banners against evictions near the Carmel's Bunkers, Turó de la Rovira. Credit: Author, 2019

Figure 13.2 Banners against evictions at the Carrer de Mallorca, next to La Sagrada Familia. Credit: Author, 2020

because the construction plans include a huge staircase leading to the main structure (Lombardi, 2020). According to local neighbourhood associations, the original plans set by Gaudí did not include this staircase (Cia & Blanchar, 2018). These associations also resorted to activism using the placement of banners on their balconies (Figure 13.2), with direct references to the 10th commandment about 'greed' regarding Joan Josep Omella i Omella (Metropolitan Archbishop of Barcelona), to evictions and the local government of Barcelona and to tourist guides who spread 'lies'.

As reported by the local newspaper *La Vanguardia* (Muñoz, 2013), there were eight possible projects for the future of the Sagrada Familia, with only one of them preserving the buildings surrounding the church. The other seven proposals involved a physical transformation of the area. In a recent news story published in *El País*, the two current projects have been challenged by the neighbours, considering that they would significantly contribute to exceeding the carrying capacity of the neighbourhood. There were also concerns about potential health risks such as exposure to asbestos (Cia, 2019). Given that Gaudí's aesthetics and architectural masterworks have been exploited for the branding of Barcelona, other conflicts in different areas of the city have been reported, including 'irregular' parking of tourist coaches near La Pedrera (Benavides, 2019). Although the tourist coaches are allowed to park in the area for 10 minutes, the local residents have accused the drivers of staying longer with the engines on with the sole purpose of avoiding fines, while exposing them to constant noise and fumes.

Moreover, in Barcelona, a sort of gentrification or touristification by art has been denounced by local social movements. In recent years, the

Figure 13.3 Banner against evictions near the MACBA, El Raval. Credit: Author, 2019

proposed plan for a Hermitage Museum near the Port of Barcelona and in proximity of La Barceloneta has been heavily contested (Burgen, 2021). According to local groups, the museum does not contribute to the needs of the local inhabitants and it would become yet another tourist destination for the city. In the past, a similar conflict took place in El Raval with the creation of the Museum of Modern Art of Barcelona (MACBA). Local inhabitants have once again placed banners in these areas to protest the excessive noise (e.g. the noise of skateboarding), the impact of Airbnb and the massive presence of tourists (Figures 13.3 and 13.4). Recently, the

Figure 13.4 Banner against Airbnb, noise, skateboarding and tourists near the MACBA, El Raval. Credit: Author, 2019

expansion plans of the MACBA museum were also contested. The local inhabitants campaigned against the concession of the building proposed for the project, demanding the expansion of local health services instead (Montañés & Blanchar, 2019).

Lack of affordable housing and a systematic expulsion of local inhabitants have been reported in other central neighbourhoods of Barcelona, from El Raval to the Gothic Quarter. In the nearby and independent municipality L'Hospitalet, local action groups such as Barris Per Viure have reported concerns about potential touristification; this is what I have labelled as 'overtourism by proxy' (Araya López, 2021a). Although L'Hospitalet has its own cultural and historical particularities, it is the proximity to Barcelona (as a European cultural city) including the popular Camp Nou, that drives more tourists to the area. These activists have reported quality of life issues linked to the unruly behaviour of tourists as well as the systematic expulsion of neighbours due to the emergence of tourist rentals (Araya López, 2021d). According to Barris Per Viure, the uniqueness of L'Hospitalet is threatened, and therefore their political activism demands that the needs of the local inhabitants should be given priority by the local government. However, as reported about the ban against new hotels in Barcelona, the local government has been eager to profit from projects linked to tourism and they have prioritised the needs of visitors. Although L'Hospitalet is its own political unit and has a distinctive identity, this exemplifies the practical challenges of the idea of the right to the city, given that decisions taken in relation to Barcelona directly and indirectly impact the living conditions of the inhabitants of nearby communities:

> The local government sees this as an opportunity, eh… They are facilitating the creation of hotels, of tourist rentals, of student accommodation for foreign students in elite universities, all these are facilities… And everything that Barcelona does not want, it's coming here… It's exactly an effect, if people in Barcelona are going to invest, and they want to put a hotel, and in Barcelona there is a suspension of licenses, and in L'Hospitalet they are eager for them to arrive… Well… They have to come here to knock on the door… and our local government… Very generous in that respect… They give them a welcome and set in place all kind of advantages… . (Informant B, Barris Per Viure, pers. comm. January 2020, translation by the author)

A similar displacement took place in the centre of Barcelona in the past, but even 'touristified' districts such as Ciutat Vella (Old City) have resisted this exclusion, for example by appropriating spaces through sporadic playful protest (Bruttomesso, 2018). Recently, the project Km-ZERO was launched collaboratively between the local inhabitants and the municipal government of Barcelona (Ajuntament de Barcelona, 2018). The explicit goal was to transform Las Ramblas and recover this space for the citizens, working to fight gentrification while protecting social and cultural heritage. In the Poblenou neighbourhood, residents have protested the project 22@, which was part of the re-invention of Barcelona as a 'tech hub' (Mansilla, 2021), but that has failed to become an alternative to the tourism revenue.

In relation to global tourism and mobilities, groups such as the Association of Neighborhoods for Tourism Degrowth (ABDT) and platforms such as Zeroport have been denouncing the projected expansion of both the Port of Barcelona and El Prat Airport. Regarding cruise tourism, Barcelona ranks in the first position of the most polluted European port-cities, as reported in a recent study by the non-government organisation (NGO) Transport and Environment (2019). This report analysed 203 cruise ships in European countries in 2017 to estimate their emissions under the assumption that the ships fully comply with existing fuel sulphur standards in the respective geographical locations. In 2019, a protest at El Prat airport formed a symbolic red line of activists to denounce the pollution and environmental degradation linked to global air transport, as part of an encounter organised by the global network Stay Grounded (Rocabert Maltas, 2019). During the pandemic, the expansion plans for El Prat Airport were heavily protested in Barcelona, particularly due to concerns about the future of the protected natural area La Ricarda in the Llobregat Delta (Blanchar & Garfella Palmer, 2021), a bird sanctuary with an adjoining lagoon.

At the city level, other groups campaign for the rights of workers in the industry, such as the collective Las Kellys (Araya López, 2021d), who denounce the physical and mental exploitation experienced by hotel housekeepers while campaigning to end the outsourcing practices of the hotel sector in Spain. The youth group Arran, a left-wing organisation, has also advocated against the touristification of Barcelona, using radical politics as a catalyst to debate the 'side effects' of the industry (Araya López, 2021a). The map provided below helps in visualising the locations of some of these social movements in Barcelona (Figure 13.5).

Not just Barcelona: Tourism Conflicts in Venice, Lisbon and Amsterdam

In this section, the local impacts of global mass tourism are briefly discussed by establishing comparisons between the social movements campaigning against mass tourism in Barcelona and those in other European cities such as Venice, Lisbon and Amsterdam. It must be mentioned that these three cities have a wide presence of local action groups and that the examples mentioned here are not exhaustive.

Similar to Barcelona, the impact of Airbnb and other online platforms that offer temporary tourist accommodation has been identified as a serious externality resulting from global mass tourism in these three cities. In Venice, the activists of Osservatorio CIvicO sulla Casa e la Residenza

Figure 13.5 Map of selected tourism-related conflicts in Barcelona. Map created by the author with uMap

(OCIO) (2021) have documented the proliferation of tourist rentals, both by collecting data and by disseminating their findings in public forums with the community. !WOON (2021), a tenant support agency in Amsterdam, has also denounced irregular tourist rentals in the city, as well as collecting data and informing the local inhabitants about their rights. In 2019, the collective Left Hand Rotation (2021) produced a documentary describing the real estate speculation in Lisbon, referencing processes of exclusion of disenfranchised inhabitants and its relation to wider processes of touristification and gentrification. In Lisbon, the work of Cocola-Gant and Gago (2019) has explored the tourism-driven displacement in the Alfama neighbourhood, including through interviews with residents.

The unruly behaviour of tourists, which in Barcelona is commonplace and has been protested against in popular areas such as La Barceloneta, has captured the attention of both social movements and local authorities in various European cities. In Amsterdam, the famous Red Light District attracts massive numbers of tourists every year and this has created a series of problems ranging from public urination to noise and illegal activities. While sex workers in the area have become a sort of tourist attraction, they have also reported dissatisfaction with tourists that photograph or film them (van der Lit, 2019). Local citizen groups such as Stop de Gekte (2021) have campaigned to create a liveable environment for the inhabitants, proposing short- and long-term solutions which include restrictions on cannabis consumption by foreign tourists in the coffee shops, limits on the sale of alcohol (for example, in minimarkets) and the creation of alternative venues for sex workers (which is a contested and highly complex subject that includes issues of structural racism, xenophobia, sexism and classism). As a result, the local government in Amsterdam has created its own strategies targeting specific subsets of international and domestic visitors, which include the We Live Here! campaign and the creation of fines for unruly visitors (Araya López, 2020).

In reference to mobilities, resistance linked to the various impacts of big infrastructure projects has been growing in these cities. The group SchipholWatch (2021) has denounced the impacts of the air transport industry in general and of Schiphol Airport in particular, focusing on diverse issues such as CO_2 emissions and aircraft noise. In 2019, the group took part in the Protestival at Schiphol Plaza, a creative form of performative resistance which involved environmental groups such as Greenpeace and Extinction Rebellion (Stil, 2019). Indeed, SchipholWatch has also produced its own app ExPlane to help people track and collect data about the real effects of aircraft noise. Throughout the pandemic, the group has also campaigned online against the inherent risks that air transport means for the dissemination of infectious diseases. Near Lisbon, a plan for a new airport in Montijo (now cancelled) was challenged by local environmental activists from Aterra (2021). Suppression of protest is of concern, as shown in the case of a protest in which a protester interrupted the speech of a local politician, while throwing paper planes from the stage; this activist was later threatened with a two-year jail term (Banha, 2022). In 2021, the plan for new parking at the Marco Polo Airport near Venice also caused resistance due to the potential felling of 220 trees in the surrounding area, with activists from CriaaVe (a citizens' collective campaigning against the environmental impact of the airport) engaging in a flash-mob. According to local media (Francesconi, 2021), airport authorities reported that the protest was unnecessary, given that the information about the project was inaccurate and the trees were not threatened. Additionally, airport authorities also emphasised the seriousness of the airport crisis, with explicit references to the pandemic.

Finally, the dissent against cruise tourism is also identifiable in these diverse contexts. While Barcelona has no specific group campaigning against big cruise ships, concerns about their impact have been voiced by groups such as the ABDT and the collective Arran. In Venice, the local group No Grandi Navi and organisations such as We Are Here Venice have pointed out the damaging effects of cruises in the Venetian lagoon, including issues such as air pollution and erosion (Araya López, 2021b). In Amsterdam, the criticism of cruise tourism is less evident, but individual activists have denounced this through social media such as Twitter. Recently in Lisbon, the organisation Zero (2021) has criticised the lack of mitigation measures in relation to cruise tourism, particularly after the long pause caused by the pandemic. Zero expressed explicit concerns about the quality of air for local inhabitants once the cruise ships return.

Indeed, there are complex solidarity networks between social movements campaigning for specific causes, and this can be perceived at the national, European and global scales. The network Sur de Europa ante la Turistización (Southern Europe against Touristification, SET), for example, is a coalition of cities in Southern Europe denouncing the touristification of their communities and cities, while demanding access to affordable housing and the safeguarding of natural spaces. Similarly, the group Stay Grounded, which campaigns for tourism and air transport degrowth in the context of the climate emergency and air transport, is an umbrella organisation with participants from all over the world.

Discussion

In their book describing the long history of tourism in Venice, Davis and Marvin explain that:

> Although each tourist may take up Venetian space for only a short while, in the aggregate they are there all the time: their transitory occupation might be called a flow, but for the Venetians themselves it appears as a continual, solid mass. (2004: 4)

This statement could describe the realities of local inhabitants in many other cities in Europe and around the world, but each of these singular 'travel destinations' have followed its own particular process of 'touristification'. Through strategies of city-branding and by identifying physical, social and cultural value that could be 'attractive' to visitors, this process of touristification seems to be extractivist in nature (Salerno, 2018), adding more and more events, venues, infrastructure, services and resources needed to attract and satisfy domestic and international tourists.

As discussed for the case of Barcelona, the colonisation linked to the global tourism industry is perceived and experienced by local inhabitants as a series of threats and harms. Returning to the aforementioned typology proposed by Colomb and Novy (2017: 18–20), it is possible to identify an economic conflict related to the lack of affordable housing, exacerbated by a physical conflict such as the excessive noise created by tourist coaches, combined with social–cultural conflict created by perceived issues of public order linked to the disorderly behaviour of visitors and these collectively can produce psychological conflicts ranging from feelings of frustration to the loss of sense of belonging. The political banners included in this chapter are proof of this lived experience of the negative effects of mass tourism and their local activism campaigns against skateboarding noise, corporations like Airbnb or the perceived role of tourist guides as 'complicit' in touristification processes are therefore not surprising. Indeed, local inhabitants are not only forced to adapt to the city-wide transformations needed to boost the tourism industry, they are also targeted by propagandistic narratives. These are what Mansilla and Hughes (2021) identify as 'discourses of coexistence', in which local inhabitants are invited to be welcoming and to appreciate tourism as a national and city project that benefits 'us' all. This narrative has been paired with a self-victimisation discourse by various tourism stakeholders in the context of the pandemic and the climate crisis (Gössling & Schweiggart, 2022: 916–917), whereby these tourism stakeholders highlight their struggles and call for greater support.

The notion of the right to the city has been used politically to express alternatives to these diverse economic and political projects linked to global tourism, with activists frequently advocating for tourism degrowth and environmental preservation. The emergence of local, regional, national and global coalitions of movements, including action groups at the city-level such as the Association of Neighborhoods for Tourism Degrowth (ABDT) or regional networks such as the network Southern Europe against Touristification (SET), demonstrate that these activists are responsive to the systemic nature of the side effects of global tourism, valuing the sharing of common experiences while also devising strategies to face the issues through a unified front. At the global level, these networks include umbrella organisations such as Stay Grounded (2021), which campaigns for aviation degrowth, and the Global Cruise Activist Network (GCAN, 2021), which was created in the context of the pandemic and advocates for a 'no return to business as usual' for the cruise industry.

The strategies used by these action groups are diverse, and although most of these activists engage in non-violent demonstrations and claim making, there are a few episodes in which a form of 'radical' politics has been chosen (for example, the symbolic 'attack' against a tourist coach by the youth group Arran in Barcelona in 2017 (Araya López, 2021a)). These acts of dissent aim to capture media attention and to communicate alternative visions of the future of the city, which are oriented toward the needs of the local inhabitants and not necessarily compatible with infinite economic growth. Instead of projects that prioritise and serve a mobile class of leisure travellers, the local inhabitants demand projects that protect their traditions and identity, that contribute to improving the quality of life of those who live and work in the city or neighbourhood, and that are compatible with the protection of natural spaces in the context of the climate emergency.

For tourism stakeholders and local and national governments, the negative impacts of global mass tourism can be readily fixed with technological advancements and policies. The enclosure of public spaces, such as Park Güell in Barcelona, is expected to be replicated at the city-level in Venice, with the recent introduction of a single-entry ticket (*contributo di accesso*) (Città di Venezia, 2022). This measure has been highly questioned by local activists, and it is considered 'ineffective' just like other plans that proposed to set up turnstiles to regulate tourist flows. In the next years, while tourism stakeholders and local and national governments rely on global tourism as part of the economic recovery from the crisis caused by the pandemic, the activism of these groups campaigning for the right to the city must be tracked, observed and studied in all of their complexity, not only because they have become an effective barrier against the total touristification of the cities, but because their political vision is in synchrony with needed global societal cha(lle)nges based on sustainable and responsible consumption.

Conclusion

This chapter has examined the local impacts of global mass tourism in Barcelona and the social movement activism of local communities this has spawned, with brief comparison to similar dynamics in Lisbon, Venice and Amsterdam. Following the notion of the right to the city, the mapping of conflicts in Barcelona demonstrates that the debate on the notion of the right to the city is considered crucial to understand these local processes of 'reclaiming the city', but the dangers of oversimplification and of all-encompassing definitions of 'the city' present a challenge for theorising this right in relation to highly complex global and local scenarios of political dissent. The question about how much of these conflicts is attributable to tourism remains open, but the main concern for residents is their systematic expulsion from or segregation within the city, particularly in terms of access to housing. Other issues such as the unruly behaviour of tourists and the replacement of local services for tourism-oriented businesses have also been identified. In this sense, the notion of the right to the city seems to fit the needs of local inhabitants, who are demanding the right to take part in the decision-making processes for their city.

These social movements are not only acting at the local scale, but they are establishing valuable networks with similar groups at the national, European and global levels. Although the protest strategies range from radical politics to digital activism to performative protest acts, these collectives and individuals characterise themselves as emerging experts in local legislation and the technical aspects of the projects threatening them, oftentimes devising strategies for data collection and for engaging the national and global communities in their cause. Future research should address this citizen-produced knowledge. Similarly, instead of focusing on a specific conflict within the city, scholars should explore the interconnections between diverse conflicts through the city, particularly when approaching the notion of the right to the city, which has become a slogan for various social movements.

Finally, while in the early days of the pandemic there was a perception that the global tourism industry would reinvent itself and that the impasse would translate into a more sustainable form of tourism (Gössling &

Schweiggart, 2022), it seems that the future of tourism will be defined by a combination of vulnerabilities, anxiety and instability. Several activist groups have been denouncing new concessions given to the industry, ranging from bailouts to more 'business-friendly' legislation. As discussed in some of the examples above, the pandemic has become a sort of excuse to justify a fast recovery of global tourism, which could imply worsening living conditions for the local inhabitants, increased environmental degradation and the postponement of strategies for tourism degrowth. It seems likely that the struggle for the right to these cities will continue far into the future.

Acknowledgements

This publication has been partially based on data collected for the project RIGHTS UP, which was hosted at Ca' Foscari, University of Venice between 2018 and 2020. The RIGHTS UP project received funding from the European Union's Horizon 2020 research and innovation programme under the Marie Skłodowska-Curie grant agreement no. 792489.

References

Ajuntament de Barcelona (2017) *Proposta Estratègica per al Park Güell 2017–2022.* November. See https://ajuntament.barcelona.cat/gracia/es/el-ayuntamiento/estrategia-y-accion-de-gobierno/planes-y-proyectos/plan-estrategico-del-park-guell (accessed 22 December 2021).

Ajuntament de Barcelona (2018) The new Rambla, a transformation at the service of people. *Ajuntament de Barcelona,* 2 November. See https://ajuntament.barcelona.cat/participaciociutadana/en/noticia/the-new-rambla-a-transformation-at-the-service-of-people_730914 (accessed 22 December 2021).

Airbnb (2018) Healthy Travel and Healthy Destinations. *Airbnb,* May. See https://press.airbnb.com/wp-content/uploads/sites/4/2018/05/Healthy-Travel-and-Healthy-Destinations.pdf (accessed 22 December 2021).

Andrés, G. (2018) Activistas de Arran se encadenan al dragón del Park Güell para protestar contra el turismo. *El País,* 19 June. See https://elpais.com/ccaa/2018/06/19/catalunya/1529394344_263530.html (accessed 22 December 2021).

Araya López, A. (2020) Policing the 'anti-social' tourist. Mass tourism and 'disorderly behaviors' in Venice, Amsterdam and Barcelona. *PACO PArtecipazione e Conflitto* 13 (2), 1190–1207. https://doi.org/10.1285/i20356609v13i2p1190.

Araya López, A. (2021a) A summer of phobias: media discourses on 'radical' acts of dissent against 'mass tourism' in Barcelona. *Open Research Europe* 1 (66), no pages. https://doi.org/10.12688/openreseurope.13253.1

Araya López, A. (2021b) Saint Mark's Square as contested political space: Protesting cruise tourism in Venice. *Shima* 15 (1), 168–196. https://doi.org/10.21463/shima.119.

Arias-Sans, A. and Russo, A. (2017) The right to Gaudí. What can we learn from the communing of Park Güell, Barcelona. In C. Colomb and J. Novy (eds) *Protest and Resistance in the Tourist City* (pp. 247–263). New York: Routledge.

Aterra (2021) See. https://aterra.info/ (accessed 22 December 2021).

Attoh, K.A. (2011) What kind of right is the right to the city? *Progress in Human Geography* 35 (5), 669–685. https://doi.org/10.1177%2F0309132510394706.

Banha, I. (2022) Ativista que interrompeu António Costa nega ter organizado o protesto. *JN*, 13 January. See https://www.jn.pt/justica/ativista-julgado-em-lisboa-por-interromper-antonio-costa-14488661.html (accessed 26 May 2022).

Benavides, L. (2019) Los vecinos de la calle Provença, hartos de la Zona Bus de la Pedrera. *El Periódico de Catalunya*, 15 May. See https://www.elperiodico.com/es/barcelona/20190515/autocares-turismo-pedrera-barcelona-7455136 (accessed 22 December 2021).

Blanchar, C. and Garfella Palmer, C. (2021) El ecologismo mide su fuerza en la calle para blindar el 'no' a la ampliación de El Prat. *El País*, 19 September. See https://elpais.com/espana/catalunya/2021-09-19/el-ecologismo-mide-su-fuerza-en-la-calle-para-blindar-el-no-a-la-ampliacion-de-el-prat.html (accessed 22 December 2021).

Burgen, S. (2021) Hermitage Museum proposal divides Barcelona authorities. *The Guardian*, May 30. See https://www.theguardian.com/world/2021/may/30/hermitage-museum-proposal-divides-barcelona-authorities (accessed 22 December 2021).

Bruttomesso, E. (2018) Making sense of the square: Facing the touristification of public space through playful protest in Barcelona. *Tourist Studies* 18 (4), 467–485. https://doi.org/10.1177/1468797618775219.

Casanova, G. (2019) La Sagrada Familia consigue la licencia de obras 137 años después. *El País*, 7 June. See https://elpais.com/ccaa/2019/06/07/catalunya/1559911273_643720.html (accessed 22 December 2021).

Cia, B. (2019) Vecinos de la Sagrada Familia rechazan la escalinata que quiere construir el templo. *El País*, 28 November. See https://elpais.com/ccaa/2019/11/28/catalunya/1574955541_576455.html (accessed 22 December 2021).

Cia, B. (2020) La Sagrada Familia renegocia con el Ayuntamiento el pago de las compensaciones a la ciudad. *El País*, 17 December. See https://elpais.com/espana/catalunya/2020-12-17/la-sagrada-familia-renegocia-con-el-ayuntamiento-el-pago-de-las-compensaciones-a-la-ciudad.html (accessed 22 December 2021).

Cia, B. and Blanchar, C. (2018) El proyecto de Gaudí no contemplaba la escalinata de la Sagrada Familia que amenaza varias viviendas. *El País*, 22 November. See https://elpais.com/ccaa/2018/11/22/catalunya/1542895436_511563.html (accessed 22 December 2021).

Città de Venezia (2022) *Contributo de accesso*. See https://www.comune.venezia.it/it/content/contributo-accesso (accessed 22 December 2021).

Cocola-Grant, A. and Gago, A. (2019) Airbnb, buy-to-let investment and tourism-driven displacement: A case study in Lisbon. *EPA: Economy and Space* 53 (7), 1671–1688. https://doi.org/10.1177%2F0308518X19869012.

Coleman, N. (2013) Utopian prospect of Henri Lefebvre. *Space and Culture* 16 (3), 349–363. http://sac.sagepub.com/content/16/3/349.

Colomb, C. and Novy, J. (2017) *Protest and Resistance in the Tourist City*. New York: Routledge.

Davis, R.C. and Marvin, G.R. (2004) *Venice, the Tourist Maze. A Cultural Critique of the World's Most Touristed City*. Berkeley: University of California Press.

Francesconi, A. (2021) Sit-in contro il taglio degli alberi, ma la Save accusa il Comune. *Il Gazzettino*, 6 March. See https://www.ilgazzettino.it/nordest/venezia/taglio_alberi_tessera_save_comune-5811640.html (accessed 22 December 2021).

Garfella, C. (2021) La laguna de la discordia de El Prat: un pulmón verde enclaustrado entre aviones y el mar. *El País*, 9 September. See https://elpais.com/espana/catalunya/2021-09-10/la-laguna-de-la-discordia-de-el-prat-un-pulmon-verde-enclaustrado-entre-buques-y-aviones.html (accessed 22 December 2021).

García Prat, S. and Segura, S. (2021) Veïns protesten contra el tancament de les bateries antiaèries del turó de la Rovira. *Betevé*, 9 September. See https://beteve.cat/societat/veins-protesten-contra-tancament-bateries-antiaeries-turo-rovira/ (accessed 22 December 2021).

GCAN (2021) See https://globalcruiseactivistnetwork.com/ (accessed 22 December 2021).

Gössling, S. and Schweiggart, N. (2022) Two years of COVID-19 and tourism: What we learned, and what we should have learned. *Journal of Sustainable Tourism* 30 (4), 915–931. https://doi.org/10.1080/09669582.2022.2029872.

Harvey, D. (2008) The right to the city. *New Left Review* 53, 23–40. See https://newleftreview.org/issues/ii53/articles/david-harvey-the-right-to-the-city (accessed 22 December 2021).

Harvey, D. (2012) *Rebel Cities: From the Right to the City to the Urban Revolution*. London: Verso.

Kuymulu, M.B. (2013) The vortex of rights: 'right to the city' at a crossroads. *International Journal of Urban and Regional Research* 37 (3), 923–940. https://doi.org/10.1111/1468-2427.12008.

Lefebvre, H. (1996) The right to the city. In E. Kofman and E. Lebas (eds) *Writings on Cities* (pp. 147–159). Cambridge, MA: Wiley-Blackwell.

Left Hand Rotation (2021) See http://www.lefthandrotation.com/ (accessed 22 December 2021).

Lenzen, M., Sun, Y-Y., Faturay, F., Ting, Y-P., Geschke, A. and Malik, A. (2018) The carbon footprint of global tourism. *Nature Climate Change* 8, 522–528. https://doi.org/10.1038/s41558-018-0141-x.

Lindblad, S. (2020) Overtourism is bad – Undertourism is worse'. *Travel Weekly*, 15 June. See https://www.travelweekly.com/Strategic-Content/Overtourism-Undertourism (accessed 22 December 2021).

Lombardi, P. (2020) Does Gaudí's Church need a big staircase? Grand vision divides Barcelona. *The Wall Street Journal*, 11 February. See https://www.wsj.com/articles/does-gaudis-church-need-a-big-staircase-grand-vision-divides-barcelona-11581440285 (accessed 22 December 2021).

Mansilla, J.A. (2021) Social movements and class struggle: Against unequal urban transformations from the neighborhood. *Journal of Urban Affairs* 1–15. https://doi.org/10.1080/07352166.2021.1947747.

Mansilla, J.A. and Hughes, N. (2021) 'En dos años no nos vamos a acordar de la pandemia' Análisis del discurso sobre el decrecimiento turístico en Barcelona. *BARATARIA. Revista Castellano-Manchega de Ciencias Sociales* 30, 30–52. https://doi.org/10.20932/barataria.v0i30.623.

Mayer, M. (2009) The 'right to the city' in the context of shifting mottos of urban social movements. *City* 13 (2–3), 362–374. https://doi.org/10.1080/13604810902982755.

Milano, C. and Mansilla, J.A. (2018) Introducción a la Ciudad de Vacaciones. Apuntes sobre el turismo y el malestar social en Barcelona. In C. Milano and J.A. Mansilla (eds) *Ciudad de Vacaciones. Conflictos Urbanos en Espacios Turísticos* (pp. 19–79). Barcelona: Pol•len Edicions.

Montañés, J.A. and Blanchar, C. (2019) El CAP Raval gana al Macba la guerra por la Capella de la Misericòrdia. *El País*, 19 November. See https://elpais.com/ccaa/2019/11/19/catalunya/1574197605_511306.html (accessed 22 December 2021).

Muñoz, O. (2013) Barcelona propone ocho maneras de acabar la Sagrada Familia. *La Vanguardia*, 20 December. See https://www.lavanguardia.com/vida/20131220/54397390429/barcelona-ocho-maneras-acabar-sagrada-familia.html (accessed 22 December 2021).

Nofre, J., Giordano, E., Eldridge, A., Martins, J.C. and Sequera, J. (2018) Tourism, nightlife and planning: Challenges and opportunities for community liveability in La Barceloneta. *Tourism Geographies* 20 (3), 377–396. https://doi.org/10.1080/14616688.2017.1375972.

Osservatorio CIvicO sulla Casa e la Residenza (OCIO) (2021) See https://ocio-venezia.it/pagine/affittanze-dati/ (accessed 22 December 2021).

Purcell, M. (2002) Excavating Lefebvre: The right to the city and its urban politics of the inhabitant. *GeoJournal* 58, 99–108. https://doi.org/10.1023/B:GEJO.0000010829.62237.8f.

Rocabert Maltas, M. (2019) Una línea roja contra la ampliación del aeropuerto de Barcelona. *El País*, 15 July. See https://elpais.com/sociedad/2019/07/14/actualidad/1563121192_483645.html (accessed 22 December 2021).

Salerno, G-M. (2018) Estrattivismo contro il comune. Venezia e l'economia turistica. *ACME: An International Journal for Critical Geographies* 17 (2), 480–505. https://www.acme-journal.org/index.php/acme/article/view/1489/1324.

SchipholWatch (2021) See https://schipholwatch.nl/ (accessed 22 December 2021).

Stay Grounded (2021) See https://stay-grounded.org/ (accessed 22 December 2021).

Stil, H. (2019) Greenpeace gaat toch actievoeren in Schiphol Plaza., Het Parool, 19 December. See https://www.parool.nl/amsterdam/greenpeace-gaat-toch-actievoeren-in-schiphol-plaza ~ b2e4045f/ (accessed 22 December 2021).

Stop de Gekte (2021) See https://stopdegekte.nl/ (accessed 22 December 2021).

Swain, J. and Spire, Z. (2020) The role of informal conversations in generating data, and the ethical and methodological issues they raise [49 paragraphs]. *Forum Qualitative Sozialforschung / Forum: Qualitative Social Research* 21 (1), Art. 10, 1–22. http://dx.doi.org/10.17169/fqs-21.1.3344.

Transport & Environment (2019) One corporation to pollute them all: luxury cruise air emissions in Europe. *Transport & Environment*, 4 June. See https://www.transport-environment.org/publications/one-corporation-pollute-them-all (accessed 22 December 2021).

van der Lit, R. (2019) Koketteren met het rode licht'. *De Groene Amsterdammer* 143 (47), 20 November. See https://www.groene.nl/artikel/koketteren-met-het-rode-licht (accessed 22 December 2021).

!WOON (2021) See https://www.wooninfo.nl/ (accessed 22 December 2021).

World Travel & Tourism Council (2021) Travel and tourism economic impact, June 2021. See https://wttc.org/Portals/0/Documents/Reports/2021/Global%20Economic%20Impact%20and%20Trends%202021.pdf?ver=2021-07-01-114957-177 (accessed 22 December 2021).

Zero (2021) 12 September. See https://zero.ong/cruzeiros-voltam-amanha-a-lisboa-e-alerta-a-zero-a-poluicao-tambem/ (accessed 22 December 2021).

Data sets available in Zenodo

Araya López, A. (2021c) RIGHTS UP – Table with references for media texts regarding protests against mass tourism in Venice, Amsterdam and Barcelona (2014–2017). [Data set]. *Zenodo*. http://www.doi.org/10.5281/zenodo.4740404.

Araya López, A. (2021d) RIGHTS UP – Interviews with activists protesting mass tourism in Barcelona [Transcripts – English translations]. [Data set]. *Zenodo*. http://www.doi.org/10.5281/zenodo.4739716.

Araya López, A. (2021e) RIGHTS UP – Photographs from the RIGHTS UP project (2018–2020) [Visual data]. *Zenodo*. https://doi.org/10.5281/zenodo.5006898.

14 Conclusion: What is to be Done?

Freya Higgins-Desbiolles and Bobbie Chew Bigby

Introduction

This book has gathered multi-layered insights into the dynamics and possibilities of the local turn in tourism, seeking the views of scholars and practitioners working in communities and places around the world. We have intentionally not clearly defined the borders and boundaries of the concepts of 'community' and 'locals' in the introduction to this work in order to leave open a gateway for multiple perspectives and interpretations. We have offered the concept of 'relatedness' as a binding force that connects people to people, people to the more-than-human and people to place. It is also this relatedness that prevents the local turn from turning hostile, isolated, insular, parochial and chauvinistic. Relatedness underpins the local, but it opens up relations between locals, locales and interlocalities.

We as editors are not unaware that these words 'what is to be done?' are commonly attributed to Vladimir Ilyich Lenin's work by that name penned in 1902. We risk the association with the tenets of communism in using these well-known words, but we also do not shy away from championing the commons, processes of recommoning and championing the community as the proper beneficiaries of their common resources. However, this phrase 'what is to be done' is also expressive of the human condition, where we constantly confront problems and contradictions of inequities and injustices and work to create social movements for emancipation from multiple oppressions and seek justice as communities. Simultaneously, we try as individuals to live lives of meaning, value and positive impact. 'What is to be done?' is a question of perennial importance but even more so as we face the complex, compounding and cascading crises caused by human-induced climate change (Morton, 2022).

In our studies of tourism, we have witnessed communities around the world become increasingly frustrated with the negative impacts of tourism, the profiting of the tourism industry at local community expense, the failure of governments at many levels to regulate tourism for the local community's interest and benefit and tourists who have brought disrespect and displacement from their tourism behaviours and choices. The recent phenomenon of overtourism represents the culmination of these dynamics and informs us that a rethinking of tourism is long overdue. These exploitative forms of tourism are labelled 'WEIRD' in tjukonai's first words opening this book, an acronym for 'Western Egocentric Indulgent Ruthlessly Dollarised' pointing to some of the values and ideologies in play. The rise of community activism and social movements as reported in some of these chapters indicates communities are activating to secure their own interests and futures in/against/through tourism.

The local turn, despite the challenges outlined in some of the chapters of this work, offers one among many pathways forward to transform tourism to serve in building better futures. In the foreword, Norberg-Hodge argued that localising tourism presents a positive agenda through which local organisations and sensitive governments can work collaboratively to 'boost both the esteem and prestige of the local and artisanal economy, as well as its economic viability'. But she cautioned this cannot be left as solely a niche approach; greater structural changes and degrowth are also essential. In this next section, we outline some of the key themes that have emerged from our reading of the chapters presented in this book. We then turn to an effort to outline the further work to be done in this thinking space before closing and leaving our readers to consider taking this agenda forward.

Themes Emerging from these Chapters

The chapters in this book are quite diverse in their locations, positionalities and focuses, but they all tell a story about how local communities interface with tourism and its development potential in communities. We the editors have worked to commission and weave the chapters together to narrate a story of the hopes, efforts and challenges for communities in creating their desired futures. Below are some of the themes we have identified as emergent from these powerful chapters.

Putting locals at the forefront of tourism

The Introduction to this book argued for the necessity to redefine tourism through a focus on the local community and to centre their rights, needs and benefits in the future development of tourism (Higgins-Desbiolles *et al.*, 2019). Many chapters in this book offered relevant insights into the possibilities and challenges to these efforts. For instance, Jungersted's Chapter 8 narrates a recognition among Destination Marketing Organisations such as Wonderful Copenhagen that local communities are central to the responsible management of tourism. This is further underscored in the brief case study that follows from Group Nao on the 'DMOcracy' project. In Chapter 7, Jones and Zarb report on a 'Meet the Locals' project for Malta that intended to centre the local community and their cultures at the heart of Malta's tourism and thus challenge the dominance of the mass tourism model. In Chapter 11, Müller, Fletcher and Blázquez-Salom explain how the community surrounding La Trapa in Mallorca, Spain demonstrated the values of 'proximity travel' during the COVID-19 pandemic as they sought to protect this public commons as a space to enjoy nature during this difficult time.

But as Ooi well explains in Chapter 4, communities are not cohesive and there are many disputes, conflicts of interest and corruptions that disrupt any romantic notions of harmonious ideals. Some of the chapters in this book provide clear insights into the tensions and pressures in involving community in tourism, particularly how some working in the tourism industry logically support growth dynamics of tourism, viewing it as in their best interest. In Everingham and Francis-Coan's Chapter 12, we gain insight into the complexities of identifying exactly who is local to the community and sense how pressures of development, including tourism, can divide once cohesive communities. Seyfi and Hall in Chapter 3 refer to a two-part concept of 'neighbourliness' which they explain reveals tensions between the recognition that everyone has the right to make a living and the recognition that everyone who is affected by tourism development has a right to be consulted. Finally, in Chapter 10, Bigby and Brown-Burdex focus on the development of the Greenwood Cultural Centre in Tulsa, Oklahoma where we see a community divided by both the difficult legacy of the 1921 Race Massacre and the ways that this history intersects with tourism. One result of this community division has been the establishment of rival cultural centres given that the narrative of truth-telling, calls for reparations and justice, and acceptable forms of tourism engagement are not approached in the same way across the community. This points to a fundamental challenge of how we might develop strategies to bridge divides, promote reconciliation and develop consensus in and across communities.

Tourism as industry, markets, realities: How real are the alternatives?

In their analysis of the interface between communities, sanctions and the ethical contexts of tourism consumption in Chapter 3, Seyfi and Hall convincingly argue that dynamics of contemporary tourism fall under capitalist market distribution systems which limit the abilities to press ethical claims on tourists or tourism. This corresponds to the framing of tourism as an industry and an understanding of it as a system built on the industry's supply serving to meet the tourists' demand (see Higgins-Desbiolles, 2006). Similarly, Ooi investigated moral limits in localising tourism which arise from its distribution through market mechanisms. However, some of the chapters in this book offer considerations of enduring forms of tourism that sit outside of marketised, privatised and commodified systems of capitalism, including forms of social tourism, cooperative networks and exchanges. There is also the emerging work on diverse economies approaches to tourism, including Buen Vivir and Gross National Happiness in Bhutan outlined in the Introduction. Little's discussion of community cryptocurrencies in Chapter 6 may be situated under this umbrella. Additionally, Tomassini and Cavagnaro's conceptualisation of 'circular *oikonomia*' could be considered a diverse economies approach.

The essential point of debate is how much capability such diverse economies approaches have to withstand the pressures of capitalism and offer real alternatives and how much they represent a small niche destined to operate on the side lines of capitalistic systems. In Chapter 11, Müller, Fletcher and Blázquez-Salom argue for a transition to a post-capitalist future and their exploration of 'convivial' tourism praxis represents one pathway towards that goal. This issue of the realities of capitalism and the possibilities of moving beyond its dominance is a point of some contention between our authors about the import and impact of diverse and alternative approaches. In the foreword, Norberg-Hodge offered us the examples from her years of work with the community of Ladakh describing exchanges between this community and communities of the developed countries such as the UK to offer opportunities to critically reflect on the impacts of capitalist growth as part of the efforts to build movements for different economics.

Circles, cycles and connections

'Tours' and 'turns' invoke movement through circles and cycles. Particularly, when we think about regenerative tourism, we begin to leave Western linear models behind and think about circles, cycles and connections (Ateljevic, 2020). This stands in contrast to sustainability thinking where we focus on maintaining consumption while working to minimise negative impacts. Regenerative approaches to tourism require net benefits for communities and ecologies and represents an important shift in our thinking in tourism. In Chapter 2, Tomassini and Cavagnaro explain to us the iconic image of the eternal circle and propose we consider viewing tourism as a 'circular *oikonomia*'. By this, they mean adapting the circular economy concept to re-engage with the philosophical basis of oikonomia, meaning management of the household. This brings in a needed rebalancing to capitalism's overemphasis on economics and markets, to bring in ethical values, human connections and obligations 'to our household… the Earth'. Little's Chapter 6 also features circular economy ideas with its engagement with doughnut economics and description of Monteverde's inclusive model of using community currencies to reduce waste, share abundance and connect the community in challenging COVID times.

In Chapter 12, Everingham and Francis-Coan use autoethnographical approaches to delve into the ideas of 'more than-tourism' to consider engagement with home as tourists and thereby building a (re)enchantment with place in the context of COVID-19 pandemic public health measures. Rather than the linear journeys of conventional tourism pre-COVID, we experience circular journeys around local place, getting to know it and its many layers with others in perhaps slower circuits, delving into the changing face of place by moving through space and time in attunement allowing for new meanings and feelings to arise. But we should take caution from tjukonai who explained that 'tour' and 'turn' are word twins but with historical associations to conquest.

Democratising tourism

Another theme found in this work is the need to democratise tourism so that it becomes answerable to the communities in which it is undertaken. The introduction outlined some tools and approaches that might be useful in such efforts, including appreciative enquiry and citizen assemblies. Democratising tourism is exemplified in the work of Chapter 8, where Signe Jungersted shares with us her experience in developing Copenhagen's 'Localhood' strategy from 2017 and in her discussion looking forward to 'Localhood 3.0 as DMOcracy' in 2022, which she defines as a shift from seeking residents' tolerance for tourism to actual resident empowerment (see also the associated case study). Jungersted also addresses the role and mandates of Destination Marketing/Management Organisations revealing how little capacity they currently have to manage tourism for the public good in their jurisdictions. Araya López explained in Chapter 13 how communities comprising Barcelona, Spain have used the capacities of social movements to assert community rights against inappropriate forms of tourism imposed on their neighbourhoods. Araya López also advocates for recognition of 'citizen-produced knowledge' as important in these efforts. This calls for citizen involvement in the data-gathering and research that underpins decision-making and planning that manages tourism in the community.

Canosa's Chapter 5 is an important reminder to think of diversity and inclusion when we take up work to democratise tourism. Her chapter presents the findings from a project to engage youth perspectives and activism to shape tourism for better outcomes in the town of Byron Bay, Australia. The Introduction provided the term 'subsidiarity' which is a key principle of empowerment of the local community in tourism; but as Canosa's chapter alerts us, focused attention on pathways for diversity and inclusion is essential to such efforts. Bridging tourism and development, Winkler in Chapter 9 provides insights into how an NGO operating in Cambodia, the

Cambodian Children's Trust, evolved away from a paternalistic desire to do good to facilitating a community-led process for development using a Freirian practice of conscientisation and learning to work together in mutuality (see also associated case study). This Village Hive Model represents a passing of power and authority to the local people and the building of firm foundations for local governing authorities to move away from external dependency on development aid towards holistic, self-determining approaches. Winkler explains this as an 'upstream approach' and it assists in our efforts to think beyond notions of community capacity-building and outsider-provided empowerment, to genuine community self-determination. Winkler's narrative also suggests how supposedly caring forms of tourism, such as volunteer tourism, can disempower local people and lead to exploitation. This takes us back to the theme of this book that local communities must be central to tourism.

Relations and relatedness through tourism

A key theme of this book is the idea that communities in tourism are defined by relatedness, using the definition of community provided by Higgins-Desbiolles and Bigby (2022). In Chapter 10, Bigby and Brown-Burdex share with us the development of truth-telling through tours emerging from the Greenwood Cultural Centre in Tulsa, Oklahoma. These are mechanisms through which the Greenwood community can present their story of the 1921 Tulsa Race Massacre and its ongoing aftermath in an effort to secure reparative justice for this Black American community. Canosa's chapter on youth also reinforces the important point that inclusion and relatedness must be intergenerational, thus engaging youth in managing and harnessing tourism is essential. In Chapter 1, Bigby, Edgar and Higgins-Desbiolles explain place-based governance in tourism as a multitude of connections in relatedness between local people and place, people and all their generations and local people and tourists working collaboratively to share an intimate connectedness that results in care and custodianship. In Chapter 9, Winkler explains how the Village Hive approach used in Battambang, Cambodia creates a community-wide network of protection and social solidarity preventing children and young people becoming vulnerable to institutionalisation in orphanages, as occurred in the past and made worse by orphanage tourism demand. In Chapter 2 Tomassini and Cavagnaro use the term 'throwntogetherness' derived from Massey (2005) to describe the way we find ourselves living with various others in states of relatedness.

As the Introduction explained, we view the idea of the local turn to be relational and referred to Latouche's concept of 'relational localisms' (2009: 47). This emphasises the point that these agendas of local empowerment of communities in tourism are not intended to advocate for isolationist strategies, but rather to encourage partnerships that help build mutual support networks both near and far. For Canosa, her work guided her to an understanding of 'relational activism' in Chapter 5, describing how youth activate for social and ecological justice within the community context. Relational activism may be evident in Chapter 13, where Araya López explains how community activists are linking up across Europe and even globally to support each other in their demands on governing authorities, tourism industry and tourists, as seen for example in the network called Southern Europe against Touristification. However, tensions exist in relatedness as well, as expressed in the case study of protecting La Trapa found in Müller, Fletcher and Blázquez-Salom's Chapter 11 where they note the local community want the commons protected and accessible, but they may not feel fully hospitable to visitors from the Spanish mainland. The relations described throughout this text are complex, multi-layered and forever changing.

In the first words to the book, vesper tjukonai reminds us that relatedness includes the temporal dimension, where past–present–future are related in important if sometimes overlooked ways. Through this word journey, we are shown the ways the practices of the past leave subtle reverberations in the present. Through these thoughtful words, we may contemplate the meanings and the way such words place us in relation to the land and to each other. We might set the task for tourism of 'listening to those who've gone before, preparing for those who come after us'. This relatedness is complex in both spatiality and temporality and requires multiple disciplines and knowledges for us to finetune our abilities to listen, learn and act.

In summary, this book argues that we must imagine a relational imperative. We cannot think of localisation as a purification process to be free from 'the Other'; it is not possible, nor is it desirable. The question is: how might we transform relations to move away from exploitative and instrumental approaches to tourism development operating currently to ones that are collaborative with communities, enhancing their well-being, regenerative of people and place and respecting planetary boundaries? This is an inspiring and needed agenda for transforming tourism that could take us on a pathway in which all may thrive.

There is More Work to be Done

This edited volume has offered critical exploratory analyses of the possibilities for the local turn in tourism. It is intended to open up dialogue, debate and deeper enquiries into these issues because we think the future will require us to be better locals and to nurture our communities and localities. We note we tried to commission chapters from a number of diverse locales, communities and campaigns, but unfortunately witnessed a few promising contributions fall through due to the demands of ongoing campaigns, the pressures of COVID and other difficulties that mark this era. In the interest of transparency, we note one chapter was withdrawn at a late stage due to a conflict that could not be mutually resolved. Ooi's words on community fractiousness are all too true and our small community of practice through this book fell vulnerable to it. There is no utopia achievable in our endeavours, only collaborations moving through issues with all of our imperfections and human foibles.

We have made a start with this edited volume, but there is much more that remains to be done. Research on localisation of tourism is essential, particularly to reveal conflicts, power dynamics and tendencies to exclusions. As Kenis and Mathijs noted 'assessment of localisation discourses from the vantage point of the (post-)political, evidently requires empirical inquiry' (2014: 176). We would argue that empirical and conceptual work, comparative work and engaged, activist work are all essential in providing needed insights into how to better shape tourism for social and ecological justice. Figure 14.1 offers some pointers and possibilities for future research on the local turn in tourism.

Local community level dynamics & worldviews	The ways stakeholder roles are transformed by local empowerment	Approaches to addressing power effects on local communities at all levels – local, regional and global	The wider structural context under which tourism occurs in local communities
• Roles of culture, religion, politics, ecology, etc. in shaping societal dynamics • Processes for inclusion of marginalised & discriminated against populations • Philosophical & cultural approaches to understand diverse worldviews	• Case studies of success & failure • Practices e.g. appreciative enquiry & co-design labs to engage stakeholders in transformative approaches • Political approaches (e.g. subsidiarity for enacting a localising approach)	• Complex, integrated, systems analyses to understand the limits & possibilities of localising approaches • Harnessing regional & international governance measures to enable local empowerment • Capacities of alternative economic approaches to support local communities (e.g. social enterprises, cooperatives & well-being economics)	• Systemic injustices inhibiting empowerment (colonialism, whiteness, patriarchy, etc.) • Global crises inhibiting thriving (e.g. climate change, pandemics, global financial crises, conflicts) • Ongoing & thriving humanistic/ communitarian/ Indigenist/ socialist social systems & supports

Localising- local to global approaches addressing power, justice & benefits

Figure 14.1 A preliminary schema to advance the local turn in tourism studies. Reprinted from *Annals of Tourism Research*, Vol 92, F. Higgins-Desbiolles and B.C. Bigby, 'A local turn in tourism', 103291, with permission of Elsevier

In Chapter 1, Bigby, Edgar and Higgins-Desbiolles presented Figure 1.4 focused on the principles of place-based approaches to tourism governance. This may be refined to develop a tentative framework for community involvement in directing tourism to community well-being, as suggested in Table 14.1 offered below.

Conclusion

This book represents an effort to start the conversation on the local turn in tourism that builds on the insights of communities working

Table 14.1 Tentative framework for community involvement

Principles	Involvement methods	Examples of indicators
Citizen-led participation in decision-making	Community representation on boards, citizen assemblies, participatory budgeting	Breakdown of board membership in the categories outlined; memberships of liaison committees, number of meetings convened, attendance
Inclusive of all diversities – inclusion of marginalised in communities	Critical mapping of community diversities, consultation with each on their involvement	Indicators co-developed with each group
Involvement of schools, youth groups	School fieldtrips, youth exchanges, citizen science and social science activation	Number of trips, exchanges, youth participation, resource support
Strengths-based approaches to all work	Citizen consultations to identify what is well-loved and visioning on positive co-created futures	Documentation of strengths, dissemination strategies
Committing to place among all stakeholders	Visitor pledges developed and implemented by communities as an ongoing practice of care	Number of participants in pledge-making workshops, numbers of visitors signing on, ongoing meetings to refresh pledges
Listening to all the generations of a community, holding a view of the connectedness of past-present-future	Collectively gathering histories of place, using multiple methods to communicate temporal engagements with a place	Metrics of community forums (number of, attendees, outcomes implemented), qualitative data in the form of stories, performances, art
Causing little to no harm through tourism and its management	Community involved in development of models of low harm tourism development	Indicators of social, ecological, economic and political impacts of tourism (as per models like the Tourism Optimisation Management Model – Jack, n.d.)
Related localisms – sister city/ community diplomacy in localising tourism	Forming networks of related localisms in tourism	Number and strength of sister relationships developed and measuring their engagement
Regenerative approaches	Community appreciative enquiry sessions on forms of tourism that regenerate communities and ecologies	Metrics and narratives of tourism developments that build community well-being, resilience and connectedness

through difficult times as they work to recover from the pandemic and also prepare themselves for future crises that we know are coming. It builds on the multitude of earlier work on community empowerment, community-based tourism and community participation in tourism planning and decision-making (e.g. Blackstock, 2005; Scheyvens, 2002; Tosun & Timothy, 2003).

However, it would seem that this time in post-COVID (or at least, pre post-COVID) is heralding a new era, the contours of which remain unclear and undetermined. Our intervention through collecting the chapters in this book is to sketch the ways communities might go about and indeed are building and guiding their own futures and working to shape tourism and visitation to their needs. It offers both conceptual and practical insights that may prove helpful in guiding various kinds of communities in taking control of their future and shaping tourism to their needs. While we situate the imperative in terms of crises, such as tourism dependency vulnerabilities, climate change and conflicts, we also envision this as a very positive and hopeful endeavour that can only contribute to building better communities and collective futures. This represents an open invitation for all of the stakeholders in tourism to embrace the local turn and discover ways they can contribute to making tourism more life-enhancing for local communities, local ecologies and all the generations that are and will be.

References

Ateljevic, I. (2020) Transforming the (tourism) world for good and (re)generating the potential 'new normal'. *Tourism Geographies* 22 (3), 467–475. https://doi.org/10.1080/14616688.2020.1759134.

Blackstock, K. (2005) A critical look at community based tourism. *Community Development Journal* 40 (1), 39–49. https://doi.org/10.1093/cdj/bsi005.

Higgins-Desbiolles, F. (2006) More than an industry: Tourism as a social force. *Tourism Management* 27 (6), 1192–1208.

Higgins-Desbiolles, F. and Bigby, B.C. (2022) A local turn in tourism studies. *Annals of Tourism Research* 92, 103291. https://doi.org/10.1016/j.annals.2021.103291.

Higgins-Desbiolles, F., Carnicelli, S., Krolikowski, C., Wijesinghe, G. and Boluk, K. (2019) Degrowing tourism: Rethinking tourism. *Journal of Sustainable Tourism* 27 (12), 1926–1944. https://doi.org/10.1080/09669582.2019.1601732.

Jack, L. (n.d.) Tourism Optimisation Management Model. Retrieved 17 March 2022. http://www.regional.org.au/au/countrytowns/options/jack.htm.

Kenis A. and Mathijs E. (2014) (De)politicising the local: The case of the Transition Towns movement in Flanders (Belgium). *Journal of Rural Studies* 34, 172–183.

Latouche, S. (2009) *Farewell to Growth*. Cambridge: Polity.

Lenin, V.I. (1902) What is to be done? In *Lenin's Selected Works*, Volume 1, pp. 119–271. See https://www.marxists.org/archive/lenin/works/1901/witbd/ (accessed 13 March 2022).

Massey, D. (2005) *For Space*. London: Sage.

Morton, A. (2022, 28 February) Climate scientists warn global heating means Australia facing more catastrophic storms and floods. The Guardian. See https://www.theguardian.com/environment/2022/feb/28/.

climate-scientists-warn-global-heating-means-australia-facing-more-catastrophic-storms-and-floods (accessed 13 March 2022).

Scheyvens, R. (2002) *Tourism for Development: Empowering Communities.* Harlow: Prentice-Hall.

Tosun, C. and Timothy, D.J. (2003) Arguments for community participation in the tourism development process. *Journal of Tourism Studies* 14 (2), 2–15.

Index

Figures are shown in *italics*, tables in **bold**.

Aboriginal cultures xxxi–xxxv, 9, 21, 32, 33, 35–46, 49
activism *see* social movements; protest movements; youth activism
Actor Network Theory 55
Adams, William M. 220
agriculture 116–117, 122–124
  see also community currency
Aikau, Hokulani K. 33
air travel xxvii, xxviii, 259, 261, 263
Airbnb 64, 88, 157, 250, 254, 257, 259–260
Aitken, David 76
Albergo Diffuso concept 16
Albrecht, Julia N. 110
Alfred, Prince, Duke of Edinburgh xxxiii–xxxiv
Amsterdam 33–34, 63, 64, 259–261
animal-based tourism 63–64
anti-tourism *see* protest movements; tourismphobia
Aotearoa/New Zealand 7–8, 21, 32, 33, 34, 110
  see also Māori culture
Appadurai, Arjun 11
appreciative enquiry (AI) 19
Araya López, Alexander 4, 14–15
Arnstein, Sherry R. 136, 137
Arran (action group) 254, 259, 263
Asia Pacific Economic Cooperation (APEC) 89
Attoh, Kafui A. 252
Aung San Suu Kyi 69, 75
Australian Volunteers International programme 183
autoethnography 234–236
  see also Newcastle, NSW

Balearic Group of Ornithology and Nature Defence (GOB) 222–228
Balearic Islands 217–218
  see also Mallorca
Barad, Karen 61
Barcelona 4, 253–259, *255*, *256*, *257*, *260*
Barinanga, Ester 119
Barris Per Viure (action group) 258
bartering systems 88
Bayley, Sam 36, 39
Becky (Newcastle NSW tour guide) 236–242
Berndt, Ronald and Catherine xxxiii–xxxiv
Besgrove, Samantha 184, 185
Bhutan 45
Bianchi, Raoul 11
Biddulph, Robin 19–20
Biden, Joseph 203, *208*
Bigby, Bobby Chew 7, 17, 55
Bitcoin 119
Black Americans *see* Greenwood District, Tulsa
Blackstock, Kirsty 5
blockchain technology 117, 119–120, 124–130, **124**, *127*
Böhm, Steffen 13
Bordeaux Tourism 8
Born, Branden 10
#BoycottMurree campaign 72
boycotts *see* travel boycotts
Boyle, David 7
Braidotti, Rosi 58, 59
Bramwell, Bill 137
Braungart, Michael 56
Briassoulis, H. 8
Britton, Tessy 21

Brown-Burdex, Michelle 195, 205–210, *208*
Brunei 75
Buen Vivir 44, 45
Burtner, Jennifer 72
Büscher, Bram 218–219, 221
Butcher, J. 10
Butler, Richard 137, 144, 145–146
'buy local' movement 4
Byron Shire, New South Wales 103, 106–110
Byron Youth Theatre 109, 110

Calvia, Mallorca 136
Calvino, Italo 57
Cambodia
  community empowerment 170, 178–189
  family-based care model 175–177
  orphanages 89–90, 170–176, 182
  poverty levels 169–170
Cambodian Children's Trust (CCT) 171–172, 175–181, 183–189, 193
Canan, Penelope 9
Caritas Canada 20
Castañeda, Quetzil E. 72
Caterina-Knorr, Tanner C. 245
Chemainus, Canada 89
Chenoweth, Erica 73
child protection
  downstream approach 169–175, *174*, 181–183
  family-based care model 175–177
  upstream approach 173, *174*, 178–181, 183–189
childism, defined 112
children
  activism projects 106–107, 108–110
  connection to place 103, 107
  definitions 105–106, 112
  and relational activism 111
Chomsky, Noam 10
Chua, B. 50
circular economy (*oikonomia*) 54–65
  definitions 56, 57
  and posthumanism 59–64, *62*
  practical examples 63–64
  theoretical framework 55, 59–60, 61–63, *62*
citizen assemblies 19
Cittaslow (slow town movement) 19

city tours 236–242
Clarke, Alan 135, 144, 148
climate crisis 5, 12, 17, 20, 54, 119, 234–235, 244, 250
Cochet, Yves 12
Cocola-Gant, Agustin 260
Cole, Mary Hill xxxiii
Colomb, Claire 253
colonialism *see* decolonial tourism; neo-colonialism
commodity-consumerist model xxvii
commoning 17
communities
  complex nature of 91–93, **93**
  framework for community involvement **277**
  holistic definitions 7, 17
community currency
  generally 17, 117, 119–120
  Verdes (Costa Rica) 117, 120, 124–130, **124**, *127*
community protests *see* protest movements
community washing 83, 162–163
community-based tourism (CBT) 5, 10, 91–92, 135–137, 149–151
computing programs, delivery of 180–181
convivial conservation 217, 218–219, 223–229
cooperative enquiry 31–32, 49
Copenhagen *see* 'Localhood for Everyone' strategy
corruption 87, 94, 171, 181
  anti-corruption training 187
Costa Rica *see* Monteverde
couchsurfing 88
COVID-19 pandemic
  approach of hospitality industry 64
  and ecology 54
  and food insecurity 116–117, 123
  and inequality 47
  and local communities 9, 15–16, 109, 185
  and proximity tourism 1–2, 216–218
  spread along tourism routes 216
  and sustainable tourism 232
  as 'tourism reset' xxvii, 105, 233–234, 250–251
critical consciousness 18
cruise tourism 2, 50, 259, 261, 263

Crutcher, Terence 202
cryptocurrencies (CCC) 117, 119–120, 124–130, **124**, *127*
Cyan Planet 63–64

Dahles, Heidi 90
dall'Ara, Giancarlo 16
Dapp, Marcus M. 119–120
Davis, Robert C. 262
decolonial tourism 33, 47–50
  see also community currency; Village Hive (child protection model)
Defensem Park Güell (action group) 254
degrowth 11, 47–49
Deleuze, Gilles 59
Denmark 64
  see also 'Localhood for Everyone' strategy
destination management organisations (DMOs)
  extent of mandate 161–162
  involvement of local residents 8, 164, 166–168
development *see* international development
Diaz-Parra, Ibán 3
Diaz-Soria, Inma 217
digital travellers 157
Diniz, Eduardo H. 119, 128
DMOcracy project 8, 164, 166–168
Dodds, Rachel 136, 137, 144, 145–146
donation and bartering systems 88
doughnut economy model 118, 120–121, *120*, 244–245
downstream development 169–175, *174*, 181–183
Drinnan, Keir 184, 185

ecology 54, 216
ecotourism 10, 217–218, 220–221
  see also convivial conservation
Edensor, Tim 232
Edgar, Uncle Joe 36, *37*, 41–42
Einstein, Charles 12
Ellen MacArthur Foundation 56
Ellsworth, Scott 196, 200
empowerment *see* community empowerment
equality 47, 91–92
Escobar, Arturo 12, 34, 47
Ethereum 119

ethical tourism 70–71, 77
  see also travel boycotts
ethnography 234–236
  see also Newcastle, NSW
European Commission 56
Everingham, Phoebe 245

Fairbnb.coop 15, 18–19
fairness *see* unfair distribution of benefits
farm tourism 88
Fiji 1–2, 21
Fletcher, Robert 11, 218–219
food insecurity 116–117, 122–124
  see also community currency
Freeman, R. Edward 86, 136, 141
Freire, Paolo 18, 20, 90, 139, 170, 186, 191–193
Frew, Elspeth 236
Friedman, Monroe 71–72
future research 275–276, *276*

Gago, Ana 260
Galvez, Alyshia 116
Gates, Eddie Faye 207
gender equality 91–92
gentrification 256–260
Gibson, Chris 246
Giridharadas, Anand 34
Global North, development approach 169, 169–175, *174*, 181–183
Global South
  poverty levels 169–170
  use of community currency 118–119
globalisation 10–12, 14–15, 33
Goble, Danny 195
Gonzalez, Vernadette Vicuña 33
Goodson, Lisa 139
governance
  market regulation 88–89
  place-based approaches 45–46, *46*
Gozo *see* 'Meet the Locals' project
Graham, Mary xxxii, 49
greenhouse gas emissions 250
Greenwood District, Tulsa, Oklahoma
  Greenwood Cultural Centre (GCC) 195, 200–211, *201*, *203*, *208*
  Greenwood Rising Museum 204–205, 209–210
  Tulsa Race Massacre (1921) 194–201, *198*, *199*, *202*, 205–210

Grok Learning 180–181
Gross National Happiness 45
Group NAO *see* DMOcracy project
Guattari, Félix 59
Guia, Jaume 60, 61

Hacket, Mr (Narrung) xxxiii
Hall, C. Michael 8, 9, 75, 76, 136–137, 146, 217, 232
Hall, Derek 135
Hallward, Maria Carter 74
Hambleton, Robin 35
Hardingham-Gill, Tamara 64
Harkin, Natalie xxxv
Harvey, David 252
Hawai'i 9, 33, 110
Helsinki Freedom Campaign 163
Hennessy, Michael 9
Henry, Jacob 191
Heron, John 31
Hickel, Jason 11, 16, 17
Higgins-Desbiolles, Freya 2, 6, 7, 17, 45, 55, 90, 105, 186, 229, 233
Hirmer, Monika 33, 48
hooks, bell 31
hospitality industry 64
Hotel Bauen, Buenos Aires 15
Howitt, A.W. xxxii
Hudson, Simon 75
Hughes, Neil 4, 263
human rights *see* travel boycotts
Hutton, Jon 220

Iceland 110
Illich, Ivan 57
inclusive tourism 19–20
indigenous tourism xxxi–xxxv, 9, 21, 32, 33, 35–46, 49
Indonesia 90–91
inequality 47, 91–92
international development
 capacity sharing approach (Frere) 191–193
 downstream approach 169–175, *174*, 181–183
 family-based care model 175–177
 upstream approach 173, *174*, 178–181, 183–189
 see also DMOcracy project; Sustainable Development Goals

international poverty line 169
Israel, boycott 72

Jackson, Sue 18
Jamal, Tazim 60
Jigme Khesar Namgyel Wangchuck, King 45, 49
Jones, Andrew 135, 148
Josefsson, Jonathan 111
Jover, Jaime 3

Kallis, Giorgos 12
Kangaroo Island, South Australia 45–46
Karajarri tourism model, Western Australia 35–44, *37*, *41*
Kay, Delta 110
Kemmis, Stephen 139
Kenis, Anneleen 13, 14, 276
Kennedy, Emily Huddart 111
Key West, Florida 2
King, Brian 91
King, Uncle Thomas 'Dooley' 37–38, 42–44
Kirby, Erin 184, 185
Komodo National Park, Indonesia 90–91
Krutwayso, Oratai 137
Kunoth-Monks, Rosalie xxxiv
Kuokkanen, Rauna 34

Ladakh, India xxvii–xxix
Lai, Kun 146
Lamers, Machiel 72
Lapadat, Judith C. 236
Las Kellys (action group) 259
La Sagrada Familia, Barcelona 255–256, *256*
Lasso, Aldi H. 90
Latouche, Serge 11, 12, 16, 17, 34, 57
Latour, Bruno 54, 55
La Trapa, Mallorca 215, 221–228, *222*, *225*
La Via Campesina 12, 13
Lawson, Rob 136
Lediard, Danielle E. 193
Lefebvre, Henri 251–252
Left Hand Rotation (artistic collective) 260
Linehan, Denis 47
Lirrwi Tourism, Northern Territory 45
Lisbon 259–260, *261*
Little, Mabel B. 201

'live like a local' tourism campaigns 10
Llurdés-Coit, Joan Carles 217
local currency *see* community currency
Local Futures xxviii, 13
local washing 83, 162–163
localhood 155–164, 157
'Localhood for Everyone' strategy 155–163
   impact beyond Copenhagen 158–159
   implementation of strategy 160–163
   importance of permanent locals 157–158
   strategy document 63, 155–158
'locals' and 'outsiders' 238, 241–242
localwashing 17–18, 162–163
Lukermann, Fred 32–33

Mair, Judith 236
Malcolm, Neill, 13th Laird of Poltalloch xxxiii
Mallorca 221–228, *222*, *225*
Malm, Andreas 216
Malta 137–138, 142
   see also 'Meet the Locals' project
Mansilla, Jose 263
Māori culture 21, 32, 33
marine pollution 106–108, *108*
market regulation 88–89
Marvin, Garry R. 262
Massey, Doreen 55, 57, 61
Mathijs, Erik 13, 14, 276
McAfee, Kathleen 221
McArthur, Simon 136–137, 146
McCreery, Cindy xxxiii
McCrohan, Petrine 38–40, 43
McDonough, William 56
McKenzie, Marcia 58
McTaggart, Robin 139
medical tourism 85
'Meet the Locals' project (Malta/Gozo) 133–135, *134*, 138–151, *143*, *148*, *149*
Messer, Cynthia 135, 144
Migrantour 63
Miles Partnership 168n
Millar, Craig 76
Mission Australia 109
Moloka'i, Hawai'i 9
Monteverde, Costa Rica
   doughnut economy model 120–121
   food insecurity and responses to 122–124

   growth of ecotourism 121–122
   Verdes cryptocurrency 117, 120, 124–130, **124**, *127*
moral limits to tourism 83–86
More, Sir Thomas xxxiv
'more-than-tourism' perspective 242–246
Morgan, Aunty Maria 36
Morrison, Scott 238–239
Moscardo, Gianna 135, 136, 144, 148
Mother Teresa 182
Movono, Apisalome 1–2, 21
Murphy, Peter E. 5, 135, 136
Murray, Alan 56
Myanmar 69, 73, 74–76

Nature Observation Tourism 217
neo-colonialism 33, 60, 181–182
neoliberalism 5, 9, 33, 60, 77–78, 116
   neoliberal conservation 221
neolocalism 4–5, 244
Netherlands 64
   see also Amsterdam
New Zealand/Aotearoa 7–8, 21, 32, 33, 34, 110
   see also Māori culture
Newcastle, New South Wales *240*, *241*
Newcastle, NSW 234, 236–242, *239*
non-governmental organisations (NGOs) 183–184
   delivery of computing programs 180
   expatriate vs local staff 183–186, 187
   focus on capacity building 182
   see also international development
Nonomura, Robert 72
Norberg-Hodge, Helena 13
Norkunas, Martha K. 142
Novy, Johannes 253

oikonomia see circular economy
Oliver, James 55
orphanage tourism 89–90, 170–176, 182
O'Shaughnessy, Sara 111
Osservatorio CIvicO (action group) 259–260
Our Home Holiday Town project 108–110
overtourism 3–4, *4*, 11, 137–138, 159, 250–251, 262
   see also touristification
Oviedo, Julie Benisty 8

Page, Sarah 197
Palau, Ol'au approach 45, 110
Panama, Guna 45
Parajuli, Pramod 34–35
Park Güell 254
Participatory Action Research (PAR) 20–21, 139
participatory budgeting 19
participatory filmmaking 106–107
participatory planning 40–41, *40*
participatory research projects 103–112
Pearce, Philip L. 135
Pedagogy of the Oppressed (Freire) 18, 20, 90, 170, 186, 191–192
Pedersen, Arthur 136, 145
Petrescu, Doina 17
Phillmore, Jenny 139
place-based governance 31–50
  forms of governance 34–35
  Karajarri tourism model 35–44, *37*, *41*
  other examples 44–46
  place and placelessness 32–34
  place-based approaches 46
planetary boundary framework 50
pledges 45, 109–110
Pollan, Michael 4
Pon Jedtha 171, 175–176, 184–185
Portugal 75
'positive counter-development' xxviii
posthumanism 59–61
poverty 47, 85, 169–170, 172–173
Pritchard, Annette 246
profit-share arrangements 180
promoted areas 219, 223–228, *227*
protected areas 216–217, 218, 220–221, 223
  see also Karajarri tourism model
protest movements 3–4, 253–265
  see also social movements
proximity tourism 1, 20, 217–218, 219, 228–229
public–private partnerships (PPP) 87, 90
Pulami, Peter xxxiii–xxxiv
Purcell, Mark 10, 252

Quilley, Stephen 14

racism 183–184, 196–201
Radzik, Linda 73–74, 76
Rastegar, Raymond 117

Raworth, Kate 120–121
Raymond, Eliza M. 110
Reed-Gilbert, Kerry xxxii
regenerative tourism 89–90
regulation *see* market regulation
relational activism 111
repugnant transactions 84, 86, 88, 93
resistance *see* protest movements
responsible tourism initiatives xxviii–xxix, 109–110
Richards, Greg 135
'right to the city' 251–253
Roosevelt, Theodore 182
Roots Guide 63
Ross, Don 200, 201
Rothman, Hal K. 10
Rowland, Dick 197
Royal Caribbean International 50

Samoa, 'beach *fale*' tourism 46
Saura, Bruno 32
Scheyvens, Regina 1–2, 5, 19–20, 21, 46
SchipholWatch (action group) 261
Schultz, Karl 89
sex trafficking 175
sex workers 174, 260–261
Sharp, Gene 74
Shovellor, Wynston 37, 41, *41*
Sibley, David 106
Simpson, Kate 191
Singh, Tej Vir 22
Sipe, Lori J. 4–5
Skift (think tank) 159
Sklair, Leslie 9, 48
slow tourism 15, 19
social media 72, 157
social movements 12–16
  see also protest movements
Somaly Mam 175
Spain 75
  see also Balearic Islands; Barcelona; Mallorca
Spinks, Rosie 10
stakeholder management 86–87, 92, 141–142, 144–146, 186–189
Stephan, Maria J. 73
Stepping Stones model 40–41, *40*
Stop de Gekte (action group) 260–261
subsidiarity 8, 20, *21*
sustainable agriculture 116–117, 122–124
  see also community currency

Sustainable Development Goals (SDGs) 56, 91–92, 105, 169–170
Sydney Alliance 15

Tasmania 85
taxation 88
temporary locals 156, 157
'three Ds' 17
Thrift, Nigel 62
Thunberg, Greta 111
Tolkach, Denis 91
tourism research, indigenising 21
tourismphobia 3–4, 250
　see also protest movements
tourist destination, as term 18
tourist rentals 259–260
　see also Airbnb
touristification 254, 256–262
Transition movement 13–14
travel boycotts 69–79
　boycott defined 71–72
　debates over effectiveness 74–76
　and destination communities 71, 76–79
　growth of 72
　and social justice activism 73–74
triple bottom line (TBL) framework 86–87, 90, 105
Tuck, Eve 58
Tulsa Race Massacre (1921) 194–201, *198*, *199*, *202*, 205–210
　see also Greenwood District, Tulsa

Uber 88
Ubuntu 44, 45
Uluru, Central Australia *4*
undertourism 7, 250
unfair distribution of benefits 85–86
United Nations
　Convention on the Rights of the Child (UNCRC) 105–106
　Sustainable Development Goals (SDGs) 56, 91–92, 105, 169–170
　World Commission on Environment and Development ('Brundtland Commission') 104
　World Tourism Organisation 3, 116, 217
upstream development 173, *174*, 178–181, 183–189

Vanuatu 50
Veal, Anthony J. 136
Venice 259–260, 261, 262, 263–264
Verdes cryptocurrency (Costa Rica) 117, 120–121, 124–130, **124**, *127*
Village Hive (child protection model) 178–181, 185, 186–189
Vimercati, Luigi 64
Vision Walks (ecotourism operator) 110
visitor pledges 45
volunteer tourism 85, 191–193

Waiheke Island, Aotearoa/New Zealand 8, 34
Walia, Sandeep Kumar 135, 144
Wall, John 111
Weaver, David B. 105
Weinberg, Adam 122
W.E.I.R.D. tourism xxxi–xxxv, 270
white saviourism 181–182
Wiedmann, Thomas 13
wildlife tourism 63–64
Williams, John 136
Wiltshier, Peter 135, 144, 148
women
　relational activism 111
　women's rights 91–92
Wonderful Copenhagen *see* 'Localhood for Everyone' strategy
Wooltorton, Sandra 31–32, 33, 49
!WOON (support agency) 260
World Childhood Foundation 187
World Tourism Organisation (UNWTO) 3, 116, 217
Worldwide Opportunities on Organic Farms (WWOOF) 19, 88
Wyler, Steve 7

Yousaf, Salman 72
youth activism 103–112
　connection to place 103, 107
　against mass tourism 254, 259, 263
　Our Home Holiday Town project 108–110
　participatory filmmaking 106–107
　relational activism 111
　Youth4Sea project 107–108, *108*

Zapata Campos, María José 76
Zapatistas, Mexico 48
Zarb, Julian 135, 136, 141, 145, 148
zoonosis 216

For Product Safety Concerns and Information please contact our EU Authorised Representative:

Easy Access System Europe

Mustamäe tee 50

10621 Tallinn

Estonia

gpsr.requests@easproject.com

www.ingramcontent.com/pod-product-compliance
Ingram Content Group UK Ltd.
Pitfield, Milton Keynes, MK11 3LW, UK
UKHW021836140426
5217IPUK00021B/1490